Managing Protected Areas

Niall Finneran · Denise Hewlett ·
Richard Clarke
Editors

Managing Protected Areas

People and Places

Editors
Niall Finneran
University of Winchester
Winchester Hampshire, UK

Richard Clarke
National Association for AONB
Marlborough, UK

Denise Hewlett
PeopleScapes Research & Knowledge
Exchange Centre, Department
of Responsible Management
University of Winchester
Winchester, UK

Bournemouth University
Dorset, UK

ISBN 978-3-031-40782-6   ISBN 978-3-031-40783-3   (eBook)
https://doi.org/10.1007/978-3-031-40783-3

Cover illustration © Sirayot Bunhlong/Alamy Stock Photo

This Palgrave Macmillan imprint is published by the registered company Springer Nature Switzerland AG
The registered company address is: Gewerbestrasse 11, 6330 Cham, Switzerland

Paper in this product is recyclable.

CONTENTS

List of Contributors

Stephanie Aburrow Dorset AONB, Dorset, UK

Sheela Agarwal Department of Tourism Management, University of Plymouth Business School, Plymouth, UK

Neil Amswych Temple Beth Shalom, Sante Fe, NM, USA

Jocelyn Belfort Université Vincennes Saint-Denis (Paris 8), Saint-Denis, France

Elise M. S. Belle WCMC Europe, Bruxelles, Belgium

Olta Braçe Health & Territory Research (HTR) and Department of Human Geography, Universidad de Sevilla, Seville, Spain

Martin Breed Flinders University, Bedford Park, SA, Australia

Danny Byrne Independent Consultant, Hampshire, UK

Zachary J. Cannizzo National Marine Protected Areas Center, National Oceanic and Atmospheric Administration Office of National Marine Sanctuaries, Washington, DC, USA

Tracey Churcher National Trust Purbeck, Dorset, UK

Richard Clarke National Association for AONB, Marlborough, UK; Environmetal Consultant, Marlborough, UK

Nigel Dudley Equilibrium Research, Bristol, UK

Mariana Napolitano Ferreira WWF-Brasil, Brasília, Brazil

Niall Finneran Historical Archaeology and Heritage Studies, University of Winchester, Winchester, UK

Simone Grassini Department of Psychosocial Science, University of Bergen, Bergen, Norway;
Cognitive and Behavioral Neuroscience Lab, University of Stavanger, Stavanger, Norway

Marco Garrido-Cumbrera Health & Territory Research (HTR) and Department of Physical Geography and Regional Geographic Analysis, Universidad de Sevilla, Seville, Spain

Rachel Golden Kroner World Wildlife Fund US, Washington, D.C, USA

Debra Gray PeopleScapes Research & Knowledge Exchange, University of Winchester, Winchester, UK

Richard Gunton Winchester Business School, University of Winchester, Winchester, UK

Julie Hammon Dorset AONB, Dorset, UK

Denise Hewlett PeopleScapes Research & Knowledge Exchange Centre, Department of Responsible Management, University of Winchester, Winchester, UK;
Bournemouth University, Dorset, UK

Oliver Hutchinson London, UK

Tara Inniss The University of the West Indies (UWI), Cave Hill Campus, Barbados

Natalia Lavrushkina Faculty of Management, Bournemouth University, Bournemouth, UK

Godson Lubrun Haitian Ministry of Tourism, Port-Au-Prince, Haiti

Tom P. Mommsen Canada University of Victoria, University of Victoria, Victoria, BC, Canada

Tom Munro Dorset AONB, Dorset, UK

Mohammad K. S. Pasha International Union for Conservation of Nature (IUCN), Asia Regional Office, Bangkok, Thailand

Malgorzata Radomska Department of Responsible Management and Leadership, Faculty of Business and Digital Technologies, University of Winchester, Winchester, UK

Fahima B. Rahman Independent Scholar, London, UK

Eleanor Ratcliffe School of Psychology, University of Surrey, Guildford, UK

Mitali Sharma Independent Consultant for Environmental Organisations, Singapore, Singapore

Chris Skelly Department of Health and Social Care, London, UK

Risa B. Smith IUCN/World Commission On Protected Areas, Gland, Switzerland

Hugues Séraphin Tourism, Hospitality and Events, Oxford Brookes University, Oxford, UK

Ainara Terradillos Universidad de Sevilla, Andalusia, Spain

Philip Weinstein School of Public Health, Faculty of Health and Medical Sciences, University of Adelaide, Adelaide, SA, Australia

Christina Welch University of Winchester, Winchester, UK

LIST OF FIGURES

LIST OF TABLES

CHAPTER 1

Managing Protected Areas: People and Places: Introduction

Niall Finneran, Denise Hewlett, and Richard Clarke

For decades, literature on the management of green and blue spaces and their uses has been founded in two distinctive sources. The academic repository, comprising primarily peer-reviewed journal papers, has tended to detract practitioners for their very nature of academic design, composition, debate and discourse traditionally deployed. Yet within these works, a critical voice on both practice and theory is developed contributing to

N. Finneran (✉)
University of Winchester, Winchester, UK
e-mail: niall.finneran@winchester.ac.uk

D. Hewlett
PeopleScapes Research & Knowledge Exchange Centre, Department of Responsible Management, University of Winchester, Winchester, UK

Bournemouth University, Dorset, UK

D. Hewlett
e-mail: denise.hewlett@winchester.ac.uk

R. Clarke
National Association for AONB, Marlborough, UK
e-mail: richard.clarke069@btinternet.com

© The Author(s) 2024   1
N. Finneran et al. (eds.), *Managing Protected Areas*,
https://doi.org/10.1007/978-3-031-40783-3_1

enhancements in the management of natural spaces. The alternative body of knowledge comprises practitioner publications and personal accounts. These are frequently descriptively formatted and articulated as prescriptions of best practice in the management of green and blue spaces and on the practical implementation of national and international policies.

Whilst each of these substantial bodies of interrelated knowledge facilitates insights into how the practice of managing green and blue spaces of cultural and environmental importance might be enhanced, the traditional two-pronged approach to literature, has, we contend, not been exploited to their fullest potential. Debates and key challenges, both new and traditional, could be enhanced by refining a model of co-working amongst practitioner and academic authors that might exploit the rigour of academic research with the in-depth expertise and insight of the practitioner across the range of rural and urban green and blue spaces found worldwide.

The importance of progressing this collaborative approach has been made even more important in recent years as a state of political, social, economic and environmental flux has developed worldwide. The results of this state are well known, and examples include at a regional level, the full socio-economic-political impacts of Brexit; at an international level a polycrisis driving the displaced and refugees; wars leading to insecurities of the most basic survival needs; extensive and increasingly intensive impacts of climate change; impacts on post-COVID-19 pandemic societies and threats of additional zoonotic sources of public health concerns.

Such events have resulted in multiple changes to environmental, political, economic, technological and social systems, impacting on how green and blue spaces worldwide are being managed. In order to contribute to their long-term and sustainable management, this information needs to be captured, reported in documentary sources and research is required to enable the potential for refreshed if not revised practice. Many of the most current challenges in the management of greenspaces and protected areas and voids in current literature will be addressed in this book '*Managing Protected Areas: People and Places*'.

As with any publication project, relying upon specially commissioned papers and the constraints of people with busy and challenging work schedules, our original concept for the book shifted over time. Initially, we had an idea to present a more structured and 'siloed' work, with chapters grouped under thematic concerns. In each case an academic would be working alongside a practitioner, and essentially would engage in a dialectic approach, working towards a synthesis of background ideas and a melding together of theory and practice. Essentially this ethos has been preserved, but the end result the reader experiences here, has evolved over the last two highly eventful years worldwide and has involved a great deal of conversation and debate. We feel the present result is actually far better, capturing many of the key issues and new ways of thinking that have evolved around the relationship between people and green and blue spaces worldwide as they have unfolded.

We have attempted to move beyond purely the confines of natural heritage and have brought together a range of authors from many different academic and practice-based backgrounds. This makes we feel, for a stronger and more diverse volume and one which we think essentially captures the key themes, issues and debates around wider approaches to the management of natural heritage and cultural heritage in the twenty-first century. This book frequently considers applied research and practice. It is therefore not only aimed at academics, practitioners and policy-makers working in the area of natural heritage and cultural heritage, it also provides for many insightful debates of universal value, demonstrates the application of theoretical frameworks in practice and addresses voids in our knowledge base, providing for alternative and practical solutions. As such there will unquestionably be utility in this work for the undergraduate and postgraduate students of social, environmental, health and political sciences. We engage with place and space on a quotidian basis and, as we emphasise here, there are many challenges and dangers which we face. Above all, this book we believe emphasises the key principle that people and places are essentially interlinked and interdependent. Through recognising these interconnections, we demonstrate importance for a sustainable approach to managing green and blue spaces that is underpinned by encouraging and maintaining the importance of peoples and places.

In the 16 specially commissioned chapters that follow, you will find a number of different voices from different backgrounds, working at the regional, national and international scales of engagement. There are some resonant names, and perhaps others that may not yet be known within the world of natural heritage management. We urged at the outset that the authors took a personal and (if they needed to) challenging and provocative (in the best sense) overview. We encouraged a personal, expressive and reflective writing style, admitting to problems where they were found and seeking ways to do things better. The tone we hope is constructive and proactive and not in any sense redolent of professional navel gazing.

The authors we have gathered here are in many cases professional colleagues and acquaintances drawn from pretty much a 360° perspective of the management of natural heritage and its relationship with the public. Naturally a good deal of the book focuses upon the disciplines of human and (to some extent) physical geography and its related concerns. So there is a core emphasis upon ecological and environmental issues, as well as elements that relate to the tourist industry and the pressures that sector entails. This is only part of the picture, however. Other themes we articulate are around health and wellbeing, educational outreach and to the far less reported context of spirituality, important themes in the concept of post-COVID-19 approaches to our relationship with natural spaces. So, in this connection the reader will be exposed to ideas drawn from psychology, archaeology, anthropology, religious studies, even Islamic and Jewish theology and to the application of information technologies and virtual reality experiences.

As we have noted above, we have attempted to give a degree of editorial free rein. The only condition that we stress is that each chapter must relate to the core ethos of the book: that the relationship between people and places is something to be celebrated, and promoted in the years after the COVID-19 pandemic and against a background of climate change. We acknowledge the fragility of our natural heritage and do not pretend to offer definitive solutions, instead we hope to provoke a constructive and ongoing debate which is illustrated where appropriate with case study examples. The chapters vary in length and are written in many cases (except for Chapters 2 and 3) from diverse and multiple viewpoints. The chapters are of course fully referenced and the additional of URL links and doi links allows for an easy follow up of the key information. The language is, we hope, accessible, and graphics where added are intended to be informative and to capture the human spirit of the chapters.

The second chapter is written by someone who will need no introduction to many of the readers of this volume. Nigel Dudley's work on broad-scale conservation, protected areas and society and environment gives him an informed, analytical and evaluative insider's view on the most current and critical issues facing protected areas in the twenty-first century. In Chapter 2 'What does the Global Biodiversity Framework (GBF) mean for protected and conserved areas?' Dudley takes a critical and insightful view on the current progress of the GBF and its ambitious attempts to meet the 30 × 30 target of conserving at least 30 per cent of land, inland water and marine ecosystems in protected and conserved areas by 2030. A lively, reporting writing style is retained highlighting the complex and nuanced political contortions of the development of the GBF and its implications for nature conservation.

Chapter 3 'landscapes of the romantic sublime: the legacy of nineteenth-century artistic visions and contributions to the development of the management of natural heritage' sets out a deep culture-historical context for the philosophy of natural heritage management in the UK. It is written by Niall Finneran who is from an archaeological-cultural heritage background demonstrates how the romantic cultural movement of the late eighteenth–nineteenth century framed our perceptions of land-scape and informed Victorian management strategies, principles of which remain with us today.

Chapter 4 focuses on a case study drawn from the insular Caribbean. Writing with Niall Finneran is Tara Inniss, a Caribbean academic and heritage practitioner based in Barbados. This chapter, 'Islandscapes: tourism, COVID, climate change and challenges to natural landscapes. A Caribbean perspective and view from Barbados' considers specifically the management of the fragile natural heritage of the Caribbean islands. Inniss' work has focused upon heritage and wellbeing, and she was instru-mental in helping frame the UNESCO World Heritage Site management plan for the Garrison site in Bridgetown. She has worked alongside Finneran in mainly an archaeological capacity, but both authors here bring their wider disciplinary and geographical perspectives together to throw light on a somewhat understudied geographical area.

In recent years, the National Trust of England and Wales has come in for some stringent political criticism from certain sectors of the national media and the political world for its attempts to make itself more inclusive and also to reveal hidden histories relating to its holdings (note we make no mention of the concept of 'rewriting history'!). The National Trust is perhaps the premier management framework for natural and cultural heritage in England and Wales, and building on the historical contexts for its establishment previously discussed in Chapter 3, Chapter 5 ('managing heritage landscapes of cultural value: a view from the National Trust portfolio in Purbeck, southern England') brings a practitioner view on the realities of managing a prime area of historical and tourist-focused real estate in Dorset, southwestern England: the Isle of Purbeck. Tracey Churcher manages this glorious portion of the National Trust estate in England, her background is in heritage management, and writing with Niall Finneran, the challenges of her daily work in the post-COVID-19 world are laid bare.

Chapter 6 'Between high and low tide. Participatory approaches to managing England's coastal and riverine natural and cultural heritage: a case study from the CITIZAN initiative' is written from a maritime-archaeology perspective. It essentially draws attention to the importance of collaborative and participatory approaches to the conservation of the fragile coastal cultural and natural heritage of the UK. The lessons learned here are equally applicable in other geographical contexts, as the authors (who both have worked with the Coastal and Intertidal Zone Archaeological Network initiative, or CITIZAN) outline. Sea level rise is a real fact, and it is threatening many coastal assets. Oliver Hutchinson has been working on the frontline of community focused heritage leadership in the Essex region, and alongside Niall Finneran outlines the rewards (and pitfalls) of a collaborative public-facing approach.

Chapter 7 brings us back to the Caribbean but a quite different social, cultural and economic context from that offered by Barbados. Haiti is one of the poorest islands in the Caribbean and a nation state that faces real developmental challenges. Unlike the rest of the Caribbean, tourism here is relatively under-developed. We have asked three Haitian scholars and practitioners, Hugues Séraphin, a tourism academic, Jocelyn Belfort, a PhD student in Paris, and Godson Lubrun, who works at the Ministry of Tourism in Haiti, to offer their thoughts in a contribution entitled 'Managing a UNESCO World Heritage Site in a post-colonial, post-conflict and post-disaster destination. The case of the *Haitian National History Park*'. They touch upon the value of local guides, and the pride in which the community take in a site with a very real social memory. Although this chapter focuses more upon a cultural heritage site, the implications for broader audience engagement are foregrounded.

In recognition that much of what constitutes best practice in protected area and spatial management of green and blue spaces relates to principles of project management, we turn in Chapter 8 to 'sustainable project management of green spaces, protected and conserved areas— opportunities, challenges and pitfalls'. Malgorzata Radomska's expertise in generic project management, anchors this chapter on academic theories in responsible management, operational excellence and leadership skills. Two of the authors, Richard Clarke and Denise Hewlett have considerable and wide-ranging experience around the design and delivery of multiple projects and programmes in the natural environment. Richard Clarke has a long track record in working alongside environmental organisations in the UK, particularly as the national policy and development manager for the National Association for Areas of Outstanding Natural Beauty (NAAONBs) . Denise Hewlett brings her international academic and practitioner expertise to bear on the issues and problems in sustainable management of projects in the planning and management of green spaces and protected areas.

We then change the tone towards a wellbeing and spiritual focus. Chapter 9, '(Re)connecting with Nature: exploring nature-based interventions for psychological health and wellbeing' is led by Debra Gray, a social psychologist at the University of Winchester with an interest in the relationship between place and participation, and on the link between collective participation, social identities and health and wellbeing. This chapter considers the work she has led alongside Denise Hewlett and two of the Stepping into Nature project leads at the Dorset AONB, Julie Hammon and Stephanie Aburrow, on the strategies and projects deployed in making nature more accessible to a wider range of stakeholders and above all the part that this plays in fostering wellbeing. This issue has come to the fore in the post-COVID-19 world.

Chapter 10, 'Significant spaces: exploring the health and wellbeing impacts of natural environments' draws upon a wide range of differing perspectives. Led by Denise Hewlett, and involving expertise drawn from social psychology (Gray), statistical analysis (Gunton), tourism (Agarwal), landscapes planning and management (Hewlett), projects in nationally important protected areas and networks (Munro), microbiome sciences (Breed, Skelly, Weinstein) GIS analysis (Terradillos) and data gathering (Lavrushkina and Byrne) this chapter picks up on some of the case study ideas laid out in Chapter 8. It offers both a wider perspective on how we can most effectively and sustainably use the natural environment to improve people's health and make a positive difference in their lives and introduces a conceptual framework that determines what it is about natural environments that impacts and enhances our health and wellbeing.

Chapters 11 and 12 are essentially complementary and take a more spiritual and religious direction. In these two chapters, Christina Welch, Reader in religious studies at the University of Winchester works with two religious studies practitioners, a Jewish Rabbi, Neil Amswych, based in the USA, but formerly based in Dorset, England (Chapter 11 'Judaism and engagements with nature: theology and practice') and Fahima Rahman, a young Muslim woman from northern England (Chapter 12 'Islam and engagements with nature; theology and practice'). In both chapters, the dialogue is established through Welch's extensive knowledge of both Jewish and Islamic theology and writings and the practical issues of religious engagements with environmental practice. In the case of Judaism, we see how theology is adapting to the realities of climate change, and in the case of Islam, through Rahman's personal experiences, we see the real issues around racism and accessibility to green spaces.

Chapter 13 'what have we learned from the impact of the Pandemic on our relationship with nature? The importance of views from home' is written by Marco Garrido-Cumbrera and Olta Braçe who are both geographers based in Spain. Their unique perspective frames how we experience the natural world through our windows. Again considering the health benefits of engaging with the natural world, we see here the importance during the pandemic of being able to view nature and see it, even if we couldn't necessarily engage with it first-hand. Drawing upon the results of the GreenCOVID survey, and subjects from Spain, the UK and Republic of Ireland, the importance of secondary experience of nature is clearly defined. It is an unusual and compelling study that draws together architectural space, natural space and human perception and wellbeing.

Chapter 14 'impacts and lessons learned from the COVID-19 Pandemic for protected and conserved area management' is authored by a team with links to the International Union for the Conservation of Nature (Sharma; Pasha), the World Wildlife Fund (Ferreira) and wider approaches to conservation policy and practice in the Americas (Golden-Kroner). The chapter considers the impact of the COVID-19 pandemic on protected and conserved areas and draws upon case studies from across the world, but particularly Latin American and African regions with long-standing economic issues, and highlights the importance of green listing these sites to best achieve a more sustainable and effective management framework.

Chapter 15, 'tourism and visitor management in protected areas post pandemic: the English context' focuses on primary research, funded by National Geographic, that evaluated what impact post-pandemic tourism was having on areas of environmental and cultural importance. Presented in this chapter is primary fieldwork and data gathering undertaken at coastal sites in southern Dorset. The relaxing of social restrictions during the Summers of 2020 and 2021 resulted in an almost unprecedented demand for visits to rural locations. Southern Dorset, within easy reach of a large number of conurbations, certainly bore the brunt of this influx of tourists and as we see in this chapter not all of the effects were positive.

Chapter 16 'climate change—protected areas as a tool to address a global crisis' is led by Zachary Cannizzo of the Climate Coordinator with the NOAA Office of National Marine Sanctuaries—National Marine Protected Areas Center, Washington DC alongside Elise Bell of the World Conservation Monitoring Centre, Risa Smith of the IUCN and Tom Mommsen of the University of Victoria in Canada. This chapter provides for a detailed and comprehensive current perspective on the impact of climate change on protected areas and seeks to find ways to use them more proactively as a means for mitigating the worst effects of climate change.

Chapter 17 is written by two psychologists, Simon Grassini and Eleanor Ratcliffe. Entitled 'the virtual wild: exploring the intersection of virtual reality (VR) and natural environments', the authors consider the powerful potential for the deployment of VR as a means of simulating human experience of the natural world as a cost-effective and easily accessed tool to improve human wellbeing. Taking a broad view of the technological developments in the area, the authors discuss issues around simulation and stimulation and what is an imaginative and hi-tech methodology for accessing natural spaces.

These chapters represent a broad and in many cases, unique approach to discussions taken on the management of natural spaces. The collection of discussions engages to varying degrees with the social, political, environmental and technological sciences and their value includes the currency of their content. The editors thank all the contributors for their work here and also thank the staff at the publisher for their patience whilst the project came to fruition. It has been a stimulating and thought-provoking experience to share several different yet complimentary perspectives. The relationship between people and places is a profound part of our existence, yet perhaps we take it for granted. It is hoped that the case studies, discussions and arguments presented in this volume will contribute if only in some small part to helping us think through the implications of this relationship. Let us round this introductory section off with a thought-provoking quotation from the Kenyan environmental activist, the late Wangarī Muta Maathai (1940–2011), which summarises our position in this book:

> We all share one planet and are one humanity; there is no escaping this reality.

https://www.forbes.com/2009/04/21/challenge-africa-excerpt-opi nions-business-visionaries-maathai.html?sh=3e59273e1b23.

CHAPTER 2

What Does the Global Biodiversity Framework Mean for Protected and Conserved Areas?

Nigel Dudley

2.1   Introduction

On a grey morning in December, my WhatsApp suddenly lit up with the message 'China just gavelled it!! We have GBF!!!'. Four years into the process, and at 3.30 am in Montreal, the Global Biodiversity Framework (GBF) had finally been agreed by signatories to the Convention on Biological Diversity (CBD). Except that it hadn't, not quite. United Nations decisions need unanimity, and the Democratic Republic of Congo (DRC) had been holding out for more aid money and a fund separate from the Global Environmental Facility, which DRC government representatives felt would respond better to their needs. China as chair,

---

A 'reporting style' of language is used here with agreement with the editors in order to capture a first-hand insight of the contextual influences on decisions and the intensity and subtle nuances of negotiation that occur in such forums

---

N. Dudley (✉)
Equilibrium Research, Bristol, UK
e-mail: nigel@equilibriumresearch.com

N. Finneran et al. (eds.), *Managing Protected Areas*,
https://doi.org/10.1007/978-3-031-40783-3_2

11

jumped the gun and overrode DRC's objections, calling the meeting to an end even before the translators had caught up and most delegates understood what was happening. DRC immediately objected, saying it would not abide by the decision, thereby massively weakening the agreement. Several African countries sided with DRC on a point of principle about the way that the decision was reached: others were bitterly opposed, splitting the continent. A tense 24 hours of negotiations followed before DRC finally agreed that the GBF should be adopted, whilst noting reservations to two of the targets and the GBF was gavelled for a second and final time.

This was a final piece of drama in what had been a long and exhausting negotiation. Even a day or so earlier, such an agreement seemed highly unlikely. I was in Montreal until shortly before the end of the CBD's fifteenth Conference of Parties (COP-15) in December 2022, and progress had been glacially slow. The meeting began with 700 phrases of the proposed Global Biodiversity Framework, itself only a fairly short statement, still in square brackets, meaning that they needed to be debated and consensus reached. Watching from the rear of the room (non-governmental organisations were not allowed to intervene at this stage), it felt like many delegates were too junior to make decisions on the spot and were simply holding a government line, even where a compromise should have been simple. Delegates argued back and forth about synonyms which meant basically the same thing. Some key issues failed to be resolved, including in particular issues of financial resources.

The GBF is the latest in a series of strategies set by members of the CBD to address biodiversity loss. The final text of the draft GBF contains 23 targets, including amongst others those relating to ecosystem restoration, pesticide reduction, food waste, gender equality, invasive species, financial resources, removal of perverse subsidies and disclosure of impact by companies (CBD, 2022). Like all UN decisions, adherence is largely voluntary and inevitably patchy. But governments look closely at such documents, so that slight changes of wording can have major implications on policies right down to the level of individual sites. Mention of a particular ecosystem, or conservation approach, or related social target, can have a major impact on funding streams, the ways that policies and laws develop and the attitudes of donor and recipient governments. Quantitative targets, with measurable objectives, are often more successful because there is something concrete to measure deliverables against. So, any numerical targets have a particular resonance.

This means that a few judicious words in a UN decision can push a government to change its policies and encourage donor governments and organisations to switch their funding priorities. Results eventually trickle down to the communities, protected area managers and other stakeholders at a very local level. Many of the recipients will have no idea about the work that has gone on to make this happen. Nor do they need to necessarily. However, there are practical reasons why it sometimes seems as if every word in a meeting like COP-15 is hard fought over and why interested governments and non-governmental organisations invest time and money in advocacy, both in the long build-up to these meetings and at such Conference of Parties (COP) itself.

Owing to the COVID-19 pandemic, COP-15 was already over two years late, making things even more complicated. The twenty-three targets together created a comprehensive and ambitious global strategy. Delegates sat in crowded rooms until late at night, with text projected onto a screen at the front, and they argued about each phrase in turn. There had already been three earlier meetings in 2022 alone to try to agree text. These were held in Geneva, Nairobi and in Montreal immediately before the COP, with little success. Session chairs either dithered or lost their temper, whole delegations walked out over points of principle on more than one occasion. China finally came out with a suggested final text and presented it at the high-level segment of the meeting towards the end of the fortnight and faced with that or nothing countries eventually agreed; such brinkmanship is far from unusual in these negotiations. But the fact that we have a GBF at all, is still a minor miracle.

One particular element of the GBF that is particularly noteworthy is how the language agreed amongst representatives from signatory states and members of the conference on a global level, will influence planning and management activities of protected area and greenspace managers on the ground. The processes by which such decisions are reached; the confusion and realpolitik involved are reported in this chapter from a personal account of the events leading up to the agreement of the GBF.

Particularly controversial, and the main subject of this chapter, is Target three of the GBF. This proposed that 30% of the world's land and ocean should be in '*protected areas and other effective area-based conservation measures*' by 2030, the so-called 30 × 30 target (CBD, 2020). The target reflects a significant change in attitudes towards natural ecosystems and their values and would have seemed to be a pipe dream even a decade earlier. Yet by the time the meeting kicked off in Montreal,

114 national governments had already signed onto a High Ambition Coalition 30 × 30 target, meaning that even if the GBF had failed to reach agreement—a real possibility—it seemed likely that many of the elements of this particular target would have been carried forward. But it wasn't an easy ride. Many governments remain deeply sceptical about this target and consensus was driven by years of hard lobbying, both by supportive governments and by NGOs, with presumably many deals and sweeteners offered along the way in terms of aid and support to help the process along. Eventually, it was agreed, with the final—albeit fairly strangulated—text which reads:

> Ensure and enable that by 2030 at least 30 per cent of terrestrial, inland water, and of coastal and marine areas, especially areas of particular importance for biodiversity and ecosystem functions and services, are effectively conserved and managed through ecologically representative, well-connected and equitably governed systems of protected areas and other effective area-based conservation measures, recognizing indigenous and traditional territories, where applicable, and integrated into wider landscapes, seascapes and the ocean, while ensuring that any sustainable use, where appropriate in such areas, is fully consistent with conservation outcomes, recognizing and respecting the rights of indigenous peoples and local communities including over their traditional territories. (CBD, 2022)

Every word and phrase of this agreement is significant, and some remaining ambiguities will continue to be debated for a long time yet. We have considered the implications of individual words and phrases elsewhere (Dudley & Stolton, 2022a). But notwithstanding from the nuances of how it will be interpreted, the GBF already represents an extraordinary set of changes in the way that governments, NGOs and others view the world. These are what will be examined below and nine key elements have been identified as representing significant changes in the philosophy of how we will progress nature conservation in future:

- The 30% target itself.
- Specific inclusion of inland waters.
- Increased ambition for coastal and marine areas, including the high seas.
- The rise of other effective area-based conservation measures (OECMs).

- Inclusion of indigenous and traditional territories in the target.
- The importance of management effectiveness.
- Coupled with equitable management.
- Integration into wider landscapes and seascapes.
- The rights of indigenous peoples and local communities.

All these points have major implications not only for the ways in which we implement the GBF, but for virtually every single other aspect of sustainable development. Some, but not all, are repeated in different ways in other GBF targets.

But first, a reality check, I was, and I still am, very excited about the possibilities opened up by the GBF, which was a rare bit of good news in an otherwise generally dismal year for international politics. We've been working on the implications of the 30 × 30 target for the last few years, looking at how it might be put into practice and the tools and approaches available to governments and others (Dudley & Stolton, 2022b). There is a very basic change in the way that civil society is viewing the natural world and alongside all the bad news are some real opportunities for positive progress. But will something like the GBF really make much of a difference? One of the first things I did after returning from Montreal to Wales was to attend our village Christmas party. We live in a biosphere reserve, on the edge of Snowdonia National Park. I was talking with one of my neighbours, a professional ecologist who works as a freelance bird surveyor and found that he scarcely knew that there even was a UN convention dealing with biological diversity. He certainly hadn't heard of the GBF or of the meeting in Montreal. Many field naturalists hold the same view and tend to sniff about their colleagues who spend their time arguing the minutiae of wording of paper agreements in conference centres around the world.

Previous CBD targets have mainly failed, including the much-vaunted Aichi targets, agreed at a similar meeting in 2010 in Japan, which also concluded with high hopes for the future. I am writing this only a few weeks after the exhilaration of the events in Montreal and the slow and laborious job of putting the conclusions in place hasn't even begun. At the end of this chapter, I consider some of the practical ways in which the GBF might make a difference. But beforehand, I will run through each of the nine elements above to discuss how the GBF will affect nature conservation in the near future.

2.2   30% OF THE PLANET

Most of the world's protected area system will have been established during the lifetime of most of the people reading this chapter. The shift from production—or more often potential production—to conservation on such a large scale must be the fastest major change in land use in human history. IUCN, the International Union for Conservation of Nature, first proposed that ten per cent of the planet's land surface might be in protected areas at its World Conservation Congress in Caracas in 1992. Then it seemed naively utopian. Thirty years later we are already at 17% of the land surface of the planet, as noted in Aichi Target 11 set in 2010, which is one of the few Aichi targets to have been at least partially fulfilled. The target was less successful in marine conservation and there were other elements in the target including governance, effectiveness and integration into the landscape/seascape, where progress was slower. Now there is a commitment to set aside 30% and the half earth theory, which argues that we should leave at least half the planet in a more-or-less natural state (Pimm et al., 2018), is gaining increasing acceptance. What has changed?

30 × 30 represents a coming together of many different strands of opinion. On the one hand, there is increasing realisation that unless natural or near-natural areas are deliberately set aside; they are likely to be transformed (Winkler et al., 2021). This is due to the actions of an apparently insatiable set of actors, particularly the food industry, which in some situations still clearly views land clearance as an easier option than the maintenance of the land already under cultivation. Awareness of the role and importance of natural systems has also been increasing. The focus has been expanding from biodiversity, (which for many is not a sufficient reason for such large commitments to conservation), to a much wider range of ecosystem services, including aspects of food and water security, disaster risk reduction and climate change mitigation (Stolton & Dudley, 2010). Finally, there is far more support for maintaining land under relatively natural ecosystems than there is for protected areas—national parks, nature reserves and wilderness areas—as being the best or only vehicles for achieving this. It is in large part the recognition of other effective area-based conservation measures that has tipped the balance towards 30 × 30 (Dudley et al., 2018).

2.3 INLAND WATERS

Freshwater biodiversity is decreasing at almost twice the rate of terrestrial and marine species, and freshwaters are incredibly diverse, with high rates of endemism and equally high levels of threat (Ramsar, 2018). Freshwater conservation biologists were convinced that lumping inland waters in with terrestrial ecosystems means that the former was often forgotten in practice. Even when rivers or lakes occurred inside larger terrestrial protected areas they were often virtually ignored by land and water managers. Adding the phrase '*inland waters*', which includes freshwaters and also those with high levels of salinity such as the Dead Sea, was seen by a group of NGOs as the clearest way to ensure that these ecosystems were adequately represented in national conservation plans (The Nature Conservancy et al., 2022).

Conservation groups are interpreting the final wording of the target to mean that 30% of terrestrial, 30% of inland waters and 30% of coastal and marine ecosystems should each be included separately within the target; implying that the target applies to 30% of inland waters as well. We will have to wait to see if national governments apply this in practice. But the change signals a wider shift in priorities. It is not so much that freshwaters have been ignored—there are many thousand people working throughout the world on their conservation—but that freshwater concerns have often been treated in isolation from wider conservation messaging and that bringing them into the mainstream is now more and more necessary (Garcia-Moreno et al., 2014).

2.4 OCEAN AND MARINE AREAS

The other significant shift in the range of ecosystems targeted was to increase the ambition for conservation of oceans and coastal waters to 30%, from the ten per cent target in Aichi 11. This was perhaps the biggest sticking point within Target three in the negotiations leading up to COP-15, with China particularly reluctant in the light of its huge fishing fleet and appetite for fish. Given China's pivotal role in chairing this session of the CBD, it seemed more than likely that a compromise might be reached for 30% on land but less for the ocean; in the event the original proposals held.

The final decision indicates two important changes. Firstly, the increased ambition of the target is very significant in itself, but secondly there has also been a gradual shift to include a greater focus on the high seas (Lewis et al., 2017). Here, protection is urgently needed to rebuild dwindling fish stocks and other impacted marine communities but is also much trickier to achieve, due to the need to reach consensus from all countries about conservation of areas beyond national jurisdiction. Given the remoteness of many areas and the existence of huge fishing fleets acting outside the law, it will also be hard to implement even if boundaries are drawn on marine charts. Exactly, how this will play out remains unclear, and we might speculate that several countries signed up without necessarily intending to follow through. There are also large questions about the effectiveness of many marine protected areas, which have been set up with so few controls that it is hard to distinguish them from the rest of the marine environment. At coastal level, there are numerous community-managed initiatives such as locally managed marine areas (LMMAs) (Newell et al., 2019), many long-tested in real conditions. Some of these fall outside the protected area network but are ideally situated to be recognised as OECMs. They are likely to become increasingly important as the 30 × 30 target starts to be implemented, but such approaches are less available on the high seas.

2.5   OTHER EFFECTIVE AREA-BASED CONSERVATION MECHANISMS (OECMs)

OECMs are not new exactly but they are newly emergent. The phrase appeared at the last minute in Aichi Target 11, at around 3.00 in the morning when negotiators were exhausted and no-one had the energy to object. And the addition caused quite a lot of consternation; it took almost eight years after the Japan meeting to agree a definition, and governments are now scrambling to catch up. There is still a huge amount of confusion about what OECMs are in practice. The definition was finally settled in late 2018 at the previous CBD Conference of Parties in Sharm el Sheik in Egypt:

> A geographically defined area other than a Protected Area, which is governed and managed in ways that achieve positive and sustained long-term outcomes for the in-situ conservation of biodiversity, with associated ecosystem functions and services and where applicable, cultural, spiritual, socio–economic, and other locally relevant values. (CBD, 2018)

This definition covers three main cases:

- Ancillary conservation—areas delivering in situ conservation as a by-product of management, even though biodiversity conservation is *not* an objective (e.g. some military training grounds).
- Secondary conservation—active conservation of an area where biodiversity outcomes are only a *secondary* management objective (e.g. some conservation corridors).
- Primary conservation—areas meeting the IUCN definition of a protected area, but where the governance authority (i.e., community, indigenous peoples' group, religious group, private landowner or company) does not wish the area to be reported as a protected area (IUCN Task Force on OECMs, 2019).

Everyone knows OECM is a horrible acronym, but it is the only term that has official government recognition and for the moment we are stuck with it. In practice, many people refer to 'protected and conserved areas' as being equivalent to 'protected areas and OECMs' although there is no real agreement as yet, on the scope of a 'conserved area' and it is used in a variety of ways.

OECMs are a beguiling prospect for many governments, worried that they don't have enough land or marine areas to set aside major new national parks or similar but wanting to meet biodiversity goals. The fact that existing management is already maintaining biodiversity means that these uses can continue, removing the usual social and economic objections that often face protected area establishment. Commitments are more about maintaining current patterns of land and water use into the future and monitoring status to make sure that biodiversity (and other) values are maintained. Huge areas of indigenous peoples' territories could become OECMs in theory, although there are complications as discussed in the next section.

Although OECMs were mentioned by name in Aichi 11, the fact that they hadn't been defined until near the end of the target period means that few governments tried to apply them in practice. The largest change in the GBF may well be the emergence of OECMs as a potentially huge land-use category within 30 × 30. But this won't be trouble-free. There are large and justifiable fears that OECMs will be used as a means of greenwashing—designating areas with little biodiversity value in order to make up the numbers—and conversely that they could be used by governments as a further excuse for dispossessing people from their traditional territories. Both conservation and human right groups, therefore, have concerns and some indigenous peoples' organisations are deeply sceptical. Additionally, and currently less discussed, is the fact that because OECMs are recognised from places that are *already* successfully conserving biodiversity, meaning that they do not lead to any actual gain in land and water under effective conservation, although they may increase the long-term security of such areas. Questions about whether OECMs could be created, for instance, through restoration policies, have hardly been discussed as yet. Over the next few years, it will become clear whether the OECM concept ends up revolutionising conservation or becomes a slightly irrelevant side-show.

2.6   Indigenous and Traditional Territories

The biggest surprise for Target three in Montreal was the success of the International Indigenous Forum on Biodiversity (IIFB) in introducing a third category into the 30 × 30 target for indigenous territories outside protected areas and OECMs. This was resisted by many developing countries including the European Union but was loudly endorsed by many developing countries, who applauded every intervention that supported the proposal. Conservation NGOs were privately concerned but mainly kept silent. The fears were: (a) that *any* indigenous territory might be recognised, even if it had no conservation value (IIFB statements suggest this was never the intention) and (b) deciding which territories 'count' could take years: it took eight years to agree the OECM definition. The phrase '*where applicable*' implies that indigenous territories will need to show effective conservation, equitable governance, etc., to meet 30 × 30.

The IIFB argued that the addition was needed because of reluctance amongst many indigenous peoples to being included in any top-down scheme, and fears that, whatever was said at international level, creation of more protected areas or OECMs in their territories will risk further loss of access to land, water and resources. There is also a belief that inclusion as a separate category in Target three will further help gain tenure rights on traditional territories, still the primary concern of many groups. However, there is also a risk that governments will simply 'recognise' existing indigenous territories as part of the 30 × 30 package without providing anything more in terms of support or legal safeguards, leaving such areas open to further incursion and degradation. It is clear that 30 × 30 is only going to be possible with the support and active participation of many indigenous peoples (e.g. Fa et al., 2020), and a strong belief amongst conservation groups that this partnership could be mutually beneficial. But it is still too early to say whether the inclusion of a third category will help or hinder this process.

2.7 EFFECTIVE CONSERVATION

Some of the changes are not so much to do with wording as with emphasis. The words 'effectively conserved' appeared identically in Aichi 11 but reappear in the GBF with a greater commitment to implementation, not least due to a constant stream of criticism that says conventional protected areas (often dismissed by critics as 'fortress conservation') is not working. A massive growth in ambition for area under protection will bring with it an equally large demand for proof that such areas are delivering the goods. This will perhaps be even more important in the case of OECMs, where the whole definition is based on a perception of effectiveness that will need to be demonstrated.

There are many practical implications. Management effectiveness assessments will need to become a standard part of management and increasingly linked with comprehensive assessment of conservation and social outcomes, i.e., it is not enough to know that areas are being managed well but also that this is having the desired result in terms of ecosystem protection. Even well managed protected areas can fail if outside pressures, including climate change, are too great. Measuring outcomes implies continual monitoring, which can prove expensive and time-consuming, although technology is expanding rapidly with access to more effective satellite data, cheaper camera trap technology, acoustic monitoring, DNA sampling and online tools for identification. Ranger-led monitoring tools such as SMART (Cronin et al., 2021) are also helping to transform monitoring options for managers. Meanwhile agreed management standards—a step beyond assessment of effectiveness—are emerging, including the IUCN Green List (Hockings et al., 2019) and Conservation Assured, a species-based approach with standards already available for tigers, jaguars and river dolphins (e.g. Dudley et al., 2020).

2.8   EQUITABLE CONSERVATION

The concept of protected area management effectiveness has been around for 25 years (see for instance Hockings et al., 2006), pioneered by the IUCN World Commission on Protected Areas (WCPA) and has developed steadily ever since. Although still far from perfect we have a whole generation of conservationists who have developed their careers with these tools and concepts already embedded in their approach. Ideas about looking critically at the equity of conservation and attendant issues of social impact and governance quality are newer (Franks & Booker, 2022), but equally important. Equity and governance are also less well supplied with tools and experience although this situation is changing fast. The International Institute for Environment and Development (IIED) has been developing and testing a series of tools to assess social impacts and governance quality, with recent efforts to amalgamate the Management Effectiveness Tracking Tool (METT) (Stolton et al., 2019) with IIED's Site-level Assessment of Governance and Equity (SAGE) (IIED, 2022).

Issues of equity go far beyond an understanding of whether the results of biodiversity conservation are fairly distributed. There is now an expectation that any new protected areas, and any recognition of OECMs, include full participation of those people living in, using and abutting the proposed area. In the case of indigenous people such participation should also include a formal process of Free, Prior and Informed Consent (FPIC). These demands have been in place for some time, but not infrequently ignored in practice by governments. This issue of human rights is now centre-stage and impossible to ignore. I'll return to this in the last of our nine points below.

2.9   INTEGRATION INTO WIDER
LANDSCAPES AND SEASCAPES

Unless they happen to be protecting an island, protected areas should not be 'islands' (McArthur & Wilson, 1967). In other words, an area that is cut off from other suitable habitat, because surrounding land or water has been transformed and will no longer support most native species, is unlikely to maintain its values in the long term. Yet that is exactly what many protected areas are in practice, a casual glance at Google Earth will show patches of green in the middle of areas of farmland or conurbation, with little opportunity for species to achieve genetic interchange with other populations. Recognition that, however well managed, many of these areas will inevitably degrade over time has prompted increased interest in connectivity (Hilty et al., 2020) and more generally in the notion that protected areas and OECMs should be integrated into wider landscape and seascape approaches (e.g. Chatterton et al., 2017) rather than being seen as separate and somehow removed.

The challenges of addressing this part of the target should not be underestimated. Many protected areas—whether they are managed by park authorities, or private individuals or groups of indigenous peoples and other local communities—were established precisely because the surrounding land or water had been transformed, overexploited or otherwise degraded from an ecological perspective. OECMs are often being suggested as tools for increasing connectivity, but as noted above this won't help in areas where isolation has already occurred. In these cases, reconnecting protected and conserved areas will require both careful negotiation with existing users and owners, including judicious land and water purchase on occasion and active restoration. This will not happen overnight and, in some cases, will require generational change stretching over many decades. Managers of isolated protected areas will need to keep a watch on impacts, such as inbreeding, and if necessary, provide stop-gap solutions such as introduction of individuals from other populations or special conservation measures for species in decline.

2.10   THE RIGHTS OF INDIGENOUS PEOPLES AND LOCAL COMMUNITIES

In this instance, the GBF is creating an additional push for a process that has been ongoing for twenty years, but which clearly hasn't yet gone far enough. There has been a long-term problem of authoritarian governments creating protected areas through processes that include forcible loss of rights, ranging from hunting controls to full-scale relocation of communities, sometimes with minimal consultation or compensation. Protected area rangers have also sometimes come under fire for human rights abuses, including violence and sexual assault; giving a poorly trained person access to a uniform and a gun can be dangerous. The role of non-governmental conservation organisations in these issues is complicated, but there is a general perception amongst human rights' groups that conservationists have been too complicit in the mistreatment of some of the poorest people on the planet. Complicity can vary from active engagement to failing to speak out against abuses in countries and areas where they operate.

This is an enormously complex issue. Any conservation change is going to affect someone. A ban on hunting the last population of a unique primate group will impact the hunters who would like to shoot them for the pot, and biodiversity conservation necessarily involves a series of compromises including negotiation, trade-offs and sometimes also compensation. But in some places, the approach has clearly been off-balance. The issue first gained global attention at the Fifth World Parks Congress in Durban, South Africa, in 2003, when various leaders of indigenous peoples' organisations called for the 'elimination' of national parks. The debate started a twenty-year process of reform. Now all reputable conservation organisations have detailed safeguarding processes that attempt to avoid human rights abuses at any stage in the creation or management of protected areas. The extent to which they are applied varies, more between the attitudes of individual project leaders and managers than between organisations; over the years we have experienced both good and bad.

As a consultant, I have worked with all the largest conservation NGOs and have seen major changes in approach over the last two decades. A new generation of conservationists is emerging, who are far more aware of social issues than conservationists have been previously and these people are now reaching a stage in their career where they can be influential in shaping decisions and policies. And yet, the accusations and anger have if anything increased rather than decreased. To some extent there may be a historical lag, with old wounds still untreated, but it is also due to the far better access to information available today, with instant communications and a huge social media network tracking these issues. Any controversial actions now are likely to be circulated to activist groups around the world within hours, whereas in the past many problems went unnoticed and unchallenged. Issues that are only peripherally linked to conservation, such as the management of private hunting reserves, often get labelled as a form of fortress conservation. Responses need to be not only about preventing bad practices but also about how responsible parties react when something bad comes to light. A few years ago, a series of sexual abuse scandals were reported in the disaster relief sector; those organisations involved that acted swiftly and transparently to admit and react to the issues, fared better in public opinion than those seeking to brush the problems aside.

A more intractable issue relates to criticising governments if abuse comes to light. Many conservation NGOs are reluctant to call out governments because they fear that doing so will lose them access to the country in question (a far from implausible response). This has left criticism to smaller activist groups inside or outside the country concerned and thus further increased splits within the conservation and human rights lobbies. Some sort of independent review process may be needed in the near future. A number of delegates at COP-15 argued that the issue of human rights should be presented as an overarching chapeau rather than listed multiple times against individual targets. But the IIFB and others countered that it was important to embed principles into the actual targets to provide a necessary and that this is particularly true in Target three, due to the long controversy about protected areas, provided a necessary emphasis. It is clear not only that expectations about participation; FPIC, governance quality, equity and social outcomes are far higher than they have been in the past, but that governments and conservation NGOs will be under far greater scrutiny than hitherto. This is going to be uncomfortable on occasions but is definitely to be welcomed overall.

2.11   IMPLEMENTING TARGET THREE

To come back to my original caveat: is this just a lot of talk? Just as the people in conservation NGOs who do the lobbying are seldom the same as those putting ideas into practice in the field, the environment ministers who sign agreements are often at odds with ministries responsible for mining, agriculture and urban development. The latter are usually more powerful. Governments change and they have multiple conflicting priorities. Unforeseen events, like the COVID-19 pandemic, can create global changes. There is a long and depressing failure of international conservation and sustainable development targets to deliver and there will be plenty of forces working hard to ensure that the GBF goes the same way. The UN Sustainable Development Goals, an ambitious attempt to address all kinds of development problems over a decade, are on track for failure (SDSN and the Bertelsmann Stiftung 2019). The CBD admits that the Aichi targets were generally a failure (Secretariat of the CBD, 2020). Aichi 11, the equivalent of GBF Target three, was in fact the most successful. It broadly met its area target on land and was at least part of the reason for creation of many marine protected areas. But other aspects, such as effectiveness, connectivity and integration into the landscape or seascape, which were also included within Aichi 11, and these were much less successful. The 30 × 30 target is far more ambitious overall, emerging at a time when many parts of the world are already in chaos and where economic recession is widespread.

2.12   CONCLUSION

So how can we make this work? Money is important; small beer compared with spending on the military or energy production, but the sums needed today are much larger than any environmental investments in the past. The largest and most bitter arguments in Montreal were around money, with developing countries walking out of the negotiating chamber on two occasions. The commitments that were made are welcome, but total commitments are still far less than is needed and whether developed governments actually pay up is far from proven.

The other challenge will be to implement this level of conservation on such a large area in so relatively short a time. A recent analysis showed that there are currently nowhere near enough trained rangers to manage the existing protected area estate, let alone a major expansion (Appleton et al., 2022). Management of OECMs may be even more complex, given that the people involved will not usually have biodiversity conservation as a priority. Evidence of past success is no guarantee that this will continue into the future. The climate is changing and many societies are also under-going rapid development. The next generation may view their traditional lands and waters in very different ways. Ensuring—as far as is possible—longevity in OECMs will be a major hurdle. Participatory approaches to conservation are proven to be more effective than top-down approaches but they take time and a balance must be found between the urgency of conservation and the need to make sure that all stakeholders are comfortable with the direction of travel. Indigenous peoples and conservationists often have near identical aims, but there is also a long history of mutual suspicion that will not fade away overnight.

But there is no need to end on a depressing note. Despite the real challenges identified above there is also a welcome feeling of cautious optimism about the future amongst the conservation lobby. Notwith-standing shortages in terms of skills and people, there is already a huge cadre of dedicated conservationists, all the way from village level to the highest reaches of government. More companies are prepared to integrate biodiversity properly into their policies, there are more sources of revenue and, perhaps most important of all, a growing public recognition of the benefits from conservation in terms of a range of ecosystem services, cultural and aesthetic benefits and potential for financial revenues. The talking is finally over, now is the time to roll up our sleeves and get to work.

REFERENCES

Appleton, M. R., Courtiol, A., Emerton, L., Slade, J. L., Tilker, A., Warr, L. C., Malvido, M. Á., Barborak, J. R., de Bruin, L., Chapple, R., & Daltry, J. C. (2022). Protected area personnel and ranger numbers are insufficient to deliver global expectation. *Nature Sustainability, 5,* 1100–1110. https://doi. org/10.1038/s41893-022-00970-0.

CBD (Convention on Biological Diversity). (2018). Decision adopted by the Conference of the Parties to the Convention on Biological Diversity 14/8 Protected areas and other effective area-based conservation measures. Fourteenth Meeting, Sharm El-Sheik, Egypt, 17–29 November 2018, Agenda item 24. CBD/COP/DEC/14/18. https://www.cbd.int/doc/decisions/ cop-14/cop-14-dec-08-en.pdf. Accessed 13 April 2023.

CBD. (2020). Zero draft of the post-2020 Global Biodiversity Framework: Note by the co-chairs. CBD/WG2020/2/3, 6 January 2020. Secretariat of the CBD, Montreal, Canada. https://www.cbd.int/doc/c/efb0/1f84/a89 2b98d2982a829962b6371/wg2020-02-03-en.pdf. Accessed 13 April 2023.

CBD. (2022). Nations adopt four goals, 23 targets for 2030 in landmark UN biodiversity agreement. Secretariat of the Convention on Biological Diversity, Montreal. https://www.cbd.int/article/cop15-cbd-press-rel ease-final-19dec2022. Accessed 13 April 2023.

Chatterton, P., Ledecq, T., & Dudley, N. (Eds.). (2017). *Landscape elements: Steps to achieving sustainable landscape management.* WWF. https://wwfint. awsassets.panda.org/downloads/final_wwf_landscape_elements_09_11_i_1. pdf. Accessed 13 April 2023.

Cronin, D., Dancer, A., Long, B., Lynam, A., Muntifering, J., Palmer, J., & Bergl, R. (2021). Application of SMART software for conservation area management. In S. Wich & A. Piel (Eds.), *Conservation technology* (pp. 203–223). Oxford.

Dudley, N., Jonas, H., Nelson, F., Parrish, J., Pyhälä, A., Stolton, S., & Watson, J. (2018). The essential role of other effective area-based conservation measures in achieving big bold conservation targets. *Global Ecology and Conservation, 15,* e00424. https://doi.org/10.1016/j.gecco.2018.e00424.

Dudley, N., Stolton, S., Pasha, M. K. S., Sharma, M., Chapman, S., Roberts, J., Baltzer, M., Yap, W. L., Long, B., Yadav, S. P., & Gopal, R. (2020). How effective area tiger conservation areas at managing their sites against the Conservation Assured | Tiger Standards (CATS)? *Parks, 26*(2), 115–128. https://doi.org/10.2305/IUCN.CH.2020.PARKS-26-2ND.en.

Dudley, N., & Stolton, S. (2022a). *Target 3 of the Global Biodiversity Framework: What does 30x30 look like in practice? Briefing.* Equilibrium Research. http://www.equilibriumconsultants.com/upload/document/GBF_Target_3_briefing_-_December_2022.pdf. Accessed 13 April 2023.

Dudley, N., & Stolton, S. (Eds.). (2022b, October). *Best practice in delivering the 30x30 target* (2nd edn.). The Nature Conservancy and Equilibrium Research. https://www.nature.org/content/dam/tnc/nature/en/documents/TNC_UKDEFRA_30x30_BestPractices_Report.pdf. Accessed 13 April 2023.

Fa, J. E., Watson, J. E., Leiper, I., Potapov, P., Evans, T. D., Burgess, N. D., Molnár, Z., Fernández-Llamazares, Á., Duncan, T., Wang, S., & Austin, B. J. (2020). The importance of indigenous peoples' lands for the conservation of intact forest landscapes. *Frontiers in Ecology and the Environment.* https://doi.org/10.1002/fee.2148.

Franks, P., & Booker, F. (2022). *IUCN WCPA Technical Note Series No. 7: Equity in conservation—What, why and how?* IUCN.

Garcia-Moreno, J., Harrison, I. J., Dudgeon, D., Clausnitzer, V., Darwall, W., Farrell, T., Savy, C., Tockner, K., & Tubbs, N. (2014). Sustaining freshwater biodiversity in the Anthropocene. The global water system in the Anthropocene. In A. Bhaduri et al. (Eds.), *The global water system in the Anthropocene: Challenges for science and governance* (pp 247–270). Springer International Publishing.

Hilty, J., Worboys, G. L., Keeley, A., Woodley, S., Lausche, B., Locke, H., Carr, M., Pulsford, I., Pittock, J., White, J. W., & Theobald, D. M. (2020). *Guidelines for conserving connectivity through ecological networks and corridors.* Best Practice Protected Area Guidelines Series No. 30. IUCN.

Hockings, M., Stolton, S., Leverington, F., Dudley, N., & Courrau, J. (2006). *Evaluating effectiveness: A framework for assessing management effectiveness of protected areas* (2nd ed.). IUCN.

Hockings, M., Hardcastle, J., Woodley, S., Sandwith, T., Wilson, J., Bammert, M., Valenzuela, S., Chataigner, B., Lefebvre, T., Leverington, F., & Lopoukhine, N. (2019). The IUCN Green List of protected and conserved areas: Setting the standard for effective area-based conservation. *Parks, 25*(2), 57–65. https://doi.org/10.2305/IUCN.CH.2019.PARKS-25-2MH.en.

IIED (International Institute for Environment and Development). (2022). *Site-level Assessment of Governance and Equity (SAGE)* https://www.iied.org/site-level-assessment-governance-equity-sage. Accessed 13 April 2023.

IUCN-WCPA Task Force on OECMs. (2019). *Recognising and reporting other effective area-based conservation measures.* IUCN.

Lewis, N. A., Day, J. C., Wilhelm, A., Wagner, D., Gaymer, C., Parks, J., Friedlander, A., White, S., Sheppard, C., Spalding, M., & San Martin, G. (2017). *Large-scale marine protected areas: Guidelines for design and management.* Best Practice Protected Area Guidelines Series, No. 26. IUCN.

McArthur, R., & Wilson, E. (1967). *The theory of island biogeography.* Princeton University Press.

Newell, S., Nidhi, L., Doubleday, N., & Bloecker, A. (2019). The potential for Locally-Managed Marine Areas (LMAAs) as a participatory strategy for coastal and marine ecosystems—The global commons. *International Journal of Sustainable Development, 12*(4), 47–62.

Pimm, S., Jenkins, C., & Li, B. (2018). How to protect half of Earth to ensure it protects sufficient biodiversity. *Science Advances, 4,* eaat2616. https://doi.org/10.1126/sciadv.aat2616.

Ramsar. (2018). *Global wetland outlook: State of the world's wetlands and their services to people.* Ramsar Convention Secretariat.

SDSN (Sustainable Development Solutions Network) & Bertelsmann Stiftung. (2019). *Sustainable Development Report 2019: Transformations to achieve the Sustainable Development Goals.* SDSN. https://s3.amazonaws.com/sustainabledevelopment.report/2019/2019_sustainable_development_report.pdf. Accessed 13 April 2023.

Secretariat of the Convention on Biological Diversity. (2020). *Global Biodiversity Outlook 5.* SCBD.

Stolton, S., & Dudley, N. (Eds.). (2010). *Arguments for protected areas.* Earthscan.

Stolton, S., Dudley, N., Belokurov, A., Deguignet, M., Burgess, N. D., Hockings, M., Leverington, F., MacKinnon, K., & Young, L. (2019). Lessons learned from 18 years of implementing the Management Effectiveness Tracking Tool (METT): A perspective from the METT developers and implementers. *Parks, 25*(2). https://doi.org/10.2305/IUCN.CH.2019.PARKS-25-2SS.en.

The Nature Conservancy, Conservation International, IUCN World Commission on Protected Areas and WWF. (2022). *A pathway for inland waters in the 30x30 target: Discussion document.* IUCN.

Winkler, K., Fuchs, R., Rounsevell, M., & Herold, M. (2021). Global land use changes are four times greater than previously estimated. *Nature Communications, 12,* 2501. https://doi.org/10.1038/s41467-021-22702-2

Landscapes of the Romantic Sublime: The Legacy of Nineteenth-Century Artistic Visions and Contributions to the Development of the Management of Natural Heritage

Niall Finneran

3.1 INTRODUCTION

This chapter focuses upon how people think about places, or to be more precise, how from the late eighteenth and early nineteenth centuries, people took more notice of the landscapes that surrounded them and were motivated to investigate them, discover them and above all *sensed* them (Hess, 2012). This is really a story about embodiment, about developing a sense of place and engaging with it through all the human senses.

N. Finneran (✉)
Historical Archaeology and Heritage Studies, University of Winchester, Winchester, UK
e-mail: Niall.Finneran@winchester.ac.uk

© The Author(s) 2024
N. Finneran et al. (eds.), *Managing Protected Areas*,
https://doi.org/10.1007/978-3-031-40783-3_3

We can adopt here a philosophical term, and label this a phenomenological approach, a break from the Enlightenment-inspired Cartesian separation of mind and body (for a classic exposition of the idea see Bachelard, 1957; Merleau-Ponty, 1945). In short (and overlooking a great deal of philosophical and theoretical nuance), this idea posits that rather than wandering around space as an automaton, space/place reacts upon the body and the body reacts to space/place. It is a two-way relationship. As we shall see the phenomenological perspective allows for an interesting and provocative perspective on the relationship between people and places, emphasising the interplay of sense, movement and emotions (e.g. Taun, 1974 [1990]). Having accepted that there is an important, subtle, complex and fluid two-way relationship between people and place, it is axiomatic that people would feel the need to conserve and protect such places. This is why we have protected areas in all their diverse iterations.

The concept of managing natural landscapes in the UK has deep historical roots, although the accent has not always been on the appreciation of landscapes for their aesthetic and environmental meanings, nor indeed the notion that they should be widely accessible (for broad overviews see *inter alia* Crane, 2016: 273ff; Davies, 1996; Hoskins, 1955; Johnson, 2007; Millman, 1975; Rackham, 2020). If we were to take a view of medieval England for a start, we would see primarily feudal economic interests at work. Examples would include medieval urban burgage plots and rural tenures, parcels of land which carried monetary or service-based obligations within the feudal system and which have their roots in Anglo-Saxon defensive systems (Baker & Brookes, 2013). Here, the landscape was seen very much in terms as a mirror of the social order and an economic resource first and foremost, but perhaps some form of elemental and numinous belief in the power of place survived. The Anglo Saxons certainly felt fear and superstition about certain places within their landscapes (Semple, 2013). Of course, the Christian world view, hierarchical and structured, was all pervading and this governed perceptions of the environment and natural world. But this presupposes an old-fashioned structural dichotomy between nature and culture. In reality the concepts elide, and the so-called natural landscape could be controlled and encultured (managed, in fact).

Royal forests (such as the New Forest) were maintained primarily (as the French meaning of the word implies) as landscapes for hunting, solely by the King and aristocracy, and access to such lands, as well as severe penalties for crimes such as poaching there could be enacted

(Hooke, 2011; Young, 1978, 1979). People who lived in these areas, however, could be granted certain rights, such as pannage, for example, (allowing pigs to graze in areas of the forest) and Verderers Courts in the New Forest (Hants. UK) and Forest of Dean (Glos. UK) preserve today these ancient forms of land management. All the while through the later medieval period agricultural technologies developed and resources such as meat, crops and wool contributed to the wealth of southern and midland England in particular. Can we speak of a medieval environmental movement? In a sense protection of the natural world was something that was very much a concern of the elites, but not for the purposes of aesthetics, or indeed free access and social justice. The British and European landscapes of this period were hierarchical and controlled. We live with these legacies of ownership and control of access today (Hayes, 2020).

We can add in too the historical issues around common land and enclosure, and a general trend to a seizure of control of the countryside from the eighteenth century, with very little attempt to address the rights of non-land owning people (Wordie, 1983). Another dimension of economic land management prevalent up until the time of the dissolution of the monasteries in the 1530s by King Henry VIII, were the vast tracts of land held by monastic houses, such as Glastonbury and Westminster, as well as land held more generally by the Church. These lands were either absorbed by the Crown or transferred into the ownership of the Church of England (Heldring et al., 2021). Early religious-educational establishments such as the Universities of Oxford and Cambridge and their constituent colleges also established large estates and land holdings, along with the aristocracy. This Enlightenment conception of control extended to landscape, not just in terms of ownership, but physical manipulation of space, evidenced, for example, in the popularity of formalised and ordered gardens which were so popular in the seventeenth and eighteenth centuries, and themselves expressions of nascent capitalist ideologies (e.g. Leone et al., 2005).

By the time we reach the start of the nineteenth-century land management in England was even more in the hands of the elites and these rich agricultural and industrial holdings (discoveries of coal, e.g., could be very lucrative at the start of the Industrial Revolution) contributed to many fortunes. Factor in land holdings elsewhere in the Empire, particularly the sugar plantations of the Caribbean, and we can see the economic importance of landscape in maintaining the position of the wealthy elites. This is a very broad brush approach to a lot of history around the management of natural landscapes in England, and of course limitations on word counts prohibit a nuanced consideration of the wider picture in the British Isles and the rest of northern Europe, but it brings us to the starting point for this chapter: the artistic and cultural collision of nostalgia, aesthetics and anti-Enlightenment thinking that defines one of the major cultural currents of the late eighteenth–early twentieth century: the Romantic movement (Blanning, 2011).

This chapter does not seek to outline the history of the Romantic movement, but rather to see how the vision of romantically influenced writers, poets and painters influenced nineteenth-century thinking about natural landscapes and how they should, in the shadow of the Industrial Revolution and increasing urbanisation, be conserved and managed for the benefit of all. Our analysis takes us primarily through England, and the background to the emergence of the National Trust in 1895, but will also touch upon the contribution of Henry Thoreau in the United States and German writers and painters of the nineteenth century whose identity was not framed in terms of a nation state (Germany did not exist as a unified political phenomenon until 1871) but in terms of *Volkish* (folk) attachment to landscape, culture and language, as well as wider European approaches (e.g. Cosgrove, 2017).

3.2   ROMANTICISM AND NATURAL PLACES: A GLOBAL VIEW

The etymology of the term 'Romantic' refers primarily to the Romance languages and inheritors of the cultures of ancient Rome (Ferber, 2010, 4–5). The word acquires its current meaning within the context of heroic medieval literary 'romances', and from the eighteenth–nineteenth centuries is broadly conceived of a movement that if not quite taking on anti-Enlightenment attitudes (Ferber, 2010, 15), certainly critiques the ideas of science, certainty and rationalism. The Romantic movement, then, in wider terms is characterised by an emphasis on human-nature relations, the power of the natural world and its ability to provoke fear through being 'sublime' (Ferber, 2010, 72). The emphasis is thus on the primeval, the atavistic, and as Simon Schama has demonstrated in his 1995 important survey *Landscape and Memory* a focus on wild places, particularly mountains, lakes and deep forests such as 'the psychology of Gothic geology…brutally jagged rock pinnacles and unfathomably deep ravines' (Schama, 1995, 453).

The Romantic ethos is thus idealistic and emphasising human sensibility, here in the context of landscape, the triumph of human emotion and sympathy over reason. Natural landscapes provoke feelings of fear and beauty in us, and the reading of landscape takes on a more phenomenological and embodied turn. The celebration of the natural world is very much part of the Romantic position. From the eighteenth century with the popularity of the Grand Tour, the celebration of raw Alpine landscapes and southwards, contemplating the ruins of Roman civilisation became popular antiquarian pursuits. The Swiss philosopher Jean-Jacques Rousseau (1712–1778) records his impressions of crossing the Alps on foot in his *Confessions* (1781–1788), and soon he was followed by other writers, poets and artists who placed the Alps as the ideal Romantic landscape: wild, uncontrollable and primal (Barsham & Hitchcock, 2012; Scaramellini, 1996).

As the Napoleonic Wars closed off access to the European Alps in the early nineteenth century, English writers such as William Wordsworth (1770–1850), Samuel Taylor Coleridge (1772–1834) and Robert Southey (1774–1843) gathered in the English Lakeland and celebrated the landscapes there through poetry. A particular segment of verse from Wordsworth's poem 'The Prelude' first published in 1799 records the act of taking a boat and captures well the feeling of the sublime. He looks up at a mountain by the lake in the darkness and:

> Growing still in stature, the huge cliff, rose up between me and the stars, and still, with measured motion, like a living thing, strode after me.

Wordsworth in particular had an almost mystical attachment to landscape, covering prodigious distances on foot (Coverley, 2012, 101–109; Solnit, 2014, 104–117). The act of walking for him was a metaphysical engagement with the natural landscape, as it was albeit in an urban context for another contemporary Romantic visionary, William Blake. We can also ascertain this fixation of walking from the writings of Coleridge and another contemporary, William Hazlitt (1778–1830). Wordsworth was also probably the first to articulate the idea of management of natural landscapes for the good of the wider public, seeing the wider Lake District as a 'sort of national property, in which every man has a right and interest who has an eye to perceive and a heart to enjoy' (Wordsworth, 1835, 88). The act of walking, an embodied engagement with place, opened up these possibilities (Hall, 2014; Nichols, 2011).

In his overview of the work of the twentieth-century landscape historian W. G. Hoskins, the historical archaeologist Matthew Johnson starts his analysis with Wordsworth and his relationship with the landscape of the Lake District. Johnson defines the following themes which characterise Wordsworth's understanding of his landscape, and by extension those of the wider Romantic circle (Johnson, 2007, 25). For Johnson, Wordsworth's landscapes are (a) experienced in solitude; (b) viewed ('gazed') from above; (c) are rooted in aesthetics; (d) are translated into text (in this case his poetry, but also his guidebook to the Lakes); (e) embrace the spirit of place, the *genius loci* as well as representing 'Englishness'; (f) the landscape was saturated with traces of the past (in Johnson's words, 'travelling backwards'); (g) he engaged with it through the act of walking (a theme we return to again in this chapter); (h) this walking was an intuitive and embodied process; (i) the familiarity with the landscape was a process of many years' of training and finally (j) there is the implication that this means of engaging with landscape is socially restricted.

It was not just writers engaging with landscape. Romantic painters too sought to visually capture the fearsome and emotional qualities of the natural landscape. The term 'landscape' derives from a Dutch word, *Landschap* (meaning region, or area), and arrives in English during the sixteenth century primarily as an artistic term (Schama, 1995, 10), but Dutch landscape paintings of the sixteenth–seventeenth century were rigidly formal and realistic topographical executions, and in any case, the vogue for painting at this time was more towards heroic and mythological scenes and portraiture (Andrews, 1999, 129). A German painter, Casper David Friedrich (1744–1840), broke this mould with his 1818 composition *Wanderer above the Sea of Fog* (*Der Wanderer über dem Nebelmeer*) showing from the back, a man standing on a mountain peak contemplating the misty sublime landscape below, a communion of individual and the natural landscape, a theme that runs through all his paintings (Amstutz, 2020; Andrews, 1999, 143–144). Later, English landscape painters such as John Constable (1776–1837) and JMW Turner (1775–1851) redefine the English landscape painting tradition. For Constable his focus was mainly on the pastoral scenes of East Anglia rather than sublime mountains, for Turner it was capturing the abstract power of nature in all its force (Rees, 1982), but somehow capturing an indefinable element of 'Englishness' there in the landscape.

The Romantic ethos even influenced the design of gardens. From the mid-late nineteenth century the vogue for formalised gardens fell out of favour. The accent was now upon trying to evoke the naturalistic and informal, yet paradoxically this attempt to make gardens reflect natural spaces meant that this could only be achieved through a good deal of human intervention. The famous Victorian garden writer Gertrude Jekyll (1843–1932), is best known for promoting this approach, yet as we have noted embodies a paradox:

> Already in the nineteenth century, ecologically sensitive narratives clashed explicitly with anthropocentric ones, while also faced with the immense difficulty of implementing relations equally conducive to cultural and biotic health. Jekyll's conflicts, clearly, mirror those of her period. (Kehler, 2007, 619)

It is also pertinent to note that a very modernist technology, photography, also emerged in the mid-nineteenth century as a means of capturing elemental landscapes (DeLuca & Demo, 2000) and cementing the idea of place and national identity, particularly in the case of the disparate German-speaking nation states (Jäger, 2021). Landscape and language were unifying factors here. The *Völkisch* Movement which emerged in the nineteenth century drew inspiration from the relativist Romantic perspective (Hare & Link, 2019), urging a rekindling of interest in German folklore and traditions started by the Brothers Grimm, a back-to-nature celebration of German cultures and landscapes, particularly the primal forests (Imort, 2001, 2005), but one which soon soured and provided a great deal of impetus later on to Nazi perceptions of Germanic racial superiority. Nationalism and attachment to place is very much a theme which emerges in the nineteenth century. The nineteenth-century relationship between people and places was very much contextualised within the idea of a national pastoral image.

In the United States of America, the philosopher and essayist Ralph Waldo Emerson (1803–1882) wrote an influential manifesto for the American Romantic movement. His 1836 essay *Nature* outlined through his philosophy of transcendentalism the importance of the beauty of the natural world as a means for understanding the human condition (Emerson, 2017). This work influenced one of the most important nature writers of the nineteenth century, Henry David Thoreau (1817–1862). In July 1845, Thoreau embarked on a two-year solitary stay at Walden

Pond, Massachusetts, using his time to reflect on the natural world and its importance to humanity, publishing his experiences in the book *Walden, or Life in the Woods* in 1854 and provoking wider debates around conservation and environmentalism (Buell, 1995; Cafaro, 2002). Thoreau's ideas were embraced by a young Scottish Immigrant, John Muir (1838–1914), and although the two did not meet, it is clear that Thoreau had a significant influence on Muir's development as a nascent environmentalist (Fleck, 1985).

In the 1860s, Muir was an enthusiastic walker and found himself, after a series of adventures, in the rugged mountains of Yosemite, California, and a place that would be his spiritual and professional nirvana. Through the later part of the nineteenth-century Muir was a prime mover in environmental preservation, fighting the logging companies and seeking a means to develop legislation and management plans for Yosemite while in the true Romantic spirit emphasising the spiritual and metaphysical associations of place (Muir, 2001 ed.). This perspective put him at odds with the leading US conservationist Gifford Pinchot (1865–1946) who rather chose to emphasise long-term sustainable conservation as opposed to the unspoilt spiritual wilderness envisaged by Muir (Meyer, 1997). His legacy though can be found in the Sierra Club, an environmental social welfare organisation he founded in 1892, and the development of Federal natural heritage legislation. Although the first US National Park was designated at Yellowstone in 1872, an umbrella controlling agency, the US National Parks Service, did not come into being until the 1916 National Park Organic Act (Fox, 1985). The foregoing section has outlined some of the key elements of 'Romantic' thinking and writing about landscape using a broadly international perspective (see also McKusick, 2000). The focus of this chapter, however, is upon natural heritage management in the UK and how it responded to these cultural currents. In the next section, we consider in more detail how active strategies and policies for the management of natural heritage here developed.

3.3   THE CULTURAL CONTEXT OF THE MANAGEMENT OF NATURAL HERITAGE IN BRITAIN IN THE LATE NINETEENTH CENTURY

An appreciation of the aesthetic and spiritual qualities of natural land-scapes was a key leitmotif of the Romantic worldview of the eighteenth–nineteenth centuries, and it influenced a number of important cultural figures during this period. In English literature, the nature writings of Wiltshire-born Richard Jefferies (1848–1887), for example, tapped into this ethos, inspiring a generation of later nature writers and drawing attention to the threats posed to the countryside by rampant Victorian industrialisation. He articulated a vision of environmentalism that framed many of the later debates in the management of natural heritage (Morris, 2006; Welshman, 2011). He did not just deal with accounts of his walks through the countryside, but also produced post-apocalyptic fiction, such as *After London, or Wild England*, that dealt with the effects of societal collapse and the rewilding of the metropolis, a theme further developed, albeit within the context of a more overt radical, nostalgic and socialist setting in William Morris's 1890 utopian novel *News from Nowhere, or an Epoch of Rest* (O'Sullivan, 2011). In Morris's imagined future, responsible and sustainable stewardship of natural heritage is an important and complimentary component of his idealised socialist political society (Macdonald, 2004).

William Morris (1834–1896), poet, socialist, radical thinker, craftsman, writer and artist, was associated with one of the most significant Romantic cultural movements in Victorian Britain, the pre-Raphaelite brother-hood (MacCarthy, 1994). The pre-Raphaelites are today primarily known for their vibrant naturalistic paintings which embraced medievalist and biblical themes (Hilton, 1985), but they were also poets, writers and designers. The pre-Raphaelites had in their formative stages coalesced around one of the most significant Victorian cultural figures, John Ruskin (1819–1900). Ruskin had grown up in comfortable middle class surroundings and been exposed at an early age to Alpine landscapes as well as closer to home the Lake District, landscapes he celebrated in poetry and painting. After leaving Oxford, he wrote extensively on architectural heritage and conservation and in the 1850s fell in with the pre-Raphaelite artistic movement which focused on qualities such as aesthetics, beauty and a celebration of nature (Ruskin was also an influence on the development of responsible historical architectural conservation and preservation movements in the late nineteenth century and was a catalyst for the formation of the influential Society for the Protection of Ancient Buildings).

Ruskin also became a key advocate of what would now be called social justice, and to this end he established in 1871 the League of St George, a series of utopian and improving communities focused upon traditional crafts and conservation. His wider interests in preserving heritage, including open spaces, contributed greatly to the impetus behind the National Trust and influenced a wider group of 'back to nature' enthusiasts in the Lake District such as Susan Beever (1805–1893), Hardwicke Rawnsley (1851–1920) and William Gershom Collingwood (1854–1932) (Albritton & Jonsson, 2016). The League of St George still exists today as a charity, which continues to place an emphasis upon sustainability and self-sufficiency, but it was at the time, just one of many Victorian societies devoted to the conservation and preservation of open-air spaces for the benefit of all. Ruskin's work certainly influenced the career of another significant Victorian environmental campaigner, Octavia Hill (1838–1912) (Hewison, 2007, 57–66).

Octavia Hill grew up in a family which had strong social convictions. Social justice was an important part of her upbringing, and in her 20s started to agitate for the improvement of slum housing in London, and at the same time developed a campaigning interest in saving London's green spaces from development, and ensuring that access to these spaces could be guaranteed for all. Hill in fact was the first person to use the term 'Green Belt' as part of her (unsuccessful) struggle to ensure that the fields at Swiss Cottage remained un-developed (Kelly, 2022, 11ff; Whelan, 2009). Hill's sister Miranda Hill (1836–1910) was also cut from a similar campaigning cloth. In 1875/1876 she established the John Kyrle Society which commemorated the philanthropic career of Ross-on-Wye's John Kyrle (1637–1724), a man who devoted himself to providing clean water and public gardens for the inhabitants of the small Herefordshire town. The John Kyrle Society in its Victorian context placed great store upon the preservation of urban green space beyond Ross-on-Wye.

The John Kyrle Society was not the only Victorian social endeavour aimed at preserving and managing green spaces in late nineteenth-century Britain (Ranlett, 1983); the Metropolitan Public Gardens Association, founded in 1882, campaigned along similar lines as the John Kyrle

Society. The Lake District Defence Society, marshalled by the redoubtable Hardwicke Rawnsley, fought the expansion of railways into the area, while the Manchester Society for the Preservation of Ancient Footpaths sought to ensure the continuity of public access to landscapes. This group was founded as early as 1826 and its ethos foreshadows the famous and ground-breaking Kinder Scout Mass Trespass of April 1932. The Society for the Checking of Abuses of Public Advertising was concerned more with the aesthetics of the countryside rather than issues around access, and they campaigned against the unsightly encroachment of advertising signage along the routes of railways as they cut through the countryside (Readman, 2001).

The Commons Preservation Society was one of the largest of the environmentalist groups and was founded in 1865 with the purpose of combatting, as its name suggests, the development of the main London urban and peri-urban commons (Cowell, 2002). It survives in the twenty-first century as the Open Spaces Society. Robert (later Sir Robert) Hunter (1844–1913) acted as the main legal advisor to the Commons Preservation Society and developed strong and enduring links with Octavia Hill, as well as Hardwicke Rawnsley in the Lake District. It soon became apparent that a more coherent approach to the preservation and management of England's natural spaces would be required, and in 1895, after a period of negotiation and planning, Hill, Hunter and Rawnsley formally established the National Trust for Places of Historic Interest or Natural Beauty.

The National Trust thus has its roots in a number of Victorian civic societies and groupings that were dedicated to the preservation of landscapes in the face of encroaching industrialisation but also placed social justice and wellbeing at the heart of their mission. In its formative years, the National Trust was very much a campaigning organisation, and over the following decades it rather lost that ethos and in the mid-twentieth century (perhaps unfairly) was regarded as being somewhat elitist and out of touch (Chapter 4 this volume). The National Trust was very much of its time as well, and must be viewed within the wider Victorian socio-cultural context too. The debt to the work of Ruskin is clear, and its mission statement and ethos mirrors other legislative and societal developments in the period which focused on the protection of natural and cultural heritage assets more widely (we have noted the establishment of the Society for the Protection of Ancient Buildings above, but this was also the period when the first meaningful legislative protection of archaeological sites, the Ancient Monuments Act of 1882 was enacted).

The first gift to the National Trust in 1895 was a parcel of landscape, five acres of clifftop at Dinas Oleu, Barmouth. In 1899 the Trust then purchased two acres of land at Wicken Fen in Cambridgeshire establishing for the first time under their aegis a nature reserve. The significance of Wicken Fen related not just to its biodiversity, but as a representative landscape of one of the last pieces of undrained fenland. Surrounded on all sides by shrunken, drained fenlands, shaped into vast agricultural fields, this small oasis of relic swampland had long attracted botanists and scientists from the University of Cambridge. Now there are approximately 629 acres under management, and the site is designated as a Site of Special Scientific Interest, a National Nature Reserve, and internationally recognised as a Ramsar Wetland Site of international importance (National Trust, 2023).

Today, the National Trust is the premier guardian of natural and cultural heritage in England and Wales. The impact of the COVID-19 pandemic on its finances in the face of falling footfall and visits by foreign tourists demonstrated what an important part of the leisure tourism landscape in the UK it is. And, as is noted in another Chapter 4 its attempts to be more socially inclusive have recently aroused the ire of a certain segment of the media and political class in the UK. Yet this is precisely what its Victorian creators envisaged: an organisation that placed social justice and access to the countryside for all at the heart of what it did. This is the theme that runs through the Victorian approaches to people and places, and as we shall see in the next section, this is not an idea that has run out of time. The Romantic spirit is still with us.

3.4   Natural Heritage and the Legacy of the Romantic Movement

> Landscape has been, and is, endlessly and thoughtfully trivialised by people who suppose it is enough to paint views....the land is the foundation on which everything stands; the ground of all action and feeling. (Neve, 2020, 12)
> The landscape commits suicide every day. (Neve, 2020, 13)

The Romantic movement of the late nineteenth–nineteenth centuries was an important artistic, cultural and social phenomenon, one which made us reassess our relationship with the natural world and how we managed it. This is not a phenomenon associated solely with Britain, but as we have seen impacted upon European and North American ways of thinking about green and blue spaces. As we have seen, this 'view from above' in the words of Mathew Johnson (2007, 22) and associated very much with Wordsworth's relationship with the landscapes of the Lake District in northern England had a profound effect on our organisation of space: 'one of the greatest achievements of the Romantics...was the development of an environmental sensibility' (Johnson, 2007, 23). We have considered here the role of the Romantic movement in influencing the management of natural (and indeed cultural) heritage from the late eighteenth century through the nineteenth century against a backdrop of radical, socialist thinking, nationalism, a desire for equality of access, an accent upon the aesthetic and some degree of nostalgia (Hill, 2021, 3). But the legacy can still be felt into the twentieth and twenty-first centuries.

In terms of environmental policy development, the influence of nineteenth-century thinking about our relationship with place has survived. More protected areas were defined under a range of different levels of legislation. The Council for the Protection of Rural England (CPRE) was established in 1926 under the presidentship of Sir Patrick Abercrombie (1879–1957), a town planner who ironically forged a reputation in his post-World War 2 British urban replanning for unsympathetic redevelopment of historical townscapes and the development of the concept of 'new towns'. The CPRE emerged with the support of among others the National Trust and Commons Preservation Society. It led the way in supporting the establishment of Green Belts and National Parks, culminating in the 1949 National Parks and Access to the Countryside Act which ultimately led to the establishment of National Parks, Areas of Outstanding Natural Beauty and National Nature Reserves. Human relations to place remained central to artistic expression.

In the twentieth century, painters still sought to capture the essence of the English landscape (Neve, 2020). Writing about place and identity became popular through the twentieth century, and not just from the perspective of historical-geographical studies like the work of W. G. Hoskins, a historian who as Johnson has demonstrated (2007) provides the foundation for historical archaeological studies of the evolution of the British landscape, but who is also in a sense carrying the romantic attachment to a sense of place and quintessential Englishness. Through the late 1950s, publishing houses such as Batsford and Robert Hale and motoring companies such as Shell and the Automobile Association produced regional touring guides capturing the essence or sense of place of discrete parcels of British landscape. Into the 1960s the outer esoteric fringes of the place writing genre (such as John Michell) sought to tap into the 'earth mysteries' and the mystical elements of the British landscape, recalling the visions of William Blake in the early nineteenth century. In the twenty-first century, writers like Robert Macfarlane, Phillip Marsden and (in a more urban environment) Iain Sinclair challenge our notions about place and placehood. We seem, thanks to the romantic perspective, to have reclaimed the sense of the genius loci, or spirit of place, in our landscape.

Writing in the aftermath of the COVID-19 pandemic, it has become clear to this author (who essentially comes from a landscape-archaeology intellectual background) that we have reclaimed a deep relationship to our familiar places. The lockdown forced us to investigate afresh our very local environs, through the medium mainly of walking. The end of lockdown saw a huge influx of visitors into the countryside (Chapter 14). We started to reconnect with our sense of place again and perhaps started to sense the more esoteric and numinous elements of our environments, which have been there all along, but perhaps something which is forgotten in capitalist Western societies (Abram, 1996). In this connection perhaps we could finish with reference to a much abused term in modern cultural and literary circles, and one which informs a great deal of contemporary place writing but which has impacted hugely on the way we react to space and place around us: psychogeography.

Psychogeography began life as a 1950s French outgrowth of the surrealist and anarchic Situationist movement. Under its 'founder' Guy Debord, it set out with a manifesto to reclaim space, in this case urban space, through the act of guerrilla walking, but as Merlin Coverley has demonstrated, the currents of psychogeography can be traced back in the Anglophone tradition as far as Daniel Defoe, arriving at Iain Sinclair via William Blake, Thomas de Quincey and Arthur Machen inter alia. We can also see reference points in the concept of the *flâneur* (urban wandered) of nineteenth-century Paris, immortalised by Charles Baudelaire and celebrated in Walther Benjamin's monumental *Arcades* project in the 1940s (see Coverley, 2010 for a comprehensive overview). It may now be a well-worn trope, but essentially psychogeography is all about the way we react emotionally to place, and place reacts emotionally to us, and it is this intangible value which ought to be at the heart of any policy development relating to the relationship between people and place. This was understood by the Romantic writers over two hundred years ago and needs to be remembered today.

References

Abram, D. (1996). *The spell of the sensuous: Perception and language in a more-than-human world*. Pantheon.

Albritton, V., & Jonsson, F. (2016). *Green Victorians: The Simple Life in John Ruskin's Lake District*. Chicago University Press.

Amstutz, N. (2020). *Caspar David Friedrich: Nature and the self*. Yale University Press.

Andrews, M. (1999). *Landscape and western art*. Thames & Hudson.

Bachelard, G. (1957). *La Poétique de l'Espace*. Presses Universitaires de France.

Baker, J., & Brookes, S. (2013). *Beyond the Burghal Hidage: Anglo-Saxon civil defence in the Viking Age*. Brill.

Barsham, D., & Hitchcock, M. (2012). 'Prophets of nature': Romantic ideals of nature and their continuing relevance for tourism today. In A. Holden & D. Fennell (Eds.), *The Routledge handbook of tourism and the environment* (pp. 76–86). Routledge.

Blanning, T. (2011). *The romantic revolution*. Weidenfeld & Nicolson.

Buell, L. (1995). *The environmental imagination: Thoreau, nature writing, and the formation of American culture*. Harvard University Press.

Cafaro, P. (2002). Thoreau's environmental ethics in 'Walden.' *The Concord Saunterer (The Journal of the Thoreau Society), 10*, 17–63.

Cosgrove, D. (2017). Cultural landscapes. In T. Unwin (Ed.), *A European geography* (pp. 65–81). Routledge.

Coverley, M. (2010). *Psychogeography*. Pocket Essentials.

Coverley, M. (2012). *The art of walking: The writer as walker*. Oldcastle Books.

Cowell, B. (2002). The Commons Preservation Society and the campaign for Berkhamsted common, 1866–70. *Rural History, 13*(2), 145–161. https://doi.org/10.1017/S0956793302000080

Crane, N. (Ed.). (2016). *The making of the British landscape*. Weidenfeld & Nicolson.

Davies, J. (1996). *The making of Wales*. Sutton.

DeLuca, K., & Demo, A. (2000). Imaging nature: Watkins, Yosemite, and the birth of environmentalism. *Critical Studies in Media Communication, 17*(3), 241–260. https://doi.org/10.1080/15295030009388395

Emerson, R., (Eds. M. Branch & C. Mohs). (2017). *The best read naturalist: Nature writings of Ralph Waldo Emerson*. University of Virginia Press.

Ferber, M. (2010). *Romanticism: A very short introduction*. Oxford University Press.

Fleck, R. (1985). *Henry Thoreau and John Muir among the Indians*. Hamden Books.

Fox, S. (1985). *The American Conservation Movement: John Muir and his legacy*. University of Wisconsin Press.

Hall, D. (2014). *Romantic naturalists, early environmentalists: An ecocritical study, 1789–1912*. Routledge.

Hare, J., & Link, F. (2019). The idea of *Volk* and the origins of *Völkisch* research, 1800–1930s. *Journal of the History of Ideas, 80*(4), 575–596. https://doi.org/10.1353/jhi.2019.0032

Hayes, N. (2020). *The book of trespass: Crossing the lines that divide us*. Bloomsbury.

Heldring, L., Robinson, J., & Vollmer, S. (2021). The long-run impact of the dissolution of the English monasteries. *The Quarterly Journal of Economics, 136*(4), 2093–2145. https://doi.org/10.1093/qje/qjab030

Hess, S. (2012). *William Wordsworth and the ecology of authorship: The roots of environmentalism in nineteenth-century culture*. University of Virginia Press.

Hewison, R. (2007). *John Ruskin*. Oxford University Press.

Hill, R. (2021). *Time's witness: History in the age of romanticism*. Allen Lane.

Hilton, T. (1985). *The pre-Raphaelites*, edition. Thames & Hudson.

Hooke, D. (2011). Royal Forests-Hunting and other forest use in medieval England. In E. Ritter & D. Dauksta (Eds.), *New perspectives on people and forests* (pp. 41–59). Springer.

Hoskins, W. (1955). *The making of the English landscape*. Hodder & Stoughton.

Imort, M. (2001). *Forestopia: The use of the forest landscape in naturalizing National Socialist ideologies of Volk, race, and Lebensraum, 1918–1945 (Germany)* (Unpublished PhD thesis). Queens University, Kingston, Canada. Available online at: https://central.bac-lac.gc.ca/.item?id=NQ54061&op=pdf&app=Library&oclc_number=1006913893. Accessed 30 June 2022.

Imort, M. (2005). A sylvan people: Wilhelmine forestry and the forest as a symbol of Germandom. In T. Lekan & T. Zeller (Eds.), *Germany's nature: Cultural landscapes and environmental history* (pp. 55–80). Rutgers University Press. https://doi.org/10.36019/9780813537702-005

Jäger, J. (2021). Picturing nations: Landscape photography and national identity in Britain and Germany in the mid-nineteenth century. In J. Schwartz & J. Ryan (Eds.), *Picturing place* (pp. 117–140). Routledge.

Johnson, M. (2007). *Ideas of landscape*. Blackwell.

Kehler, G. (2007). Gertrude Jekyll and the late-Victorian garden book: Representing nature-culture relations. *Victorian Literature and Culture, 35*, 617–633. https://doi.org/10.1017/S1060150307051674

Kelly, M. (2022). *The women who saved the English countryside*. Yale University Press.

Leone, M., Harmon, J., & Neuwirth, J. (2005). Perspective and surveillance in eighteenth-century Maryland gardens, including William Paca's Garden on Wye Island. *Historical Archaeology, 39*, 138–158. https://doi.org/10.1007/BF03376708

Macdonald, B. (2004). William Morris and the vision of ecosocialism. *Contemporary Justice Review, 7*(3), 287–304. https://doi.org/10.1080/1028258042000266013

MacCarthy, F. (1994). *William Morris: A life for our time*. Faber.

McKusick, J. (2000). *Green writing: Romanticism and ecology*. Palgrave/St. Martin's Press.

Merleau-Ponty, M. (1945). *Phénoménologie de la Perception*. Editions Gallimard.

Meyer, J. (1997). Gifford Pinchot, John Muir, and the boundaries of politics in American thought. *Polity, 30*(2), 267–284. https://doi.org/10.2307/3235219

Millman, R. (1975). *The making of the Scottish landscape*. Collins.

Morris, B. (2006). *Richard Jefferies and the ecological vision*. Trafford Books.

Muir, J. (2001). *The wilderness world of John Muir*. Houghton Mifflin Harcourt.

National Trust. (2023). *Wicken Fen*. https://www.nationaltrust.org.uk/visit/cambridgeshire/wicken-fen-national-nature-reserve. Accessed 13 April 2023.

Neve, C. (2020). *Unquiet landscape: Places and ideas in 20th-Century British Painting*. Thames & Hudson.

Nichols, A. (2011). *Beyond romantic ecocriticism: Toward urbanatural roosting (nineteenth-century major lives and letters)*. Palgrave Macmillan.

O'Sullivan, P. (2011). ¡ Homenaje a Aragón!: News from Nowhere, collectivisation, and the sustainable future. *Journal of William Morris Studies, 19*(3), 93–111.

Rackham, O. (Ed.). (2020). *The history of the countryside*. Weidenfeld & Nicolson.

Ranlett, J. (1983). 'Checking nature's desecration': Late-Victorian environmental organization. *Victorian Studies, 26*(2), 197–222.

Readman, P. (2001). Landscape preservation, 'advertising disfigurement', and English National Identity c. 1890–1914. *Rural History, 12*(1), 61–83. https://doi.org/10.1017/S0956793300002272.

Rees, R. (1982). Constable, Turner, and views of nature in the nineteenth century. *Geographical Review, 72*(3), 253–269. https://doi.org/10.2307/214526

Scaramellini, G. (1996). The picturesque and the sublime in nature and the landscape: Writing and iconography in the romantic voyaging in the Alps. *GeoJournal, 38*(1), 49–57. https://doi.org/10.1007/BF00209119

Schama, S. (1995). *Landscape and memory*. Fontana Press.

Semple, S. (2013). *Perceptions of the prehistoric in Anglo-Saxon England: Religion, ritual, and rulership in the landscape*. Oxford University Press.

Solnit, R. (2014). *Wanderlust: A history of walking.* Granta.

Taun, Y-F. (1974 [1990]). *Topophilia: A study of environmental perception, attitudes, and values.* Columbia University Press.

Welshman, R. (2011). Literature and the ecological Imagination: Richard Jefferies and DH Lawrence. *Victorian Network, 3*(1), 51–63. https://doi.org/10.5283/vn.20

Whelan, R. (2009). Octavia Hill and the environmental movement. *Civitas Review, 6*(1), 1–8. https://www.civitas.org.uk/pdf/CivitasReviewApril2009.pdf. Accessed 18 April 2023.

Wordie, J. (1983). The chronology of English enclosure, 1500–1914. *Economic History Review, 36*(4), 483–505. https://doi.org/10.1111/j.1468-0289.1983.tb01244.x

Wordsworth, W. (1835). *A guide through the district of the lakes in the North of England with a description of the scenery & c. for the use of tourists and residents* (5th ed.). Hudson & Nicholson.

Young, C. (1978). Conservation policies in the royal forests of medieval England. *Albion, 10*(2), 95–103.

Young, C. (1979). *The royal forests of medieval England.* University of Pennsylvania Press.

CHAPTER 4

Islandscapes: Tourism, COVID-19, Climate Change and Challenges to Natural Landscapes. A Caribbean Perspective and View from Barbados

Niall Finneran and Tara Inniss

4.1 Introduction

The Martinique-born Francophone author Édouard Glissant (1928–2011) in his 1958 work *La Lézard* (*the Ripening*) evokes a rich and powerful portrait of his Caribbean island home. For him his island is a feminine form, and replete with natural environmental metaphors detached from the colonial European plantation system (Heller, 1996). Islands lend themselves, as bounded physical entities, to characterisation in this manner (Hay, 2006). In the Caribbean, there has always been a tension between the idealised innocent 'pre-colonial' tropical Eden and

N. Finneran (✉)
University of Winchester, Winchester, UK
e-mail: Niall.Finneran@winchester.ac.uk

T. Inniss
The University of the West Indies (UWI), Cave Hill Campus, Barbados
e-mail: tara.inniss@cavehill.uwi.edu

© The Author(s) 2024
N. Finneran et al. (eds.), *Managing Protected Areas*,
https://doi.org/10.1007/978-3-031-40783-3_4

then the rigidly controlled industrial plantation landscapes of the sugar monocultural systems (Hollsten, 2008). Then, in the post-plantation and post-colonial Caribbean world, island economies re-orientated themselves towards a new industry: tourism. With this came a reorganisation of island landscapes yet again, with the coastlines now the focus of hotel development and the countryside, historically the place of the plantation, largely becoming a rural backwater. With this move towards a tourist-based economy came other impacts on the island: infrastructure to service the beach resorts, more roads, bigger airports and leisure places—and a host of associated problematic social and economic issues that even in the post-colonial Caribbean evidence the survival of dependent colonial attitudes (Pattullo, 2005).

Another impact of this new reorientation of island space was the commodification of the remaining green spaces on the islands, preserved now as heritage zones. In Dominica, for example, this is done as a nod to the politics of the traditional land rights of the indigenous Kalinago ('Carib') people (Hudepohl, 2008). In the Windward Carib territory of Dominica, visitors can come and see traditional Kalinago architectural forms, taste traditional food and buy traditional and authentic Kalinago crafts. The sense that this is an elemental and ancestral landscape is stressed clearly, and it fits well with the way that Dominica markets itself as an eco-tourism destination: a place of waterfalls and tropical rain forests rather than a traditional Caribbean beach holiday (Patterson et al., 2004; Slinger-Friedman, 2009).

Similarly, Dominica is home to one of the few UNESCO Natural World Heritage properties in the Caribbean region, Morne Trois Pitons, which was inscribed in 1997. The Morne Trois Pitons National Park protects 6,857 hectares which is roughly nine per cent of Dominica's land area. Its Outstanding Universal Value (OUV) rests in the conservation of its exceptional volcanic landscape including Boiling Lake, fumaroles, hot springs and mud holes. The National Park is also known for its biodiversity and as the only remaining refuge for the endemic critically endangered Imperial or Sisserou Parrot and vulnerable Red-Necked Parrot. A number of trails in the park are maintained and provide locals and visitors with a wealth of hiking options on the island. However, Dominica is prone to natural disaster with hurricane and volcanic hazards. In 2017, category five hurricanes Irma and Maria affected the Leewards and northern Windward Islands within days of one another. With 60 per cent of the island covered in forest, many areas were left with completely defoliated trees.

Although the natural environment has recovered, there are estimates that it will take a decade for the forests to fully recover. The threat of natural disaster is yet another challenge in the protection and conservation of the region's natural heritage, especially given the extreme weather events and disasters associated with climate change (ECPA, 2021).

Other islands have taken more international approaches to the management of their natural heritage. There are 19 UNESCO World Heritage Sites in the insular Caribbean. These sites have been defined as being having importance due to their natural heritage, cultural heritage or exhibit mixed values that make them significant. Cuba leads the way in terms of designations with nine, of which seven are cultural WHS (designated mainly by dint of their rich colonial architecture) and two are natural WHS, the National Parks of Desembarco del Granma (designated in 1999) and of Alejandro Humboldt (designated in 2001; see https://whc.unesco.org/fr/etatsparties/cu) which represent distinctive geological landscapes and eco-diversity. The important wetland site of Ciénaga de Zapata National Park and the inshore Cuban reef systems are both on the current tentative list for future inscription. More widely in the Caribbean UNESCO natural world heritage properties are represented by the Pitons in St Lucia and Morne Trois Pitons in Dominica, and tentative sites for future inscription include Inagua National Park, Bahamas, Scotland District in Barbados, Morne Diablotin National Park and Scotts Head-Soufriere Marine Park in Dominica, the Volcanic Areas of Martinique, the National Marine park of Bonaire, the La Brea Pitch Lake and the Main Ridge Forest Reserve in Trinidad and Tobago and the entirety of the British-administered islands of the Turks and Caicos. There is a single mixed natural and cultural heritage WHS in the region: the Blue and John Crow Mountains in Jamaica, inscribed in 2015.

One of the main attractions of UNESCO World Heritage property designations is the perceived potential positive impact it has on tourism. Inscription marks out a natural, cultural or mixed site as being globally significant and thus a place to visit and experience (Adie, 2017; Adie et al., 2018). The well-known branding of the designation can be used to promote the property, but also introduces stringent management demands and carries no direct financial input from UNESCO for site management (e.g. Ryan & Silvanto, 2010; Smith, 2002). There are, for example, implications for planning control in World Heritage properties and continuing debates around conservation strategies and development. In the context of the Caribbean islands, most of the focus has been on designating cultural heritage sites, but increasingly there is also pressure to inscribe natural heritage sites. Another important factor to note has been the gradual reorientation of Caribbean tourist focus away from beach holidays to more cultural or natural heritage experiences (Scher, 2011). This shift in tourist expectations/supply became increasingly evident during a workshop led by one of the present authors (NF) for selected heads of Caribbean tourism organisations in April 2020. This event centred around recognition for and promotion of heritage assets and was further reinforced by the results of a recent study commissioned by EUDiF (the European Union Global Diaspora Facility) in 2022 which examined attitudes towards cultural tourism by the Barbadian diaspora (Dickinson et al., 2022).

Heritage tourism then has more than implications for leisure and enjoyment; there are also significant educational benefits as well as offering potential to confront topical historical and political issues such as slavery, violence and colonialism (Fortenberry, 2021). Heritage tourism also diversifies the Caribbean tourism portfolio, which will be an important issue in terms of economic recovery post-COVID-19 pandemic. Island tourist boards will need to develop new, more exciting and stimulating products to sell, and natural heritage attractions will play a key part in this. Yet paradoxically, as has been reported in many other regions worldwide, the success of these strategies may cause damage to the very assets being promoted through increased development and footfall, and in addition the biggest issue of all which comes with increased visitation: impacts of climate change. Having sketched in the broad contextual picture, we now consider a series of case studies drawn from the natural heritage portfolio of the Caribbean island of Barbados.

Heritage Tourism and Barbados
Barbados has recognised the potential to develop its heritage tourism sector to help diversify the island's tourism product. This was even more apparent during the COVID-19 pandemic when access to beach spaces and other public green spaces were at a premium with the pandemic protocols which sometimes limited their use. Moreover, 'Welcome Stampers' (long-stay visas for persons seeking to relocate to Barbados to live while working remotely) also desired more access to green space and the promise of more active lifestyles. The heritage sector has a long history in Barbados with several sites valued for their natural heritage stretching

back into the colonial era with a nascent tourism industry as early as the eighteenth century. Certainly, a number of sites in the island have been leisure spaces for locals and visitors for quite some time including the island's cave systems, gullies, East Coast and Animal Flower Cave. Some of these sites are protected under the Barbados National Trust founded in 1961, while others are either owned or operated privately or by government. Increasingly, UNESCO WHS branding is playing a more important part in the development of the heritage sector. Barbados currently has one UNESCO (cultural) WHS, Historic Bridgetown and the Garrison, which was inscribed in 2011. Essentially, this comprises a core of historic colonial buildings in the island's capital city linked to the Garrison which fortified the city along the popular and scenic coastal Bay Street corridor. Two other properties are on the tentative list include the Industrial Heritage of Barbados: The Story of Sugar and Rum which is a serial nomination of plantation sites relating to the development of the sugar and rum industries; the second is a natural heritage site, Scotland District on the east side of the island. Here we find the main mountainous area of the island, it is relatively sparsely settled and not a great draw for tourists who tend to stay on the south and west coasts of the island, although there is a growing niche segment of the tourism market that is attracted to the eco-tourism attractions on the east coast of the island in boutique hotels specialising in surf vacations and health and wellness tourism.

4.2   The Tourist and Barbados

The island of Barbados is situated in the eastern Caribbean on the Atlantic side of the Windward chain (Fig. 4.1). With a surface area of 430 km^2, it is one of the larger islands of the Lesser Antilles and differs from its mountainous volcanic neighbours to the west in the Windwards in having a flatter limestone topography with less forest coverage. Over 300,000 people live on the island, the vast majority being descendants of enslaved Africans, and in Caribbean terms is relatively densely populated. Ruled by England since the seventeenth century, independent from the United Kingdom since 1966, and a Republic since 2021, the island still retains a distinctive cultural identity within the wider Caribbean context. A service-based economy, relying mostly on international tourism, makes up 89 per cent of Barbados' economic output according to the latest available 2017 figures, but the COVID-19 pandemic (2020–2022) had a negative

Fig. 4.1 Map of the study region (Niall Finneran)

impact on this sector and the island is now in a state of economic recovery (source for all statistics World Factbook, 2022).

Historically, Barbados has always drawn large numbers of British tourists based on the island's historic connections with the UK. Prior to 2022 there were at least two daily direct flights from the UK. The island has been perceived to offer a familiar, safe experience for the British visitor. North American visitors have also made up a significant segment of Barbados' tourist cohort. Increasingly, the value of the Barbadian Diaspora has been recognised as a source market for tourism as emigrant Bajans and their descendants travel to and from the island regularly. Emerging tourism markets such as Africa, South America and Asia are also being explored. The regional tourism market, however, has not recovered post-COVID-19 with the failure of a number of regional air carriers and the high cost of regional travel. The island has been traditionally marketed for leisure, fun and as an ideal beach holiday destination (Jönsson & Devonish, 2008). Post-COVID-19 the island was marketed as a destination for digital nomads and remote workers under its 'Welcome Stamp' programme which issued long-stay visas for persons seeking to relocate to Barbados to live while working remotely. Barbados had maintained relatively low infection rates due to its border restrictions and pandemic protocols. The programme was considered a success especially at the height of the pandemic when various northern jurisdictions were still undergoing lockdowns.

In comparison with other Caribbean islands, in terms of heritage, Barbados has a wealth of historic cultural heritage which reflects the colonial plantation experience: fortifications, large plantation houses, cemeteries and also remains of enslaved Africans' dwellings. In terms of natural heritage, however, there are no dramatic volcanic mountain ranges here, nor are there tropical rain forests, and for the most part (with the exception of the eastern Atlantic coasts) the western and southern coastlines are heavily developed. However, Barbados does have a unique limestone karst topography and is well known in geological circles for its cave systems and gullies found mostly in the interior of the island and for the unique geological formations on the East Coast. With the exception of Welchman Hall Gully and Harrison Cave, Barbados' geology is not a huge singular attraction for visitors.

The coastline with its pristine white sand beaches is perhaps Barbados's biggest potential draw, but there are a number of important issues with which to contend. There are huge pressures on beaches as leisure resources. Unlike in some other Caribbean territories, beaches are public and accessible to all (NCC Barbados, n.d.). As such, hotel developers cannot claim areas for the exclusive use of their own guests (Davis, 2010) beyond what is commonly known as 'the high water mark' (Allahar, 2015). However, across the Caribbean privatisation of beaches and private control of public land have led to accusations of high-handed neo-colonial behaviour by exclusive resorts which do not typically cater to the average Caribbean citizen (Caribbean Council, n.d.). The work of the Barbadian artist Annalee Davis (cited above) is part of a wider initiative around race and class in Barbados. Here the beach has become a politicised battle-ground, where issues of public access open up deep wounds from the colonial period.

Historically, beaches were considered 'rab' or marginal land on coastal plantations (Davis, 2010). Beaches were spaces used by fishers for boat-yards. They were places where raw sewage and refuse were disposed and animals were washed. Often latrines and abattoirs were sited in close prox-imity to the sea so beaches were by no means considered high-value real estate. A number of cemeteries dating to the 1854 cholera epidemic are also located on or near beach accesses. In fact, beachfront land especially on the south and west coasts were the preserve of tenantries and working-class housing since workers' access to land in Barbados was often limited due to the strict land-labour policies that followed emancipation. Planters and the elite remained in control of the majority of land resources on the island for most of the nineteenth century until the mid-twentieth century. However, with the advent of beach tourism by the 1950s and 1960s, beachfront land started to become highly prized and eventually land speculation for this resource priced many Barbadians out of the beachfront real estate market (Allahar, 2015).

Public beach access remains a 'hot-button' issue for Barbadians. Further privatisation of sea frontage represents their continued alienation from prime real estate which may now be owned by the local elite and foreign investors (Allahar, 2015). It is defended vociferously and if attempts are made to limit beach access they are met with widespread public protest. The outrage expressed over threats to public access of the island's beaches are best memorialised in the popular 1982 calypso 'Jack' by The Mighty Gabby which recounts when the then Chair of the Barbados Board of Tourism Jack Dear wanted to initiate beach privatisation policies. Defiantly, The Mighty Gabby sang in opposition to the measures saying:

> Tourism vital I can't deny but can't mean more than I an' I
> My navel string bury right here but a tourist own could be anywhere
> Jack don't want me to bathe on my beach
> Jack tell them to keep me out of reach
> Jack tell them I would never make the grade
> Strength and security build barricade
> That can't happen here in this country
> I want Jack to know the beach belong to we.

The song has become a protest anthem and is often incanted when public beach access is threatened. In March 2018, The Mighty Gabby performed the song on the popular Crane Beach when the Crane Beach Resort was accused of limiting vendor access to the beach (Evanson, 2018).

Increasingly, managing beach access creates conflict among several different users. Not only does the public advocate for their right to access, so do vendors and tourism operators for commercial purposes. Pre-COVID-19, there have been conflicts over beach chair rentals and commercial concessions that take up beach space creating conflicts among all users (Evanson, 2018). During the height of the COVID-19 lockdowns, the beaches were almost entirely cleared of tourist activity and associated vendor activities (Graham, 2021). Beaches, once again, became the preserve of Barbadians and local communities who used them for daily recreation within stipulated hours of use. They became a highly valued respite from stringent lockdown orders which confined most Barbadians to their homes. Post-COVID-19, tourism activity has returned, to some extent, and so have the conflicts.

Beyond beaches, there are other issues facing the seascapes of Barbados. At Folkestone Marine Reserve near Holetown on the north-western coast, the once bright and colourful coral reefs have become bleached owing to pollution and poor water quality and warming seas (Oxenford et al., 2008). Discharge of sewerage into the sea causes development of toxic algal blooms which are damaging to coral. This is an important site as it was the first marine park to be established in Barbados in 1976 under the aegis of the *Marine Areas Preservation and Enhancement Act* (1976). From the outset the key aims of the marine reserve were to: 'maintain coastal and marine ecosystems in their natural state; act as an area where marine species can breed undisturbed; provide educational recreation for residents and tourist; provide a protected area where scientist and students can engage in research' (IABIN, 2010, 7). Canada's McGill University maintains a marine research centre at Bellairs, near to Folkestone, so the research element is still very much to the fore, but sadly the coastal and marine ecosystems have been degraded. Collaborative and community-focused work begun in 2009 to undertake long-term monitoring of the state of the reefs, but the overall picture remains concerning (University of the West Indies, 2009).

Integrated policies for the management of coastal and marine protected areas remain elusive. For a number of years, there have been signals to introduce a Barbados Marine Management Area which would see the management of Barbados' marine resources and its users (fishers, divers, tour operators, pleasure craft owners and even beach users), but an integrated policy and authority has not yet been implemented leaving coastal areas vulnerable to overfishing and lack of conservation (Biopama, 2020). Tourism development along the island's coastline continues to be intensive especially at sites with existing building footprints. Prior to COVID-19, coastal tourism developments were announced with much anticipation, but at the height of the COVID-19 pandemic, a number of these projects stalled through a combination of the lengthy government permission process and some tourism investment disquiet as the tourism industry took the greatest hit.

It was only in late 2022 that a number of these projects are now underway. But a number of these multi-million dollar multi-storey projects are located on the vulnerable coastline potentially posing a risk to investors and the island's environment and economy if global sea levels rise and storm surges associated with more intensive storm seasons increase. At least two of these projects have been announced in Historic Bridgetown and its Garrison—one at the site of the old Harbour Police Station (which was the island's Lock Hospital in the nineteenth century) in Bay Street and the nearby Pier Head development which is slated to transform the eighteenth- and nineteenth-century warehouse district into a modern upscale residential condominium development targeted to local and foreign young professionals and other investors. Both have been cited for posing challenges to the unique and vulnerable coastline in those areas which could compromise nearby cultural and natural heritage (Barbados Today, 2022; Joseph, 2019).

As an island, Barbados' coastal natural heritage is never far away from the island's terrestrial natural heritage, and in many cases the island's cave and gully systems are connected to coastal basins and the nearshore environment. The Graeme Hall Nature Sanctuary, for example, is the site of one of the island's oldest intact natural wetlands and is a Ramsar protected site. As an important coastal watershed, protection of the coastal wetland system has continuously been under threat from development since it was purchased and transformed into a private conservation area by the late Peter Allard, a wealthy Canadian lawyer and philanthropist. In 2019, the wetlands faced its most urgent threat when the South Coast Sewerage Project failed sending raw sewage into the main highway along the South Coast. The overflow from the nearby primary sewage plant was directed into the swamp endangering the delicate wetlands ecosystem (Barbados Today, 2019). In a country with very few intact natural wetlands, the Graeme Hall Nature Sanctuary is in dire need of comprehensive and integrated protection and management especially as it such a key artery for the health and wellness of human populations along the coast and the health of the nearshore area. Recent announcements about the development of the nearby and related Chapman's Swamp in St. Lawrence Gap also raises the spectre of more tourism development threats of coastal wetlands (Mahon, 2022).

The island's few remaining wetlands and coastal mangroves have been under continuous threat from development of the island's coastline for tourism development. In addition to coastal wetlands such as Chancery

Lane and Graeme Hall Nature Sanctuary, there are also several natural and human made 'swamps' or ponds that have been used historically for bird-shooting. Through activism and advocacy over the last few decades, some of these swamps namely Fosters in St. Lucy and Woodbourne Shorebird Refuge in Christ Church have now been transformed into private/ non-governmental bird sanctuaries where local and migratory species can drink and feed. Mounting pressure from within and outside of the bird-shooting community has forced participants to recognise the value of these spaces for avian conservation with voluntary bag limits for shooting and even the cessation of the practice altogether. But instead of allowing these habitats to dry up due to disuse, participants are being encouraged to transform them into sanctuaries.

Much of Barbados' natural native forest cover was destroyed to make way for intensive sugar plantation development by the 1650s. Turner's Hall Wood in St. Andrew in the Scotland District is a notable exception and is an ecological preserve with some of the best examples of endemic species on the island. The Barbados National Trust founded in 1961 operates at least two natural heritage sites: Welchman Hall Gully in St. Thomas which is part of the Harrison Cave system and Andromeda Gardens in St. Joseph which was gifted to the Trust in 1988 in the will of well-known horticulturalist Iris Bannochie. Both sites provide locals and visitors an accessible trail through the island's forested and in the case of Andromeda, landscaped, natural heritage (Carrington, 2011). Other similar sites on the island are owned/operated privately such as the Flower Forest and Coco Hill in St. Thomas and Hunte's Gardens in St. Joseph. These sites are heavily dependent on tourism for their maintenance and during the COVID-19 pandemic many struggled to remain open (Madden, 2022). In addition to managing the island's beaches and parks, the National Conservation Commission (NCC) also operates a number of parks and open spaces such as Farley Hill National Park and Barclay's Park which is part of the larger Barbados National Park System on the East Coast.

Although there have been announcements endorsing and even launching the Barbados National Park system, little has been done to implement the policies required to manage the system (Nation News, 2010). Through a combination of both private and public initiatives to conserve the island's terrestrial natural heritage, there is still a lack of comprehensive and integrated public policy to help govern its use and future development. Land planning policies may be in place through the

Fig. 4.2 Graeme Hall Swamp, southern Barbados: relic wetland, Ramsar site, bird sanctuary and tourist draw (Niall Finneran)

Barbados Physical Development Plan (PDP) and other planning instruments. But more can be done to protect these resources. The Barbados National Park system and possible protection and management through the UNESCO World Heritage Tentative List nomination for the Scotland District could provide some of the properties sited in these areas to achieve integrated protection and management (Fig. 4.2).

4.3   CONCLUSION

The foregoing discussion has outlined some of the key issues facing the management of the natural landscape one of the major Caribbean islands. Barbados is, in context, one of the wealthier Caribbean islands with a relatively good public infrastructure and a stable democracy. This is not the case for a number of other often smaller islands. In these smaller islands, economic pressures have translated into overdevelopment of tourist infrastructure without attending to the basic needs of sustainability and impact on the environment. While a large hotel development may provide many direct and indirect jobs, there is always the possibility of the downside: increased traffic, pollution and strain on already overstretched services. Tourism strategies need to be sustainable, cost-effective and environmentally friendly. Tourism, always a mainstay of the Caribbean economy, is close to recovering to pre-pandemic levels, and many islands are seeking to diversify their tourism profile.

One of the current authors (NF) devised and delivered a training session for a number of representatives of the Caribbean island tourism boards in 2020 on behalf of the Caribbean tourism organisation, and this message came through clearly. Heritage tourism, natural and cultural, is the way forward. The argument was however that this vision has to be thought through carefully, integratively considered in policy-making, and above all, be community led. Already we are seeing the impact of overdevelopment and rampant tourism on a number of Caribbean islands, and any measures taken of course have to balance the social, economic and cultural need of the island community and the fragile landscapes they inhabit. Leadership here from academics, both of local and international influence, is important. Given that there is a dearth of what we would term professional local practitioners working on the management of natural heritage in the Caribbean, the involvement of local community groups as practitioners in their own right also becomes important. The emphasis should be upon cross-fertilisation of ideas, sustainability and teamwork and an awareness of the emotional attachment to place and space in these fragile and exploited Caribbean island settings.

REFERENCES

Adie, B. (2017). Franchising our heritage: The UNESCO world heritage brand. *Tourism Management Perspectives, 24*, 48–53. https://doi.org/10.1016/j.tmp.2017.07.002

Adie, B., Hall, C., & Prayag, G. (2018). World Heritage as a placebo brand: A comparative analysis of three sites and marketing implications. *Journal of Sustainable Tourism, 26*(3), 399–415. https://doi.org/10.1080/09669582.2017.1359277

Allahar, C. (2015). De Beach Belong to We!: Socio-economic Disparity and Islanders' Rights of Access to the Coast in a Tourist Paradise. *Oñati Socio-legal Series, 5*(1): 298–317. file:///Users/a22727/Downloads/cultural-hosting-20150203-toppin-alahar-osls-rev-cta-revcr.pdf. Accessed 28 June 2022.

Barbados Today. (2019, February 2). *South coast sewage mend*. Barbados Today. https://barbadostoday.bb/2019/02/02/south-coast-sewage-mend/. Accessed 23 June 2022.

Barbados Today. (2022, May 17). *#BTEditorial—Is the Pierhead Project still a fanciful dream?* Barbados Today. https://barbadostoday.bb/2022/05/17/bteditorial-is-the-pierhead-project-still-a-fanciful-dream/. Accessed 23 June 2022.

Biopama. (2020, October 28). First-ever protected area management effectiveness assessment conducted for the Barbados Marine Reserve. *Biopama: from knowledge to action for a protected planet.* https://biopama.org/first-ever-protected-area-management-effectiveness-assessment-conducted-for-the-barbados-marine-reserve/. Accessed 25 May 2023.

Caribbean Council. (n.d.). *Beaches and public access.* https://www.caribbean-council.org/beaches-public-access/. Accessed 23 June 2022.

Carrington, S. (2011). *Preserving Paradise: A series of lectures to commemorate the life and work of the late Dr. Colin Hudson.* Barbados Museum and Historical Society.

Davis, A. (2010). *Public beach access.* https://annaleedavis.com/archive/public-beach-access. Accessed 28 June 2022.

Dickinson, J., Finneran, N., Smart, N. & Uwiminazi, N. (2022). *Youth entrepreneurship and heritage tourism: Long-term thinking for diaspora engagement: Lessons from Rwanda, Brazil and Barbados.* EUDiF (European Union Global Diaspora Facility). https://diasporafordevelopment.eu/wp-content/uploads/2022/06/Heritage-Tourism-Case-Study_Short-Version_EN-.pdf. Accessed 13 January 2023.

ECPA (Energy Climate Change Partnership of the Americas). (2021, July 14). Road to Recovery for a National Park in Dominica. *ECPA Newsletter.* https://ecpamericas.org/newsletters/road-to-recovery-for-a-national-park-in-dominica/. Accessed 23 June 2022.

Evanson, H. (2018, March 31). Doyle warned. *Nation News.* https://www.nationnews.com/2018/03/31/doyle-warned/. Accessed 26 June 2022.

Fortenberry, B. (2021). Heritage justice, conservation, and tourism in the greater Caribbean. *Journal of Sustainable Tourism, 29*(2–3), 253–276. https://doi.org/10.1080/09669582.2020.1757684

Graham, K. (2021, April 1). Government suspends beach chair vending. *Nation News.* https://www.nationnews.com/2021/01/04/govt-suspends-beach-chair-vending/. Accessed 23 June 2022.

Hay, P. (2006). A phenomenology of islands. *Island Studies Journal, 1*(1), 19–42. https://doi.org/10.24043/isj.186

Heller, B. (1996). Landscape, femininity and Caribbean discourse. *MLN, 111*(2), 391–416. https://doi.org/10.1353/mln.1996.0024

Hollsten, L. (2008). Controlling nature and transforming landscapes in the early modern Caribbean. *Global Environment, 1,* 80–113. https://doi.org/10.3197/ge.2008.010104

Hudepohl, K. (2008). Community agency and tourism initiatives in Carib Territory, Dominica. *Journal of Heritage Tourism, 3*(4), 231–241. https://doi.org/10.1080/17438730802366532

IABIN (Inter American Biodiversity Information Network). (2010). *Barbados ReeFix Exercise Draft Report.* http://www.oas.org/dsd/IABIN/Com ponent1/ReefFix/Barbados2010/Gill_%20BDS%20Reeffix%20Report.pdf. Accessed 28 June 2022.

Jönsson, C., & Devonish, D. (2008). Does nationality, gender, and age affect travel motivation? A case of visitors to the Caribbean island of Barbados. *Journal of Travel and Tourism Marketing, 25*(3–4), 398–408. https://doi. org/10.1080/10548400802508499

Joseph, E. (2019, November 28). Heritage body: 'Developers misled us'. *Nation News.* https://barbadostoday.bb/2019/11/28/heritage-body-develo pers-misled-us/. Accessed 23 June 2022.

Madden, M. (2022, May 27). *Flower attractions look forward to better business.* Barbados Today. https://barbadostoday.bb/2022/05/27/flower-attractions-look-forward-to-better-business/. Accessed 27 June 2022.

Mahon, R. (2022, October 15). *Site of the proposed development in St. Lawrence Gap.* Barbados Today. https://barbadostoday.bb/2022/10/15/site-of-the-proposed-development-in-st-lawrence-gap/. Accessed 26 October 2022.

Major, B. (2021, November 3). Barbados welcome stamp program wins for visitors and destination. *Travel Pulse.* https://www.travelpulse.com/news/ destinations/barbados-welcome-stamp-program-wins-for-visitors-and-destin ation.html. Accessed 26 June 2022.

Nation News. (2010, June 3). *National park 'major priority'.* Nation News. https://www.nationnews.com/2010/06/03/national-park-major-pri ority/. Accessed 23 June 2022.

NCC Barbados (National Conservation Commission). (n.d). *Beach access.* https://www.nccbarbados.com/beach-access/. Accessed 13 January 2023.

Oxenford, H., Roach, R., Brathwaite, A., Nurse, L., et al. (2008). Quantitative observations of a major coral bleaching event in Barbados, South-eastern Caribbean. *Climatic Change, 87*(3), 435–449. https://doi.org/10.1007/s10 584-007-9311-y

Patterson, T., Gulden, T., Cousins, K., & Kraev, E. (2004). Integrating environmental, social and economic systems: A dynamic model of tourism in Dominica. *Ecological Modelling, 175*(2), 121–136. https://doi.org/10. 1016/j.ecolmodel.2003.09.033

Pattullo, P. (2005). *Last resorts: The cost of tourism in the Caribbean.* Latin America Bureau.

Ryan, J., & Silvanto, S. (2010). A brand for all the nations: Development of the World Heritage Brand in emerging markets. *Marketing Intelligence and Planning, 29*(3), 305–318. https://doi.org/10.1108/02634501111129266

Scher, P. (2011). Heritage tourism in the Caribbean: The politics of culture after neoliberalism. *Bulletin of Latin American Research, 30*(1), 7–20. https://doi. org/10.1111/j.1470-9856.2010.00451.x

Slinger-Friedman, V. (2009). Ecotourism in Dominica: Studying the potential for economic development, environmental protection and cultural conservation. *Island Studies Journal, 4*(1), 3–24.

Smith, M. (2002). A critical evaluation of the global accolade: The significance of World Heritage Site status for Maritime Greenwich. *International Journal of Heritage Studies, 18*(2), 137–151. https://doi.org/10.1080/135272502 20143922

University of the West Indies. (2009). *Community-based coral reef monitoring and management.* https://www.cavehill.uwi.edu/cermes/projects/folkes tone-project/pub/baseline_survey_of_folkestone_reefs_2009_10_20.aspx. Accessed 22 June 2022.

World Factbook. (2022). *Barbados.* https://www.cia.gov/the-world-factbook/countries/barbados/#economy. Accessed 28 June 2022.

CHAPTER 5

Managing Heritage Landscapes of Cultural Value: A View from the National Trust Portfolio in Purbeck, Southern England

Tracey Churcher and Niall Finneran

5.1 INTRODUCTION

In this section, we will consider the broad issues faced in the management of the National Trust's natural/cultural landscapes and will illustrate the contemporary challenges framed by climate change, economic problems and post-COVID-19 recovery with reference to their portfolio of heritage assets on the Isle of Purbeck in Dorset, southern England. It is pertinent to start this analysis by revisiting the UNESCO (United Nations Educational, Scientific and Cultural Organization) definition of the World Heritage Site (WHS) as defined by their 1972 convention (UNESCO, 1972) which enshrines what we now recognise as being a false dichotomy between cultural sites and natural sites (Chapter 3). In

T. Churcher
National Trust Purbeck, Dorset, UK
e-mail: tracey.churcher@nationaltrust.org.uk

N. Finneran (✉)
University of Winchester, Winchester, UK
e-mail: Niall.Finneran@winchester.ac.uk

© The Author(s) 2024
N. Finneran et al. (eds.), *Managing Protected Areas*,
https://doi.org/10.1007/978-3-031-40783-3_5

the light of encounters with indigenous perceptions of heritage meaning in the landscape, particularly in the case of Australia, heritage practitioners and specialists working in the field today, rightly recognise the value of a nuanced 'mixed' categorisation (Bickertsteth et al., 2020). It is impossible to disentangle the natural from the cultural, and in any case 'natural' as a designation is a subjective cultural construct. As such we take the analysis of the management National Trust's Isle of Purbeck portfolio in holistic terms, as a landscape of cultural value, embodying natural, historical and aesthetic significances.

UNESCO is one global framework for the formal management of heritage landscapes but of course in many countries there also exists national, regional and local configurations. Informal management and interpretation can also be provided by other NGOs (non-governmental organisations), such as preservation societies and even local tourist boards. In the United Kingdom, each constituent country has its own national heritage management framework for natural and cultural landscapes and also individual sites. We chose England to illustrate how these management systems interlock not out of any perceived sense of superiority that this is the best system, but it is the system with which both authors are most familiar. Site management is provided by the National Trust (also in Northern Ireland and Wales, there is a separate National Trust in Scotland) and English Heritage. We will focus on the National Trust below, but it is important to note that it is an independent charity staffed extensively by volunteers. English Heritage was originally formed in 1983 as a QUANGO (Quasi-Autonomous Non-Governmental Organisation) to take charge of England's archaeological asset portfolio (i.e. monuments such as Stonehenge and Tintagel). In 1999, English Heritage merged with the RCHME (Royal Commission on the Historical Monuments of England) and the NMR (National Monuments Record). In 2015, English Heritage was broken up. A policymaking advisory arm known as Historic England took on the role of maintaining records and developing heritage guidance, whilst the new iteration of English Heritage became a charity which actually operated the historical properties (English Heritage, 2022).

When we talk about the definition of heritage landscapes from the point of view of English policy and practice, it is important to stress that overwhelmingly Historic England tends to focus on the management of buildings and archaeological sites. The National Trust is more landscape orientated, in terms of both natural and cultural assets, as we shall see below. England of course has UNESCO World Heritage (cultural) Sites which can be bounded entities (such as the City of Bath) or disparate areas of landscape grouped under a common theme, such as the extensive patchwork of industrial sites that make up the Cornwall and West Devon mining landscape, or indeed the significant group of Neolithic and Bronze Age monuments of north-central Wiltshire (Stonehenge, Avebury and Associated Sites). UNESCO, the National Trust and Historic England are the overarching international-national management frameworks of English cultural landscapes, but there are also localised and informal frameworks too.

Conservation areas are designated by local authorities. They group together buildings and features that have a distinctive historical and architectural interest or character. Mainly found in urban and often suburban contexts, conservation areas seek to maintain the physical and intangible character of an area (quite how the latter is quantified is debated by planners and academics!). Conservation areas grew out of the context of post-World War Two planning and redevelopment issues. They were introduced in 1967 and carry significant implications around planning, development and even tree preservation (Larkham, 2003; Skea, 1996). Management of heritage landscapes also comes under the aegis of National Parks as well as Areas of Outstanding Natural Beauty (AONBs), although here natural heritage (the aesthetics of landscape rather than the human imprint) take priority. Sometimes cultural landscapes can be conjured from intangible elements, linked themes or concepts. Long-distance public footpaths, for example, link to a long linear cultural feature that links and unites landscapes over a long distance (such as Offa's Dyke Path, or the Ridgeway; Busby, 1996). Tourist boards too can create cultural landscapes based on personifications. Think, for example, of Thomas Hardy's 'Wessex', an imagined literary topography mapped mainly upon the historic county of Dorset (Keith, 1969). Arguably, however, it is the National Trust that in England at least is the primary organisation for the management of heritage landscapes. We now consider how this role developed.

The Isle of Purbeck

The Isle of Purbeck is a discrete and defined area of landscape in the eastern part of the English county of Dorset. Not an island in the strictest sense of the word (more a peninsula), it is bounded on the south by the English Channel and to the east and northeast by Poole Harbour. To the north the River Frome forms another natural boundary. The only significant urban settlement on Purbeck is the seaside resort of Swanage. The main tourist focus is upon the extensive beaches at Studland Bay and the Historic village of Corfe and its ruined castle. The Isle is bisected by the Purbeck Hills, and geology has historically informed Purbeck's prosperity, initially in quarried stone and latterly through onshore oil exploitation. The National Trust is a major landowner here, through the terms of a bequest from the Bankes Family of Kingston Lacey, and formerly owners of Corfe Castle. The whole of the 'isle' is in the eastern portion of the Dorset AONB. On the heathlands to the north, bordering Poole Harbour, there are two national nature reserves (Hartland Moor/ Studland and Godlingston Heath) and a Royal Society for the Protection of Birds (RSPB) nature reserve (Arne). The western coast around Lulworth impinges upon the UNESCO natural World Heritage Site of the Jurassic Coast, and on the south-eastern tip of Purbeck is the Durlston Country Park and National Nature Reserve. It is an unspoilt, scenic and diverse landscape of great individual character, which draws in a large tourist population across the year.

5.2   Managing Heritage Landscapes in the UK: The National Trust

The National Trust is the primary charity in England, Wales and Northern Island that deals with the conservation of heritage landscapes, cultural, natural and mixed. For many readers, it will require little introduction, but it is an organisation that has recently found itself at the centre of political controversy (an issue to which we return later as it has implications for some of the issues analysed in this chapter). The Trust was founded in 1895 as the National Trust for Places of Historic Interest or Natural Beauty and was from the outset an organisation that chimed with wider currents and thinking of the time (Chapter 2). The three founder members Octavia Hill, Robert Hunter and Canon Hardwicke Rawnsley all shared a vision to democratise access to the open spaces of England

and Wales and to ensure the preservation of architectural works of merit (the Trust fossilised in its DNA a number of predecessor organisations concerned with philanthropy and what might now be termed widening participation) (for an overview see Hall, 2003).

During the mid-twentieth century the portfolio expanded towards the management and curation of large country house estates, gifted by their owners to the National Trust in lieu of death duties (a process vividly recorded by one of the main protagonists, James Lees-Milne, (1908–1997) Lees-Milne, 1975, 2001). This acquisition pattern was later widely criticised, ensuring the portfolio was skewed towards a certain type of property which became the archetypal National Trust visitor experience (house and garden of landed gentry fallen on hard times; Smedley, 2010). This had the effect of de-emphasising the natural heritage sites and also impacting upon the Trust's finances as the upkeep and maintenance of these period properties proved to be prohibitively expensive. This is a mistake that the Trust would make again with its purchase of the Victorian gothic manor at Tyntesfield near Bristol in 2001 (Bailey, 2004).

Today, the National Trust manages over 500 historic buildings and gardens and archaeological sites. Access to these properties is free for members, and non-members have to pay an entry fee. It also owns almost 250,000 hectares of land, with significant stretches of coastline and many nature reserves. Access to open spaces is free. Income is mainly derived from footfall (particularly via catering and commercial outlets), subscriptions, legacies, donations and grants from a range of governmental and charity sources. These funds have to support not just the upkeep of the properties but also the wages of c. 10,000 members of staff (who themselves managed some 53,000 volunteers; National Trust, 2022a). Financial overreach is of course a key issue faced by many charities working with historic properties, and in the historic environment, but recent travails for the National Trust have included reactions to their attempts to widen participation further and recognise 'hidden heritages'.

For the National Trust, 'hidden heritages' represent stories of individuals who come from backgrounds that have never been recognised in the heritage 'canon'. This inclusive approach though has led to criticism. There was a backlash, for example, against the Trust asking volunteers in 2017 to wear rainbow ribbons at their house at Felbrigg Hall, Norfolk, in recognition of the role in LGBTQ+ history played by its last owner, Robert Ketton-Cremer (Grierson, 2017). Flames were fanned by popular media further with the publication in 2020 (e.g. Starkey, 2021), against a wider backdrop of agitation around the Black Lives Matter movement and issues of contested heritage, of the report on the legacies of slavery in the National Trust's historic properties (Huxtable et al., 2020). And of course even greater global events of 2020, namely the COVID-19 pandemic, also impacted severely upon the footfall of National Trust properties. Where business models were focused upon maximising profits from visits by non-members, and particularly tourists from China, the USA, Europe and Japan, the impact was shattering for National Trust finances (National Trust, 2020a).

The foregoing discussion has sketched in some of the key issues that the National Trust has faced in its almost 130 years of caring for our cultural and natural landscapes. Some are long-lived issues, others are part of the cultural moment. In broad terms, against a worsening economic picture in 2022–2023, recovery from the COVID-19 pandemic, inflation, changes in consumer behaviour and also wider cultural currents (aka the 'culture war') this guardian of national heritage finds itself in a complex and precarious position. And these are only the human elements. The other elephant in the room is climate change. The impacts of climate change are already being felt in the National Trust's coastal portfolio where erosion generated by storms of increasing strength and frequency and sea level rise are increasingly evident (Hancock, 2021). At Brownsea Island in Poole Harbour, Dorset, for example, National Trust coastal engineers took the decision to remove sea defences with the express intention of not working against nature (a fruitless and ultimately unsustainable approach), but allowing a degree of managed decline. The maritime community volunteer CITiZAN initiative (Coastal and Intertidal Zone Archaeology Network; Chapter 5) undertook 3D photogrammetry of an old lime kiln on Brownsea to preserve it via recording rather than undertake active physical interventions on the structure (CITiZAN, 2022). This is undoubtedly the most cost-effective and sustainable approach to mitigating the effects of climate change on coastal heritage assets. We will now sharpen our geographical focus to consider a case study of cultural landscape management centred on the Isle of Purbeck in eastern Dorset, southern England.

5.3   THE NATIONAL TRUST ISLE OF PURBECK PORTFOLIO

The National Trust Isle of Purbeck Estate, including Corfe Castle, Studland Bay and Kingston Lacy house and gardens, forms part of a 16,000-acre bequest to the National Trust in 1982 by the late Ralph Bankes (the 'Lord High Admiral of Purbeck', 1902–1981) whose family owned the lands for many hundreds of years. This is the largest bequest ever made to the Trust. Located on the southern shore of Poole harbour in Dorset, England (Fig. 5.1), the Purbeck property is now managed separately from Kingston Lacy. Although, the majority of the 8500-acre Purbeck estate came through the 1982, Bankes bequest, there have been subsequent acquisitions which support the delivery of the National Trust aims to protect and care for places of nature, beauty and history. The most recent acquisition was in April 2022, Weston Farm, a small coastal farm on the coast at Worth Matravers.

Despite its seventeenth-century ruination, Corfe Castle remains one of the great medieval castles of Britain (Fig. 5.2). Visited by over a quarter a million visitors a year, it occupies a central part of the Purbeck ridge which forms a natural rampart across the centre of the island. It is virtually impossible to visit Purbeck without driving past it and it attracts a wide range of visitors, notably during school holidays, when younger families and international tourists predominate. The wider estate sits in a nationally and internationally recognised landscape area with the estate located within the Dorset AONB and all of the coastline within the Dorset and East Devon Coast UNESCO WHS. It also includes the key landscape features of Old Harry Rocks, Dancing Ledge and Winspit Quarry, some of the most photographed and iconic landscape features in the United Kingdom. The Isle of Purbeck, although not a physical island in the strictest sense, thus retains a very distinctive rural landscape character, the main routes in being via the Sandbanks Ferry from Poole to the east and via the A351 road from Wareham to the north.

The National Trust is also one of the major stakeholders in the Purbeck Heaths Super National Nature Reserve. Declared in 2020, adhering to the Lawton principles of 'Bigger, Better and more Joined', (Lawton, 2010) the 8650 acres area covers Hartland Moor, Stoborough Heath and Studland and Godlingston Heaths. With seven different landowners, including public, private and NGO's, the richly bio-diverse, lowland heath is managed at landscape scale to common principles. The four miles of beaches that make up Studland Bay are the most intensively visited, with

Fig. 5.1 Map of the Isle of Purbeck (Niall Finneran/https://www.openstree tmap.org/#map=12/50.6398/-2.0582&layers=O)

Fig. 5.2 Corfe Castle (Niall Finneran)

an estimated 1.5 million visitors each year. Just across the harbour from Poole and Bournemouth, private ownership by the Bankes family and its use as their summer retreat, prevented the development of Studland into a commercial resort. The Bay has probably the best-known naturist beach in the country with a history dating back over 100 years. The natural dune systems behind the beach are home to all six native reptile species and carry a full range of designations of national and international importance.

The area was requisitioned by the military in both World Wars. Most notably Studland was used as the location for Operation Smash in April 1944. A full-scale live-fire, dress rehearsal for D-Day, the Allied invasion of Europe, took place here. The legacy is of this event is still with us today: the ongoing discovery of unexploded ordnance in the dunes remains a significant issue. Although the visitor business is perhaps the best-known aspect of the National Trust, the responsibilities of the Purbeck estate include over 500 tenants and licensees, comprising residential properties, holiday cottages, farms, pubs, shops, activity operators, stone quarries, a golf course, boat posts and beach huts. There are also 26-scheduled monuments and 617 archaeological features in the National Trust's care. It is a historically rich landscape in terms of archaeological sites and historic monuments, yet also has significant economic value through tourism.

The staffing of the estate, therefore, includes not only the visitor-facing teams familiar to many, but estate management professionals, building and rural surveyors, ecologists and a ranger team. These are supported by staff from the internal National Trust 'Consultancy Service', which include curators, archaeologists, project managers, commercial specialists, historic environment professionals and land management professionals. Overall, the business is a careful balance of provision of public access with, in some cases revenue generating opportunities. There is also an obligation to care and improve the condition of the cultural and natural landscapes. As a property with high visitor numbers, Purbeck produces an overall surplus of income, some of which is internally 'taxed' to support properties which might have very high costs, but less opportunities to offset them.

The COVID-19 pandemic across 2020 and 2021 produced some real challenges for heritage management as well as the wider national economy. The UK Government Coronavirus Job Retention Scheme ultimately protected many jobs. On a practical level, the repeated opening and closing of properties called for an agile approach, retaining the bare minimum of staffing to keep the estate safe and compliant with insurance conditions. Volunteers were asked to not attend during the closed periods and staff had to adapt to do what was needed from a much smaller team. National Trust projects were paused and expenditure reduced where possible. Although ultimately the number of compulsory redundancies were reduced to half that initially planned (National Trust, 2020a), there was still a challenging impact upon the morale of the organisation. Whereas the cuts were applied proportionally across visitor-facing teams nationally, the impact was disproportionately felt in those areas that saw huge surges in visitor numbers. In 2020 and 2021, when staycation was king, Dorset drew huge numbers of visitors (Mulcahey, 2020). Studland saw an increase in visitor numbers of 25% over 2019, but had a visitor-facing team reduced by 15%, due to no seasonal recruitment in 2020 and voluntary redundancies in 2021.

At the same time, cleaning regimes were increased and social distancing applied to all activities, which were moved outside wherever possible.

Food and beverage outlets moved to a takeaway model and shop visitors were carefully managed on a one in, one out basis. Staff were concerned not only by their personal vulnerability to contracting COVID-19, particularly before the vaccination programme started in late 2020, but also by challenging public behaviour. Confrontation and outdoor defaecation were common. At the end of this period, both staff in work and those on furlough were affected by the experience. Some were understandably very tired and others found the lack of work routine difficult to manage and the return to work mostly welcome but exhausting. Close contact was kept with all staff during the period and staff were moved in and out of furlough to try to keep a healthy balance.

Early in the pandemic, in March 2020, the National Trust had determined that the access it could provide to outdoor spaces was essential for the well-being of the nation and announced that it would open its parklands and gardens for free. As visitors flocked to the sites, the Trust had to reverse this decision, stating a desire to limit the spread of coronavirus and the need to protect its staff, volunteers and visitors (ITV, 2020). In Purbeck, countryside and coastal sites were left open and were well used by locals throughout the closed period. The pandemic brought a new audience to Studland, being within easy reach of London and the M3 and M4 corridors it brought diverse visitors who could comfortably manage a day visit before overnight accommodation re-opened. Octavia Hill, one of the founders of the National Trust, was driven by the belief that everyone should have access to art and nature (National Trust, 2020b) and being able to do just that at a time of national crises, felt positive to staff.

The economic impact of the COVID-19 pandemic clearly demonstrated the importance of membership to the organisation. Approaching the pandemic, the organisation was optimistic of reaching 6 million members. The numbers speak volumes for the relevance of the organisation to a large sector of UK society. Although the ability to promote membership was lost with properties closed, membership remained fairly robust. The 2022 Annual General Meeting booklet shows that membership fees account for the largest source of income. In 2021/2022 bringing in £280.1 million (National Trust, 2022b). This contrasts to admission fees which nationally brought in £21.2 million. It is this bedrock that has permitted the ambitious programme of conservation works to be picked back up at pace during the post-pandemic recovery period.

The current 'culture wars' being played out both in the UK media and also at the National Trust 2022 Annual General Meeting have brought into sharp focus a wider understanding of the core purpose of the National Trust (the term 'culture wars' is widely recognised but is here taken to mean opposing populist positions over climate change, identity politics and decolonisation *inter alia*). An organisation named 'Restore Trust' has fielded a number of candidates and put forward resolutions that are not supported by the Board of Trustees. 'Welcome Everyone' and 'Climate Action' are key themes that underpin National Trust work programmes which are well supported by staff and volunteer teams in Purbeck. The desire to inclusively welcome all people to National Trust Purbeck is reflected in work with the adjacent Bournemouth, Christchurch and Poole conurbation (BCP) to understand potential barriers for ethnic minority engagement with the Purbeck estate. Forthcoming initiatives include an extension of the BCP, 'Beryl' public bike and scooter hire scheme to Studland and a 'Bollywood on the Beach' event in Spring 2023, working with the BCP South Asian Society. In Purbeck, the engagement team work with local schools to enable every school child to have an experience of the coast and countryside in their home area. It is critical for local children to be able to identify the area as a core part of their own sense of place, and not just an area for visitors to enjoy.

Climate action has also been key in local stakeholder work during lockdown, where Planet Purbeck, a local environmental organisation was catalysed on-line by the National Trust working with other NGO's and community leaders locally. Two years later, with over 1500 members and an annual festival, the grass roots work carried out during the pandemic has firmly positioned the National Trust as a key member of our community at a time when climate issues are becoming ever more centre stage (Planet Purbeck nd). Climate change and its impact upon coastal change is an area of focus for the Purbeck property. Three of the four most heavily used beach areas are at risk of loss to sea level rise and changing weather patterns. Response to this challenge has provided grounds for disagreements with the local community, most notably at the 2016 National Trust's AGM, where a group of local Studland residents brought forward a resolution requesting that a local beachside café, vulnerable to loss due to coastal change, should not be demolished. Although the motion was not carried, it led to several years of intense engagement trying to identify a sweet spot where compromise could be reached driven fundamentally by different views of how to respond to climate change.

The National Trust, although owning nearly 900 miles of coastline, agrees to management approaches (known as Shoreline Management Plans) as part of a wider 'Coastal Group' made up of the Environment Agency and Lead Local Flood authorities, typically based within local authorities (Gov.co.uk, 2022). The plan for the majority of Studland Bay is for 'No Active Intervention', which means that the coast will be allowed to adapt naturally to changing conditions. In Studland, this puts a wide range of facilities at threat including, cafes, car parks, toilets, shops, beach huts, boat parks and visitor welcome areas. In some cases, it will also change how the beach areas can be accessed. Before this policy was adopted, some coastal defences were put in at two of the beach locations. Whilst this has literally 'held the line', it has led to significant lowering of the beach height, as the hard stone gabion defences have prevented the natural action of the waves from releasing sandy substrate from the cliffs, maintaining the beaches to a constant level. The hard defences also create a pattern of scour in front of them as the waves bounce of a hard surface with retained energy.

Whilst opinions still vary as to appropriate responses to climate change, the recent changing weather patterns locally played out by two significant fires at Studland and cliff erosion directly caused by heavy rainfall have undoubtedly led to a greater acceptance that change is coming (BBC, 2021, 2022). The generally advocated approach of a roll back to higher and safer ground will be adopted, drawn from a master planning process considering the needs for public access to this area over the next twenty to fifty years. The decision over how to manage loss is altogether more complex in the case of Fort Henry, a World War Two observation bunker located at Redend Point and overlooking Studland Bay. Described by Rodney Legg as one of Britain's most important relics of World War Two (Legg, 1990), Fort Henry is a reinforced concrete observation post built by Canadian engineers for April 1944, where VIPs watched 'Operation Smash' the live-fire rehearsals for D-Day, were conducted at Studland Beach. The VIPs reputedly included King George VI, Eisenhower, Churchill and Montgomery. The coastal change in this area is rapid and a no active intervention approach has been adopted. A 2021 detailed survey has shown that some of the structure is at risk of loss due to erosion within the next twenty five years. The actual timescale is much shorter, as if the structure is to be moved or demolished, the surrounding ground will need to be robust enough to take the heavy plant that would be required for such an operation.

Approaches to managing the loss of historic coastal structures due to climate change is a developing area of expertise. Whilst the process of seeking to mitigate the issues with Fort Henry is at an early stage as of the time of writing (Autumn 2022), the next steps will be informed by a survey seeking to understand the heritage significance using the Historic England Conservation Principles (Historic England, 2008). These include considering concepts such as evidential value, historical value, aesthetic value and communal value. The concept of anticipatory history, exploring ways to respond to climate-based landscape change, using history and storytelling to explore an altered future is also providing some excellent thinking to shape the next steps (DeSilvey et al., 2011).

The impacts of climate change and the pandemic in the early 2020's have demonstrated that practical management of heritage assets is far more than ensuring their physical preservation for the next generation. How to respond is always complex. Whilst the National Trust is a conservation organisation, not all tangible culture can be conserved in the same state forever. Proposals for a future that involves change or loss are often challenging for those who value the item at risk. Engagement with local campaigns and resulting complaints, typically stoked by social media, are a part of the National Trust's regular communications work. At a time of increasing ideological conflict being aired in the cultural sector, challenge is also received from some aspects of the media when the National Trust works in a culturally progressive manner on elements of history or intangible heritage such as LGBTQ+ Pride events or the 2020 report on colonialism.

5.4   Conclusion

A dry academic analysis of the role of the National Trust in managing heritage landscapes in southern England over the period c. 2020–2022 cannot capture the nuances of the human elements at play. In the short term, the COVID-19 epidemic forced a rethink on the way the sites operated on a day to day basis in response to different patterns in visitor engagement. The bigger economic impact is yet to be felt, surely, even if visitor numbers have returned to pre-pandemic levels. Climate change is making itself felt, not just in terms of threats to coastal heritage assets through sea level rise but also through hotter drier summers, which as we have seen has impacted upon the heath lands of Purbeck in the exceptional weather of Summer 2022. As a practitioner, at the sharp end of the management experience, these are ongoing challenges. There is also a wider cultural and ideological context at play which has seen the National Trust become a target for elements of the media and certain populist politicians. External consultation over change is imperative, which by its very nature will reflect the current zeitgeist. The National Trust, however, is 'For Everyone, Forever', and that is the context in which operational decisions must always be made. This is a message that both authors agree should not be lost.

REFERENCES

Bailey, M. (2004, January 1). National Trust's total cost of saving Tyntesfield may be £50 million. *The Arts Newspaper.* https://www.theartnewspaper. com/2004/01/01/national-trusts-total-cost-of-saving-tyntesfield-may-be-pound50-million. Accessed 29 June 2022.

BBC. (2021, December 1). *Large fire breaks out on Studland Heath on Dorset coast.* BBC News. https://www.bbc.co.uk/news/uk-england-dorset-59485222. Accessed 23 September 2022.

BBC. (2022, August 12). *Studland fire: Ninety firefighters tackle heath blaze near beach.* BBC News. https://www.bbc.co.uk/news/uk-england-dorset-62522888. Accessed 23 September 2022.

Bickersteth, J., West, D. & Wallis, D. (2020). Returning Uluru. *Studies in Conservation, 65*(Suppl. 1), 9–17. https://doi.org/10.1080/00393630. 2020.1774102

Busby, G. (1996). Long distance paths as rural tourism resources. *Footprint, 2*(2), 10–13.

CITiZAN (Coastal and Intertidal Zone Archaeology Network) (2022). *Brownsea Island brick kiln record.* https://www.citizan.org.uk/interactive-coastal-map/78380/. Accessed 29 June 2022.

DeSilvey, C., Naylor, S., & Sackett, C. (2011). *Anticipatory history.* Uniform Books.

English Heritage. (2022). *Our history.* https://www.english-heritage.org.uk/about-us/our-history/. Accessed 29 June 2022.

Gov.co.uk. (2022). *Shoreline management plans (SMPs).* https://www.gov.uk/government/publications/shoreline-management-plans-smps. Accessed 10 September 2022.

Grierson, J. (2017, August 5). National Trust reverses decision enforcing gay pride badges. *Guardian.* https://www.theguardian.com/uk-news/2017/aug/05/national-trust-reverses-decision-on-gay-pride-badges. Accessed 19 April 2023.

Hall, M. (2003). The politics of collecting: The early aspirations of the National Trust, 1883–1913. *Transactions of the Royal Historical Society, 13*, 345–357. Io.IoI7/Soo8044oIo3ooo2o4.

Hancock, M. (2021, March 8). *National Trust maps out landslide and coastal erosion threat.* Ground Engineering. https://www.geplus.co.uk/news/national-trust-maps-out-landslide-and-coastal-erosion-threat-08-03-2021/. Accessed 29 June 2022.

Historic England. (2008). *Conservation principles, policies and guidance.* https://historicengland.org.uk/advice/constructive-conservation/conservation-principles. Accessed 25 September 2022.

Huxtable, S., Fowler, C., Kefalas, C. & Slocombe, E. (2020). *Interim Report on the Connections between Colonialism and Properties Now in the Care of the National Trust, including links with Historic Slavery.* https://www.nationaltrust.org.uk/felbrigg-hall-gardens-and-estate/news/heard-about-felbriggs-lanyards-and-badges. Accessed 29 June 2022.

ITV. (2020, March 21). *National Trust to close parks from midnight due to coronavirus.* ITV News. https://www.itv.com/news/2020-03-21/national-trust-to-close-parks-from-midnight-due-to-coronavirus. Accessed 25 September 2022.

Keith, W. (1969). Thomas Hardy and the literary pilgrims. *Nineteenth-Century Fiction, 24*(1), 80–92.

Larkham, P. (2003). The place of urban conservation in the UK reconstruction plans of 1942–1952. *Planning Perspectives, 18*(3), 295–324. https://doi.org/10.1080/02665430307975

Lawton, J. (2010, September 24). *Making space for nature; a review of England's wildlife sites published today.* https://www.gov.uk/government/news/making-space-for-nature-a-review-of-englands-wildlife-sites-published-today. Accessed 10 September 2022.

Lees-Milne, J. (1975). *Ancestral voices.* Chatto and Windus.

Lees-Milne, J. (2001). *People and places: Country house donors and the national trust.* John Murray.

Legg, R. (1990). *Dorset at war, diary of WW2.* Halsgrove.

Mulcahey, J. (2020, June 26). *Plea to respect Dorset's beaches after chaotic scenes.* https://www.dorsetecho.co.uk/news/18542769.plea-respect-dorsets-beaches-chaotic-scenes/. Accessed 10 September 2022.

National Trust. (2020a). *Our spending cut plans in response to Coronavirus losses.* https://www.nationaltrust.org.uk/news/our-spending-cut-plans-in-response-to-coronavirus-losses. Accessed 29 June 2022.

National Trust. (2020b). *Octavia Hill: her life and legacy.* https://www.nationaltrust.org.uk/features/octavia-hill-her-life-and-legacy. Accessed 10 September 2022.

National Trust. (2022a). *Information for journalists.* https://www.nationaltrust.org.uk/features/information-to-journalists. Accessed 29 June 2022.

National Trust. (2022b). *Members annual general meeting.* https://doc uments.nationaltrust.org.uk/story/agm-booklet-2022/page/1. Accessed 10 September 2022.

Planet Purbeck. (n.d.). *Working for a cleaner safer Purbeck.* https://planetpurbec k.org. Accessed 13 January 2023.

Skea, R. (1996). The strengths and weaknesses of conservation areas. *Journal of Urban Design, 1*(2), 215–228. https://doi.org/10.1080/135748096087 24382

Smedley, B. (2010). *On James Lees-Milne and the National Trust.* https://fug itiveink.wordpress.com/2010/03/29/on-james-lees-milne-and-the-national-trust/. Accessed 29 June 2022.

Starkey, D. (2021, March). We can't trust the National Trust's History. *The Critic.* https://thecritic.co.uk/issues/march-2021/we-cant-trust-the-nat ional-trusts-history/. Accessed 29 June 2022.

UNESCO. (1972). *Convention concerning the Protection of the World Cultural and Natural Heritage.* https://whc.unesco.org/en/conventiontext. Accessed 29 June 2022.

CHAPTER 6

Between High and Low Tide. Participatory Approaches to Managing England's Coastal and Riverine Natural and Cultural Heritage: A Case Study from the CITiZAN Initiative

Oliver Hutchinson and Niall Finneran

6.1   INTRODUCTION

With each low tide, England's largest open air archaeological site opens to the public. Ranging from rugged cliffs to sprawling sand dunes, mudflats and estuaries, England's 4,400+ km of coastline is as topographically varied as the archaeological remains found along it. The intertidal zone, the area of land exposed between low and high tides, reveals a significant, well preserved and arguably overlooked archaeological and natural heritage resource (Bailey et al., 2020). It provides a unique insight into responses to climatic and environmental change throughout human history on the British Isles. Many of these remains are, by terrestrial

O. Hutchinson
London, UK

N. Finneran (✉)
University of Winchester, Winchester, UK
e-mail: Niall.Finneran@winchester.ac.uk

© The Author(s) 2024
N. Finneran et al. (eds.), *Managing Protected Areas*,
https://doi.org/10.1007/978-3-031-40783-3_6

archaeology standards, remarkably accessible to anyone visiting the coast. Indeed, for those living in coastal communities, the curious timber structures, wrecks, submerged forests, prehistoric human footprints and myriad other finds and features exposed on the beaches have provided a lifetime's interest, posing ongoing questions of the what, why, who and how of their origins. In addition, visitors to these coasts can also gain a snapshot of the long maritime heritage of this country. It is an unrivalled cultural and natural heritage repository.

For those living in coastal communities, intertidal archaeology is a visible proxy of our changing coastline. The impacts of wave, wind and tide combined with the effects of human interaction with the landscape are written on the fragile archaeological remains they expose and the spaces they once occupied. They highlight the immediate impact of the growing number and intensity of storms that batter the coastline, an observed factor of human-driven changes in global climate (IPCC, 2022). Many intertidal archaeological remains and natural features (such as submerged prehistoric forests) cannot withstand such erosive forces indefinitely and almost all, once exposed, are at risk of destruction. Studying and protecting this valuable resource by more traditional archaeological means (or indeed by any other on the ground survey methodology) is a challenging task at best, both logistically and financially. Tidal windows rarely permit more than a few hours recording and the conditions on many sites prohibit the use of methods and machinery that would, on land, make the task relatively straightforward. Alternative approaches are therefore required that address the challenges of conducting archaeological survey in coastal environments, namely the scale of the archaeological resource in question, its fragility, accessibility, the movement of the tides, recording techniques, preservation and so on. We stress again that these are not solely archaeological considerations but problems faced by anybody seeking to survey the intertidal zone in a useful level of detail.

This chapter examines the work of the Coastal and Intertidal Zone Archaeological Network (CITiZAN) and its participatory approach to preserving by record England's valuable coastal heritage. It begins with a brief account of coastal archaeology in the UK, setting CITiZAN in context with those projects that preceded (and guided) it. The planning and design of both phases of the project are then addressed, establishing the methodology that would define the project and its legacy. There follows a case study of the Mersea Island Discovery Programme and the impact of the project on the community and, most importantly, vice versa, before concluding with some personal reflections of one authors' experience of the project. An additional case study is briefly presented in the box text, and this relates to another CITiZAN sub-project, the South Devon Rivers Project which was developed by NF. We stress again (and both authors have been involved in varying degrees in the work of CITiZAN) that although the central focus is on an essentially cultural (archaeological) coastal project, the lessons we have learned since 2015 have implications for project management of similar participatory community endeavours on the foreshore.

6.2   The CITiZAN Project in Context

CITiZAN was launched in 2015 with the principle aim of preserving by record England's fragile, at-risk coastal heritage. It was formed in response to the growing threats to the coast posed by a changing climate, particularly more powerful and frequent storm events that could damage or destroy exposed archaeological remains. Major storms in 2013/14 provided a clear impetus to develop a new, national response to the issue given the severe damage and flooding they wrought to large swathes of the coastline (BBC, 2014; Masselink et al., 2016). Through regular monitoring of these exposed sites, the project aimed to develop the most detailed, publicly accessible map and photographic record of England's coastal heritage to date, the results of which could inform future research in the archaeological discipline and beyond.

The scale of England's coastal archaeological resource had already been documented by two earlier projects. The National Mapping Project (NMP; Cattermole & Hoggett, 2020) in the early 1990s focussed on identifying archaeological sites from maps and aerial imagery, giving priority to those areas that were poorly understood or at greater risk of destruction. The NMP laid the foundations for the follow-up Rapid Coastal Zone Assessment Surveys (RCZAs), commissioned in the early 2000's by, at that time, English Heritage (Archaeology Data Service nd). These surveys would examine in greater detail the sites and features noted in the NMP by ground truthing them where possible. Crucially, it also sought to characterise the nature of a variety of risks to each site (natural coastal erosion, deposition of sediments, coastal redevelopment, etc.). The RCZAs were delivered by several different organisations, often on a regional basis, with the last stretches of coast in the southwest completed in 2020. They further detailed the sheer scale and often well-preserved nature of intertidal archaeological remains, but in doing so would raise questions about how this resource should and could be protected for further research, an all but impossible task for any single county archaeological service or national organisation to undertake.

From a technical perspective, the methodology used to record intertidal sites in England would be showcased by Wilkinson and Murphy during the Hullbridge Basin Survey (Wilkinson & Murphy, 1995). Focussed initially on the river of the same name, the study aptly demonstrated the value of archaeological research within an environmental framework in relation to coastal change. It developed techniques required to maximise the value of data it was possible to capture within limited tidal windows. The results married well-preserved archaeological and environmental evidence with climatic data to provide one of the most detailed chronologies of coastal and sea level change in England. Based on the success of the initial phase, the study was quickly expanded to include all major estuaries in Essex and set the benchmark for the techniques and potential of intertidal archaeological survey in England (Murphy, 2014), a benchmark upon which the CITiZAN project was founded.

Data from the NMP, RCZAS, Historic Environment Records (HERs) and several other national, regional and local datasets was consolidated during the development phase of CITiZAN to create a database containing over 17,000 archaeological sites and features—an impossible number for any single organisation to record and monitor effectively. A broader approach was, therefore, required, one that could respond effectively to high rates of change around the coast. One solution lay in developing a network of professional archaeologists and academics, bringing their expertise and support to volunteers around the country via a central project team. As with any successful citizen science project, an easily replicable, simple and scalable methodology was required that would allow anyone to participate irrespective of prior experience. CITiZAN was, therefore, always conceived as a citizen science project with grand ambition on a grander scale. Key project partners reflected this ambition and included Historic England, The Crown Estate, National Trust, Council for British Archaeology and Nautical Archaeology Society, each with established national networks of specialists and volunteers to support the project. Museum of London Archaeology (MOLA) led the project which was headquartered at their offices in London.

6.3   A CITIZEN SCIENCE MODEL
FOR MANAGING COASTAL HERITAGE

CITiZAN's participatory framework was built upon the successful models of two preceding coastal community archaeology projects, Scotland's Coastal Heritage at Risk Project, started in 2012 and ongoing and based at the University of St Andrews (SCHARP, n.d.) and the Thames Discovery Programme, started in 2010 and ongoing and based at Museum of London Archaeology, London (TDP, n.d.). Both projects developed successful participatory models for volunteer led recording and monitoring programmes on the coast. Facing a similar issue of scale, SCHARP developed an app that allowed anyone with a suitable phone to rapidly and simply survey intertidal archaeological features. Designed to provide a few key pieces of data to the project team, volunteers were able to upload photographs, a GPS-derived location and description of the feature in question. The data was used to build a map of coastal archaeology in Scotland, with volunteers able to monitor these sites by providing photographic updates throughout the year, highlighting those sites that were most at risk or were actively eroding. SCHARP's app proved that, given the right tools, volunteers could contribute meaningful, quality data remotely using technology familiar to them, and to interact with the team remotely for advice and feedback. The app was designed to supplement fieldwork projects around the Scottish coast, providing an opportunity for those that couldn't attend in person to contribute to the project.

The TDP also operate a successful framework for in person survey sessions on the Thames foreshore. Typically, volunteers attend two-day training events to develop a basic understanding of Thames archaeology and history, and the key recording techniques used on the foreshore. A session of foreshore recording follows before volunteers are formally recognised as member of the Foreshore Recording and Observation Group (a FROG), ready to survey and monitor the foreshore autonomously, sending their results to the TDP team. The two-day events are just as much as an opportunity to meet people with a similar interest and to form, or become part of established, FROG groups already operating on the Thames and providing a social network for volunteers, with many forming new friendships on the back of the experience. The TDP also offers a range of events for the public beyond registering as a formal volunteer. Guided walks, lectures, open foreshore days, and partnerships with other Thames-centric organisations ensured a range of means for people to participate in the project, particularly if archaeology wasn't their primary interest.

6.4   The CITiZAN Structure

CITiZAN was designed to combine and build upon the successful approaches of both SCHARP and the TDP, and was delivered in two phases, the first between 2015 and 2018 and the second between 2019 and 2022. Phase one focussed on establishing the projects' identity, aims and app-based methodology on a national scale. Key goals were to promote use of the CITiZAN app, train volunteer teams in coastal communities to autonomously record and monitor the archaeological heritage of England's foreshore; raising public awareness of the issues addressed by the project and developing a network of professionals and academics to support the long-term sustainability of the initiative. Six community archaeologists (CAs) were divided geographically in the southwest, north and southeast, with a project manager and project lead in London, made up the core team.

A bespoke app was commissioned as the primary means for the public to participate in the project. A modified version of that developed by SCHARP, the data captured by the CITiZAN app was amended to support the database structure of the Historic Environment Records (HERs), in theory allowing a seamless transmission of data from the project to the archive. The app required users to register with the project in order to submit data which was moderated by the project team before being put live on a publicly accessible interactive map of England (CITiZAN, 2022). A feedback loop to the contributor was designed to support their development as foreshore surveyors and provide a line of communication with the project team. In principle, the app was a simple, effective means of public participation. In practice, that wasn't always the case, a point addressed towards the end of this chapter.

The CA's remit was to create teams of volunteers capable of autonomously monitoring the coastline. A programme of events, including but not limited to, talks, seminars and guided walks, was developed to support a series of volunteer training events that mirrored the TDP approach. Opportunities to attend were promoted via local social media channels, through partnerships with local groups, and with national bodies with an established local volunteer base. Taking place at weekends, day one focussed on skills development and included basic archaeological concepts, photography, basic finds identification and rapid foreshore survey techniques. Day two was typically dedicated to foreshore survey at a key site nearby, putting skills into practice and building volunteer confidence identifying and recording archaeological features. Each region aimed to deliver the same programme to ensure consistency in learning outcomes for volunteers, a key requirement of the funding body.

A second phase was grant funded by the NHLF to follow in 2019–2022. Funding was awarded based on the high engagement figures achieved during phase one and a well-received TV programme, Britain at Low Tide, that showcased CITiZAN and the issues surrounding coastal heritage. The second phase allowed for a revision of the methodology, primarily to scale down foreshore survey to six geographically discreet Discovery Programmes: Liverpool Bay, South Devon Rivers, East Kent Coast, Humber estuary, Solent Harbours and Mersea Island. Additional archaeologists were hired to support this new structure, and, amongst other things, new funding ensured important revisions to the app could be made. A driving factor behind these changes was to focus efforts and resources to bring more depth to foreshore survey and study. A revised training programme was established with greater variability for each Discovery Programme to best fit the communities and archaeology they were focussed on. A detailed appraisal of the second phase accompanies site reports and evaluations of the project, accessible via an interactive map linked to the CITiZAN website (https://citizan.org.uk) (Fig. 6.1).

6.5   The Mersea Island Discovery Programme—A Case Study

Each Discovery Programme was managed by a single Project Officer (PO) with part-time support from a community archaeologist, therefore each was unique in terms of its structure and delivery. Each DP had its own successes and challenges, defined by the geography and archaeology of the area in which they were operated. The author managed the Mersea Island Discovery Programme between 2018 and 2021 and is therefore why it features as a case study, but it is only one of five other examples of how the project delivered a participatory model for recording and protecting coastal heritage against a background of climatic change. Details of all DPs can, for comparison, be found via CITiZAN's website.

Mersea Island sits just off the Essex coast at the mouth of the Blackwater estuary. Home to c.6,000 people, Mersea is a true island community. At the highest tides each month, it is separated from the mainland when the Pyefleet channel breaches the Strood causeway, the only point of access to the island. It is famed for its oysters, an industry

Fig. 6.1 Location of the two study areas mentioned in the chapter: South Devon Rivers Discovery Project (L), Mersea Island, Essex (R) (Niall Finneran)

around which the community has thrived for centuries and that many lifelong residents and families have been a part of. Along its southern shores, at the lowest of tides, sprawling mudflats stretch over two kilometres from the modern shoreline, exposing a beguiling landscape rich with archaeological evidence of the islands' prehistoric, Roman, Saxon, Medieval and more recent past. It is here that over 200 archaeological features have been recorded by members of the public and a dedicated team of CITiZAN volunteers since 2016 (Northall, 2019).

It is important to note that CITiZAN's presence on Mersea Island is a direct result of an already bubbling undercurrent of archaeological intrigue. It was a Mersea resident who first contacted the project team via the website. They were looking for help identifying features on the local foreshore that he had grown up observing and had been monitoring for decades. Learning of the CITiZAN project through a Google search, a meeting was arranged at which the form and tone for the project would be set. CITiZAN's study on Mersea Island was almost exclusively led by the residents of the island, centred on a core group of well-connected individuals within the community. Such was the scale of the exposed archaeological landscape that, without this intimate local knowledge and guidance, much of import would have gone unobserved. Areas of the mudflats were suggested and scouted, fish and oysterman were probed about their recent observations on the estuary, the local museum catalogues and finds consulted, aerial surveys conducted and much more, all by a committed group of volunteer residents. This framework would prove vital to the popularity of the project, making it truly a co-developed initiative. Members of the community anecdotally felt more able to participate because friends and trusted members of the community were already invested and were recognised on the island as being linked to the project. This self-seeded network was also an important conduit through which other members of the community could pass on information easily and more socially. With so many people having such direct and frequent access to the foreshore for a variety of reasons, a conversational approach to sharing sightings and finds proved invaluable.

A more formal structure of free talks, visiting lecturers, exhibitions, guided walks and finds identification events ran alongside frequent foreshore walks, surveys and chats at the local cafés. Larger in scale, they served to share the work of the volunteer team and to inspire people to participate in the project. These events were relatively successful with many in attendance as a result of word of mouth and general conversation about the project on the island. Between 2016 and 2022 over 700 people attended a total of 21 CITiZAN events on Mersea Island, with several evening talks 'selling out' and audiences of over 80 people gathering in school halls and community centres, eager to learn more of their island past. A further 72 people attended training sessions (capped at eight–ten people to ensure safety on the foreshore) and foreshore survey work over the two phases of the project. Between them they have surveyed five key sites of some significance, including a Bronze Age timber trackway, the conserved timbers now form part of a permanent display in the Mersea Island Museum. Funding for the display raised was locally to the tune of £6,000 (Hutchinson, 2022a). Another site surveyed was a mid-Saxon, likely fishing complex, with a high level of structural preservation (Hutchinson & Newman, 2022) and an as yet, unidentified but large mid-Saxon timber feature stretching over 160 m from beach to low water (Hutchinson, 2022b). Surveying these sites brought the volunteers into contact with a range of experts from MOLA, Historic England and Colchester Council, and it was these opportunities that they treasured the most:

> A Mersea volunteer: 'These projects are not about individuals; they are about whole communities. The future lies with and relies upon the generations to follow within those communities. Local people with local knowledge, not people who are here today and gone tomorrow with merely a passing interest, not one day trippers who come to snatch a brief moment to add to the list of places they have visited. The value of such projects lies in capturing the very hearts and minds of those people who have made their homes in those places and for whom moving away would be a last resort, many families have lived in, or return for holidays to these locations for generations, whose pride in that continued association is embedded in their everyday lives'.

The organically developed structure of the project gave it a truly social aspect, the value of which cannot be emphasised enough by the author. Where participants were given the freedom to lead their own investigation, with friends and people they trusted, the confidence of those participants to engage with the aims of the project, its findings and its future was remarkable. Of course, this is not the first archaeology project to embed itself in a community, but it is one of a small number in a coastal setting. The coast is, by and large, an attractive place to spend some time. As several volunteers noted, particularly those who had travelled from further afield to Mersea their interest in the project was as much for access to the coast and to meet with people as it was for the archaeology. The delightful setting of this community study also surely played a significant role in its popularity. The CITiZAN project as a whole and the Mersea Island DP were able to capitalise on an already considerable knowledge base that exists in coastal communities around the English coast. Its relatively continual presence successfully galvanised the community, supporting them to play an active and valuable role in protecting their coastal heritage.

The South Devon Rivers Discovery Project

Another sub-project within the 2015–21 CITiZAN management programme was the South Devon Rivers Discovery Project which was initially devised by one of the present authors (NF). Work focussed on recording the foreshore heritage of the Rivers Exe, Teign and Dart in South Devon. Each river has its own distinctive environmental and cultural character: the Dart is very much a focus of leisure and tourism, picturesque and with a large and devoted yachting and boating community (and in terms of socio-economic profile, relatively wealthy). The Teign is very much an industrial river, with a historic focus on granite, clay and metals exploitation. The Exe, the longest of the rivers and with the widest estuary, in contrast, is lesser exploited. Small settlements are located along the river estuary, and there is a significant National Nature Reserve at Dawlish Warren. During the COVID-19 pandemic, it was impossible to arrange face-to-face training sessions (these had proved to be very popular, taking in foreshore recording on the Exe and work at the coastal site of Berry Head in Torbay on the Torbay coastal Geopark). Project archaeologist Grant Bettinson used this time to develop self-guided walks along the estuaries (low-tide trails) and historic mapping (Storymaps) using crowd-sourced content. In addition regular blogs and webinars from guest speakers keep the project very much in the public eye during a time when any form of social interaction was impossible. This is one of the main legacies of the CITiZAN work here (see https://citizan.org.uk/discovery-pro grammes/south-devon-rivers/).

6.6 What Could CITiZAN Have Done Better?

Participatory projects operating on a national scale are no easy feat to deliver. Borrowing from the successes of other, similar project models, whilst helpful in the planning stages, was not always a guaranteed route to success. In planning CITiZAN it was difficult to estimate the level of public interest in the project, and perhaps a conservative estimate during that phase resulted in too few project staff to accommodate and develop this enthusiasm sustainably. In order to try and sustain the support of some local groups for example, the project arguably tried to accommodate too many ideas that weren't directly relevant to its core aims. Rather, on occasion it tried to satisfy a range of views of what the project *should* be doing rather than what it set out to achieve, resulting in a drawn-out case of mission creep. By aligning itself with the challenge of a rapidly eroding coastline it ran the risk of being viewed as a traditional archaeological rescue project but without the financial and archival means to be so, a situation that on occasion put it at odds with partner organisations and concerned members of the public who were frustrated at the project's inability to act quickly to rescue certain features.

In terms of the participatory model, a sustained focus on the CITiZAN app across both phases of the project was unfortunately difficult to maintain. As the central means of public participation, the app was not supported as diligently as it could have been. The app could (still) be conceived of as a standalone project that would have benefitted from constant refinement, creating what could be a useful tool for professional archaeologists and the public beyond the life of the project. The power-fully simple ideas behind the app were lost behind an occasionally difficult and drawn-out user experience made heavy by over technical language and detail that many found too clunky and specialist. It suffered from dated design throughout the second phase (by that time it was five years old) and ultimately became a side line to the more detailed foreshore surveys, perhaps alienating those users around the coast who had supplied a stream of data over the first three years. As the public facing element of the project there needed to be much more attention paid to functionality and feedback. Without a budget for consistent refinement and develop-ment this opportunity was lost, along with the chance to offer a means of continued participation for those unable to attend public events.

Participatory projects such as CITiZAN can provide unique oppor-tunities to work with an enormously passionate, knowledgeable and experienced community. The varied interests and careers of the dedi-cated volunteers and members of the public who took part in the Mersea Island Discovery Programme were central to making it a success. They provided fresh perspectives on the archaeological evidence they so routinely observed, perspectives that were born of an intimate knowledge of the landscape in which many had lived most of their lives. Beyond the issues the project faced, both self-inflicted and encountered, public enthu-siasm for coastal archaeology was always evident. The CITiZAN model works best when practitioners have the time and freedom within a project to embed themselves in the communities with which they're working. In an environment as active and changeable as Mersea, a regular presence is vital to capture and act upon the observations of the community, retaining their interest and commitment to the project.

Tailored versions of the model have obvious applications in coastal communities around the world (Milne et al., 2023), and for a variety of subjects beyond archaeology such as foreshore habitat loss and change and coastal erosion, to name but two. A small pilot project funded by the Natural Environment Research Council (NERC) titled *Changing Minds Changing Coasts,* established a methodology for the study of coastal change on Mersea Island over the last century. Delivered by CITiZAN's Mersea team during the COVID-19 pandemic, the project remotely gathered old postcards, photographs, maps and oral histories. The resulting database was used to create a GIS exploring how the islands coast had been impacted by factors including estuarine pollution and changes in marine flora and fauna (accessed via https://citizan.org.uk/low-tide-tra ils/mersea-island-changing-minds-changing-coasts/). The project found that the combination of un-seen materials it was able to obtain from residents coupled with oral histories allowed for a considerable degree of accuracy in plotting changes over the century, including a level of detail that is otherwise difficult to attain through more traditional scientific models (Hutchinson et al., 2021). The project provided an alternative way to develop the CITiZAN model to gather valuable data that would otherwise likely never been encountered by those studying the subject matter. It offered residents a unique way to engage with the project using their own resource and memories and has since been developed into a new Heritage Lottery Funded Project, again being led by MOLA.

The challenge of protecting our coastal heritage is certainly one that can be met by the public where accessible tools and participatory frameworks can sit alongside adaptability to local needs and interests. So varied is the archaeological evidence around our coastline that the interests of a wider proportion of the general public can be focussed on the protection of their coastal heritage. The scale of the CITiZAN project, its achievements, number of people involved in its six discovery programmes are testament to this potential (see Morel et al., 2023).

6.7   Conclusion

The foregoing discussion has focussed upon a participatory community heritage/archaeological project. It offers a good example of interplay between academics, specialists and the public. One of the present authors (OH) has a background within archaeological fieldwork and survey, giving a strong and credible practical underpinning to the delivery of the training. The other author (NF) has worked in an advisory academic capacity on the South Devon Rivers Discovery Project (see box; one of CITiZAN's sub-projects). This has involved helping develop training materials and underpinning research materials. Both authors have in their different ways engaged with the public during training and outreach events and as noted above although there have been obvious logistical difficulties, it is clear that well-thought-out participatory initiatives such as this have the potential to seize the imagination of the non-specialist (some of whom may in fact have better local knowledge than the so-called experts). It is this balance that makes this participatory approach so refreshing.

In the UK our coastal heritage assets, cultural and natural, are at risk from climate change. A CITiZAN group working in Poole Harbour in conjunction with the National Trust undertook photogrammetric recording of an eighteenth-century brick limekiln on Brownsea Island (see https://www.citizan.org.uk/interactive-coastal-map/78380/). Using free-source commercially available imaging software, it was possible to build a three-dimensional virtual model of the kiln and its surroundings. This is preservation via recording; the National Trust coastal management strategy is now clear: don't work against nature (National Trust nd). Manage the natural process of erosion due to rising sea levels and more frequent storm events and accept that coastal heritage assets are going to disappear. To professionally record these assets would be a hugely time consuming and expensive exercise. This is where community-focussed action comes into its own. Furthermore, one of the present authors (NF) who works on maritime cultures in the Caribbean is also exploring ways to adapt the CITiZAN model for use in insular Caribbean coastal settings. CITiZAN has now completed its two tranches of funded work yet its legacy remains and will remain an important model going forward. The third iteration of the project will focus more explicitly on the collection of oral histories around England's changing coastline, maintaining the strong personal and participatory ethos which can, as we have shown, underpin similar conservation and management projects, not just in the UK and not just on coastal archaeology.

REFERENCES

Archaeology Data Service. (n.d.). *Rapid Coastal Zone Assessment Surveys.* https://archaeologydataservice.ac.uk/archives/view/rczas/. Accessed 1 January 2023.

Bailey, G., Momber, G., Bell, M., Tizzard, L., Hardy, K., et al. (2020). Great Britain: The intertidal and underwater archaeology of Britain's submerged landscapes. In G. Bailey, N. Galanidou, H. Peeters, H. Jöns, & M. Mennenga (Eds.), *The archaeology of Europe's drowned landscapes* (pp. 189–219). Springer.

BBC. (2014, February 17). *10 key moments of the Winter storms.* BBC News. https://www.bbc.co.uk/news/uk-26170904. Accessed 12 October 2022.

Cattermole, A., & Hoggett, R. (2020). *Review of the Rapid Coastal Zone Assessment Survey (RCZAS) Programme, Project No. 7693.* https://historice ngland.org.uk/research/results/reports/8022/ReviewoftheRapidCoastalZ oneAssessmentSurvey(RCZAS)Programme. Accessed 5 May 2022.

CITiZAN. (2022). *Reports and research outputs: An interactive map.* https://sto rymaps.arcgis.com/stories/02aabaf1362a4c94a39ff63634acb73c. Accessed 12 October 2022.

Hutchinson, O. (2022a). *A Late Bronze Age timber trackway in the intertidal zone off Coopers Beach, Mersea Island, Essex.* https://www.citizan. org.uk/media/medialibrary/2022/11/Coopers_Beach_Report_2022_1.pdf. Accessed 27 February 2023.

Hutchinson, O. (2022b). *A large mid- Saxon timber structure in the intertidal zone Sea View, Mersea Island, Essex.* https://www.citizan.org.uk/ media/medialibrary/2022/11/Sea_View_Report_2022_1.pdf. Accessed 27 February 2023.

Hutchinson, O., & Newman, D. (2022). *A mid-Saxon structure in the intertidal zone at Point Clear, River Colne, Essex.* https://www.citizan.org.uk/ media/medialibrary/2022/11/Point_Clear_Report_2022_1.pdf. Accessed 27 February 2023.

Hutchinson, O., Newman, D. & Northall, L. (2021). *Changing Minds Changing Coasts: 100 years of coastal change on Mersea Island, Essex. A CITiZAN interim report.* https://citizan.org.uk/media/medialibrary/2021/ 11/CMCC_NERC_Paper_V3.pdf. Accessed 27 February 2023.

IPCC (Intergovernmental Panel on Climate Change). (2022). *Climate Change 2022: Impacts Adaptation and Versatility Summary for Policy Makers.* https://www.ipcc.ch/report/ar6/wg2/downloads/report/IPCC_AR6_ WGII_SummaryForPolicymakers.pdf. Accessed 12 October 2022.

Masselink, G., Scott, T., Poate, T., Russell, P., Davidson, M., & Conley, D. (2016). The extreme 2013/2014 winter storms: Hydrodynamic forcing and coastal response along the southwest coast of England. *Earth Surface Processes and Landforms, 41*(3), 378–391. https://doi.org/10.1002/esp.3836

Milne, G., Newman, D., Hutchinson, O., & Northall, L. (Forthcoming 2023). Citizen Science in Coastal Archaeology: CITiZAN's Community-based Research in England, UK. In D. Scott-Ireton, E. Jones, & T. Raupp (Eds.), *Citizen science in maritime archaeology: The power of public engagement.* University Press of Florida.

Morel, H., Band, L., Barrie-Smith, C., Bettinson, G., Griffths, S, et al. (Forthcoming 2023). Water Heritage and the Importance of Local Knowledge in Climate Action. *Journal of Historical Archaeology Special Issue: Historical Archaeology's Response to the Climate Crises.*

Murphy, P. (2014). *England's Coastal Heritage: A review of progress since 1997.* English Heritage Publications.

National Trust. (n.d.). *Living with change: Our shifting shores.* https://www.nationaltrust.org.uk/our-cause/nature-climate/climate-change-sustainability/living-with-change-our-shifting-shores. Accessed 13 January 2023.

Northall, L. (2019). Searching Mersea: Coastal archaeology, oral history and rising sea levels. *The Archaeologist, 108,* 16–17.

SCHARP (Scotland's Coastal Heritage at Risk Project). (n.d.). *Coastal heritage at risk.* https://scapetrust.org/coastal-heritage-at-risk/. Accessed 13 January 2023.

TDP (Thames Discovery Programme). (n.d.). *The Thames Discovery Programme.* http://www.thamesdiscovery.org. Accessed 10 January 2023.

Wilkinson, T., & Murphy, P. (1995). *Archaeology of the Essex Coast, Volume I: The Hullbridge Survey.* East Anglian Archaeology 71. Bury St Edmonds: Suffolk County Council.

CHAPTER 7

Managing a UNESCO World Heritage Site in a Post-colonial, Post-conflict and Post-disaster Destination. The Case of the *Haitian National History Park*

Jocelyn Belfort, Hugues Séraphin, and Godson Lubrun

7.1 Introduction

Heritage is constructed in the present for the present (Park, 2014). Sites of historical and social significance, albeit not originally designed as tourist attractions, have been 'heritagized' and integrated into destinations' promotional strategies (Da Cunha, 2019). For example, in Rio de Janeiro (Brazil), *favelas* (so-called slum areas), have become key aspects

J. Belfort
Université Vincennes Saint-Denis (Paris 8), Saint-Denis, France

H. Séraphin (✉)
Tourism, Hospitality and Events, Oxford Brookes University, Oxford, UK
e-mail: hseraphin@brookes.ac.uk

G. Lubrun
Haitian Ministry of Tourism, Port-Au-Prince, Haiti
e-mail: godson.lubrun@tourisme.gouv.ht

© The Author(s) 2024
N. Finneran et al. (eds.), *Managing Protected Areas*,
https://doi.org/10.1007/978-3-031-40783-3_7

99

of official sightseeing tours and government promotion strategies, along-side more 'traditional' heritage sites such as churches and museums (Da Cunha, 2019). The values which define heritage are constantly shifting in response to external social, cultural and economic factors. The growing interest visitors have for heritage has positive economic impacts for destinations, but also as has long been recognised, a downside, as the overflow of visitors is impacting negatively on the preservation of many world heritage sites. Venice (Italy) epitomises this situation (Nolan & Séraphin, 2019): its heritage as a key visitor attraction, has become a victim of its own success. These are but some of the delicate issues faced by heritage tourism professionals.

This chapter focuses on heritage tourism management in a post-colonial, post-conflict and post-disaster destination focusing on the western Caribbean island of Haiti. Haiti has a single UNESCO World Heritage Site, The Haitian National History Park, which comprises three significant historical built sites, Citadel, Sans-Souci and Ramiers (https:// whc.unesco.org/en/list/180). These nineteenth-century structures are significant and unusual in a Caribbean context because they were built by freed enslaved African-Caribbean peoples (much of the nineteenth-century built Caribbean heritage, for example Barbados' UNESCO World Heritage Site at the Garrison is largely a product of European colonial activity). Although the case study is a cultural site, we raise wider issues that speak to Haiti's ability to effectively manage and promote its natural heritage too (Fig. 7.1).

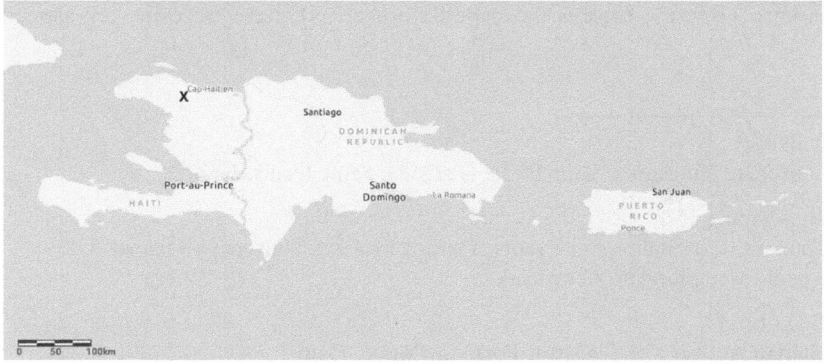

Fig. 7.1 Location of the site (Niall Finneran, basemap free source ArcGIS)

This study is of importance because published literature (see, for example, Da Cunha, 2019; Dodds & Butler, 2019; Séraphin et al., 2020) discusses the impacts of over-tourism on heritage, but little is known about the management of heritage sites in under-visited destinations such as Haiti. This study is addressing this gap in knowledge. Haiti is one of the poorest countries in the Western Hemisphere (c. US$ 1800 per capita in late 2022 according to the World Bank [World Bank, 2022]). Unlike other Caribbean islands, Haiti struggles to attract tourists—900,000 according to the 2019 estimates (*source* World of Data, 2019)—and ranks 21st in the Caribbean. Tourism as a whole contributes only 3 per cent of the Gross Domestic Product. The central question of this chapter is how can the Haitian National History Park play a greater role in driving the tourist economy here and what steps need to be taken to ensure long-term and sustainable success?

7.2   A Wider Context

In this section, we unpack key theoretical and operational concepts that underpin the analysis in this chapter. If we take the UNESCO framework for defining World Heritage Sites as a starting point (UNESCO, n.d.), heritage can be defined as related to cultures or to natural environments. In reality, UNESCO world heritage policy recognises that sites cannot be neatly pigeonholed (Uluru in Australia is an example of a mixed site). UNESCO also defines another sub-category of cultural heritage, intangible heritage, which relates to non-physical expressions of human culture such as food, dance, storytelling and craft-making traditions (Smith, 2006) which have become increasingly popular over recent decades contributing to the basis for experience-focused tourism.

Heritage is often used as a key element of destination marketing strategies in tourism (Park, 2014). This is typically the case of destinations promoting or branding themselves as cultural and 'authentic' destinations (Davidas, 1997). It is important to highlight the strong potential that history, culture and heritage tourism have in terms of capacity to attract visitors. This is mainly due to the fact that anything related to heritage tends to be connected with positive emotions and feelings such as nostalgia, trust, attachment, commitment, etc. Within the context of the Caribbean in particular, heritage tourism is used by destinations to stand out from other destinations that are mainly focusing on assets such as the sun, sand and sea. Heritage tourism is also addressing a consumers' needs and demands, especially for those who want to be engaged in a learning process while also having fun (Urry, 1990). Heritage tourism in this way contributes to a sense of wellbeing and fulfilment.

A negative characteristic of heritage tourism is commodification of a site (Urry, 1990). Here visitor expectations and practical concerns override the need for heritage managers to retain the authentic character of a site. Economic development and preservation of heritage are thus sometimes conflicting (Chakravarty, 2001). When the character of a heritage tourist destination is turned into a mere product or service, the destination is said to offer a 'staged authenticity' (Andrews & Leopold, 2013). Over-tourism is therefore a real threat to heritage sites; the ingress of tourists to the same place often at the same time subsequently impacts negatively on the environment of the destination, and negatively affects the quality of life of local residents (as well as the actual tourist experience itself). Sites can thus become a victim of their own success (see, for example, Sanchez & Adams, 2008; Dodds & Butler, 2019; Séraphin & Yallop, 2020; Séraphin et al., 2020).

In Haiti tourism in general, and cultural tourism in particular, remains relatively lightly developed in comparison with other Caribbean islands. The popular conception of Haiti remains stereotypically focused upon the role of Vodou in Haitian life, and although this has developed into something of a trope, Vodou is socially significant as it played a pivotal role for the survival of Haitians during their enslavement (Séraphin & Nolan, 2014). It enabled the enslaved not only to maintain a connection with their continent of origin (Africa), but also provided them with a mechanism to resist forced conversion to Christianity (Dalembert & Damoison, 2003; Saint-Louis, 2000). Vodou is undoubtedly an important component of Haitian identity, it is a vibrant and dynamic African-Caribbean religious expression yet it has also been commodified and packaged for tourist consumption (Séraphin & Nolan, 2014). Indeed, the original logo of the Destination Marketing Organisation (DMO) of Haiti designed in 1939 was a black mambo or Houngan (a Vodou priest). In 2012, it was changed to a more neutral logo depicting sun, sea and sand thus reflecting a move away from a stereotyped vodou theme (Bonnardel et al., 2020). Séraphin et al. (2016b) have argued that a balance should be kept between the marketing or commercial aspect of the branding strategy and the identity/cultural dimension of the destination. This ambidextrous management approach is important for the success of any management strategy (Vo-Thanh et al., 2020). We now turn to a more specific consideration of the local conditions on the ground in Haiti. We have noted above that in comparison with other Caribbean tourism destinations, Haiti is less developed and this is a consequence of colonial history.

7.3   Haiti as a Post-Colonial, Post-Conflict and Post-Disaster Destination

During the colonial period, Haiti was the richest French sugar colony in the Caribbean and was popularly known as the 'pearl of the Antilles' (Deneault, 2005). Under the military leadership of the rebel Toussaint L' Ouverture (1743–1803), a successful slave rebellion resulted in the overthrow of the French colonial system (Roc, 2008). Under the leadership of Dessalines, Haiti became in 1804 the first African republic in the world (Gilles, 2012). Haiti today though is associated with political instability, poverty and violence, all of which results from its historical French colonial exploitation (Eve di Chiara, 1988; Higate & Henry, 2009; Girault, 2010). The negative image of the destination represents a major barrier to tourism development of the destination (Séraphin, 2014a, 2014b). To appreciate the potential of Haiti as a tourist destination, a chronological overview of the tourist development of the island is required.

During the period c. 1939–1960, Haiti was the most popular tourist destination of the Caribbean and it traded extensively on its heritage as a selling point. Major events such as the Port-au-Prince (capital of Haiti) International Exhibition of 1949 helped to put Haiti on the map of world cultural tourism destinations (Clammer, 2012; Séraphin, 2010; Tardieu, 2014; Thomson, 2014). Haiti really reached its peak between 1950 and 1956, a period known as the Golden Age of Tourism in Haiti (Destin, 2014). From 1960 to 1986 Haiti was ruled by the Duvalier family, Francois and Jean-Claude Duvalier, respectively known as 'Papa Doc' and 'Baby Doc'. This was a period of political repression and violence. The violence orchestrated by the 'Tonton Macoute' (after the army in the service of the Duvalier) effectively put an end to the tourism industry in Haiti during this period (Séraphin, 2014a, 2014b; Thomson, 2014).

The period from 1986 to 2016 is remembered for two significant events. Firstly, the 2010 earthquake which destroyed the country, and subsequently decimated the tourism industry. Haiti became known during this period as the Republic of Non-Governmental Organisations (Séraphin et al., 2018). Secondly the administration of President Michel

Martelly (2011–2016) saw an influx of international hotels and Haiti was rebranded as a 'sun, sand and sea destination'. Businesses were encouraged to invest in the country as part of the scheme 'Haiti is open for business' (Séraphin, 2014a, 2014b). Since 2016, the country has been plunged again into political, social and economic turmoil. As a result of these events, the very limited number of tourists who visited the country were being targeted and assaulted by rioters. In addition, the COVID-19 pandemic struck in March 2020 and restrictions were placed on public travel worldwide (Jamal & Budke, 2020; Séraphin, 2020). These local and global issues have combined to make Haiti's journey to becoming a competitive, desirable heritage tourist destination even more difficult.

The Haitian National History Park (HNHP) could play an important role in the recovery of Haiti's tourism industry. The number of visitors to the park, reached its peak (80,238 visitors) in 2013, under the administration of President Martelly, whose objective was to turn the northern region of Haiti (where the park is located) into a tourism destination (Banque Mondiale, 2018; Belfort, 2018). The site is closely associated with Henry Christophe, a self-proclaimed king who came to power in 1807, and who was responsible for the construction of the key structures in 1811. Nowadays, the HNHP features in most of the promotional materials of the Haitian Destination Marketing Organisation (Belfort, 2018). This said, the HNHP is more than a tourism or UNESCO World Heritage Site. It represents one of the most widely recognised symbols of the Haitian heritage, as many fights for freedom against the French colonial system started from Citadelle Laferriere (Belfort, 2018). As such the HNHP occupies an important place in Haitian social memory. The HNHP only really became regarded as an important part of the Haitian heritage canon from 1932 during the administration of President Stenio Vincent (Belfort, 2018), and during this decade (1939) the first Haitian Destination Management Organisation (DMO) appeared (Séraphin, 2014a, 2014b). From the 1930s, the HNHP began to play an important role in the development of heritage tourism in Haiti.

7.4   MANAGING THE HAITIAN NATIONAL HERITAGE PARK

In this section, we examine the management of the HNHP in more detail. The work presented here is based upon first-hand fieldwork undertaken by the authors at the HNHP. Underpinning our research design are two interlinked contentions. Firstly, we recognise that the site is an important international heritage site under a global management framework (UNESCO) and it is also an economically important tourism site too. Secondly, any management plan needs to take into account these competing demands, as well as those of all the stakeholders involved (foreign and domestic tourists, local people living near the site, businesses and heritage managers in Haiti).

Our starting point is that the HNHP is very important for locals from a symbolic perspective, and it is an important part of local and national social memory. Yet owing to weaknesses in the management of the site, few local people benefit economically from the HNHP. Secondly, it is important to recognise that the potential pull factor of the HNHP is important at pre-visit stage (through marketing). However, in terms of the actual visitor experience, surveys have shown that visitors expectations are not met. The overall hypothesis is that there is a dissonance between the potential of the HNHP and what it actually delivers to both locals and to visitors: all of which is due to a poor management approach, mainly caused by the political, economic and social instabilities of the destination.

Our research study is first and foremost based on an ethnographic approach which demands the research is immersed in the quotidian life of the target groups (Hammond & Wellington, 2013, 62; (Fox et al., 2014, 64). Participant observation is a core part of the ethnographic approach, and fieldwork is a prerequisite for this method (Brunt et al., 2017).

Research was conducted over four days in the north of Haiti meeting with people in charge of tourism development in the area, observing and interviewing three tour guides and three tourists. The tour guides were asked questions about their perception of the role of tour guide, their qualifications, the content of their presentation when tour guiding and their perception of the HNHP (Séraphin, 2013). As for the tourists, one female, and two males were selected. They were asked about their motivation, experience of and perceptions of the park. Further interviews were carried out in Port-au-Prince with three experts: the DMO manager, a former Haitian tourism minister and finally the Project Manager for the Haitian National History Park in charge of the scheme *Projet de Préservation du Patrimoine et Appui au Secteur Touristique*. They were each asked about the Government plan for tourism overall as well as the importance given to heritage and the development of the *Haitian National History Park*.

A selection of excerpts from the interviews conducted are presented below. An indicative translation is provided beneath each section for English speakers.

Tour guides emphasised that tourists were keen to know about the history of Haiti:

> Quand vous faites visiter aux touristes et que vous commencez à les expliquer les lieux, ils commencent à nous poser des questions sur des lieux (souvent nos discours sont en fonction des questions posées par des visiteurs). Tout autant que la personne pose des questions, cela rend l'histoire plus belle. Cependant, on part d'une histoire construite avec la vie de Christophe, son palais, sa femme et pour terminer avec une visite de plus d'une heure à la Citadelle. C'est surtout à la Citadelle que les touristes posent beaucoup plus de questions.
>
> When you show tourists around and start explaining places to them, they start asking us questions about them (often our guided talks are based on questions asked by visitors). When a person asks questions, it makes the story more beautiful. However, we begin from a story built around the life of Christophe, his palace, his wife and we end with a visit of more than an hour to the Citadel. It is especially at the Citadel that tourists ask many more questions.

Continued visitor access is being impeded due to the site not being well maintained:

> On ne peut pas continuer à faire visiter ce monument car rares sont les espaces qui sont ouverts au public. Nous sommes obligés de continuer ainsi, mais nous savons bien qu'on peut faire mieux. On a tellement de choses à raconter sur la Citadelle.
>
> We can't continue to visit this monument because few spaces are open to the public. We have to continue like this, but we know we can do better. We have so many things to tell about the Citadel.

Some of the guides are not trained as tour guides. A result of this lack of training, and lack of focus on quality, means that the narratives conveyed by tourist guides to visitors frequently lack any sense of consistency in terms of content:

> Les guides ne racontent pas les mêmes histoires. Il y en a ceux qui racontent les vraies histoires du site et d'autres qui racontent n'importe quoi. Nous ne sommes pas responsables des dérives quant à l'histoire racontée

sur les sites. Mais les guides qui sont formés racontent la vraie histoire de la Citadelle.

Guides don't tell the same stories. There are those who tell the real stories of the site and others who tell made up stories. We are not responsible for any deviations from the story told on the sites. But the guides who are trained tell the real story of the Citadel.

In some families, being employed as a tour guide has become a tradition:

Quand mon père faisait visiter le Parc aux touristes, j'avais l'habitude de l'accompagner. Après, j'ai pris la relève, et je fais ce métier depuis plus de 40 ans.

When my father showed tourists around the Park, I got into the habit of accompanying him. Afterwards, I took over and I have been doing this job for more than 40 years.

For Cohen (1985), the role of the tour guide has four main functions: first as pathfinders, ensuring that tourists visit the right place, safely and have a good understanding of the places visited. The second role is the animator. As part of this function, tour guides need to socialise with visitors, and be able to respond well to their questions. The third role is as the tour leader. As part of this role, the guide encourages visitors to interact with each other, with local people and with their surrounding environment. Last but not least, is the professional role. As part of this function, the tour guide needs to provide detailed, accurate and consistent information to visitors.

Séraphin (2013) and Thomson (2014) note that in Haiti tour guides are not performing any of these roles because of their lack of knowledge and training. Their motivation is not always to share a passion of their knowledge about a site, but to obtain money from visitors. The lack of political stability and the insecure climate in Haiti makes it difficult for the guide to ensure the safety of visitors. It is important for local authorities to regulate the profession of tour guide, as this role is important in terms of image building and image changing of a destination (Dahles, 2002). Professional practice is a very important element in destination planning too (Meliou & Maroudas, 2010). Any strategy promoting cultural and natural heritage sites in Haiti needs to take these issues into account.

The reactions of tourists are summed up in the following brief excerpts. What is particularly evident is that those tourists, representing the Haitian diaspora, convey a particularly positive emotional importance to the HNHP:

> Dès que j'ai mis mes pieds sur la route de la citadelle, j'ai senti une émotion à la fois de tristesse, de joie et d'incompréhension. De la joie c'est quand même mon peuple qui a bâti ce moment (sic) sans ciments. Et aujourd'hui en France c'est du jamais vu. Un sentiment de tristesse parce que je me suis dit qu'ils ont souffert pour construire ce bâtiment sur cette hauteur.
>
> As soon as I set foot on the road to the citadel, I felt an emotion of sadness, joy and incomprehension at the same time. Joy that it is still my people who built this monument without cement. And today in France it is unheard of. A feeling of sadness because I said to myself that they suffered to build this building.

Haitians are proud to have a site listed by UNESCO. The park makes them proud to be Haitian (even if not born in the country):

> C'est l'une des merveilles du monde d'après l'UNESCO. Le roi Henri Christophe également, c'est quelqu'un qui est connu à travers le monde.
>
> It is one of the wonders of the world according to UNESCO. King Henri Christophe too, he is someone who is known throughout the world.
>
> Permettra de faire une rétrospective sur nous-mêmes en vue de comprendre comment les ainés avaient travaillé durs pour nous donner cette nation haïtienne, et ce que nous devons faire aujourd'hui pour garder haut ce flambeau. Quand on regarde ce qu'ils ont fait à de maigres moyens, c'est comme si nous ne faisons rien aujourd'hui. Connaître ces monuments nous permettra de mieux connaitre où nous sommes aujourd'hui.
>
> This will allow us to reflect on ourselves in order to understand how the elders had worked hard to give us this Haitian nation, and what we must do today to keep this flame alive. When you look at what they have done with meager means, it is as if we are doing nothing today. Knowing these monuments will allow us to better know where we are today.
>
> Pour moi qui ne suis pas né en Haïti, une visite à la Citadelle est une fierté. Lorsqu'on regarde ce qu'un petit pays comme Haïti avait fait il y a plus de deux-cents ans, on doit être fier de notre origine.
>
> For me, who was not born in Haiti, a visit to the Citadel is a source of pride. When we look at what a small country like Haiti did more than two hundred years ago, we must be proud of our origins.

The Haitian diaspora has played a significant role in the rebuilding of the hospitality sector in Haiti. Indeed, after the 2010 earthquake, many international hotels opened in Haiti, either owned or managed by Haitians from the diaspora (Séraphin & Paul, 2015). This diaspora is contributing to the development of significant skillsets in Haiti today (Minto-Coy & Séraphin, 2017). Paul and Séraphin (2014), even argued that the Haitian diaspora have the social responsibility to develop the country. This makes the diaspora links so crucial. Indeed, Meylon-Reinette (2010) explained that the connection between Haiti and Haitians born in Haiti and Haitians born overseas is not the same. She noticed a 'de-diasporisation' of the Haitians born abroad, in other words, a lesser interest for the home-country or country or origin.

Finally we turn to issues expressed during our interviews with tourism leaders. Key themes which arose here included: visitors receive limited if not poor understanding/knowledge of (the history of) Haiti; Park management is difficult owing to balancing activities required to meet tourism needs with expectations of residents in the Park; importance for community engagement in tourism development projects is emphasised; there is a lack of vision in terms of developing the Park as a tourism attraction; and in turn, a lack of tourism promotion and marketing of the area.

Existing literature on tourism in Haiti primarily highlights political, economic, social factors experienced in the country, conveying a negative image of the destination, impeding the development of tourism. However, little or nothing has been said about the lack of vision and leadership in tourism in Haiti. The key issue with Haiti is that there is not an educated elite (Roc, 2008). The classic role of the elite is usually one of investing in the country and generating economic activity, wealth, jobs and influencing the Government to increase trade, promote productivity, ensure stability and protect investments. In Haiti, the elite does not assume this role. Its basic activity is trade, with minor investments in case of crisis in the country. The weak governance and absence of accountability has facilitated the creation of parallel economies and patronage patterns. The Government and representatives of the private sector (the elite, local and international companies in Haiti) have done very little to encourage the sustainable economic development of the country (Barreau, 2012; Séraphin, 2012).

7.5   Conclusion

The HNHP is an important site for all Haitians (both diaspora and non-diaspora). The HNHP can contribute to fostering a sense of community among Haitians. Equally importantly, it can help to make them proud of their roots despite the negative image of the destination conveyed by the media. All attempts to turn the Park into a viable tourist product have failed, despite its huge potential. On top of the political, economic and social context, the lack of strategic political leadership is playing an important role in the current poor state of this UNESCO site. Difficulties of managing UNESCO sites in post-colonial, post-conflict and post-disaster destinations include: the wider socio-political and economic context of Haiti as a whole, a lack of professionalism of tourism actors, lack of leadership, absence of vision and lack of data insights on visitors and poor marketing images being conveyed of the destination. Clearly the political and economic climate in Haiti needs to improve before the situation might improve, but the value of this site for helping guide improvements in socio-economic terms is huge. This we would argue is also the case for other cultural and natural heritage sites on the island.

This study has highlighted the fact that for post-colonial, post-conflict and post-disaster destinations, heritage sites are more than 'just' heritage sites. They can also represent an important centralising force for good and a well of social memory and they provoke an emotional and sensory attachment (Andrews & Leopold, 2013; Chapter 2). This is a force for good for both local Haitians as well as member of the diaspora who have the potential to bring in important economic capital and educational skills to the community. Heritage can make these connections stronger (Foley et al., 2012; Kikuchi-Uehara et al., 2016; Nunkoo, 2017). Last but not least, vision and leadership are as important as a context in the success and performance of a tourism destination. Although our analysis has considered the role of a single cultural heritage site in Haiti, we argue that all heritage, cultural and natural, has an important part to play in the healthy social, cultural and economic life of the nation.

References

Andrews, H., & Leopold, T. (2013). *Events and the social sciences*. Routledge.
Banque Mondiale. (2018, April 3). *Quel tourisme pour Haïti? Renforcer la résilience et développer une nouvelle destination touristique dans les Caraïbes.*

La Banque Mondiale. https://www.banquemondiale.org/fr/news/feature/2018/04/03/what-is-next-for-haitis-tourism. Accessed 13 February 2023.

Barreau, J. (2012, August 28). *Investissements directs étrangers : La difficile équation haïtienne.* Le Nouvelliste 28 August 2012. https://www.lenouvelliste.com/article/108334/investissements-directs-etrangers-la-difficile-equation-haitienne. Accessed 13 February 2023.

Belfort, J. (2018). *La valorisation touristique du patrimoine: Outil de développement territorial et enjeux mémoriels. Le cas du Parc national historique (Citadelle Laferrière, Sans-Souci et Ramiers).* Unpublished Post Graduate thesis, University Paris 1 – Pantheon. Sorbonne.

Bonnardel, V., Séraphin, H., Gowreesunkar, G. & Ambaye, M. (2020). Empirical evaluation of the new Haiti DMO logo: Visual aesthetics, identity and communication implications, *Journal of Destination Marketing & Management.* https://doi.org/10.1016/j.jdmm.2019.100393

Brunt, P., Horner, S., & Semley, N. (2017). *Research methods in tourism, hospitality and events management.* Sage.

Chakravarty, I. (2001). Tourism development in small islands: The case of Elephanta Island, Maharashtra, India. *Tourism Recreation Research, 26*(3), 15–23. https://doi.org/10.1080/02508281.2001.11081195

Clammer, P. (2012), *Haiti.* Guildford: Bradt.

Cohen, E. (1985). The Tourist Guide. *Annals of Tourism Research, 12*(1), 5–29.

Da Cunha, N. (2019). Public policies and tourist saturation in the favelas of Rio de Janeiro. In C. Milano, J. Cheer & M. Novelli (Eds.), *Overtourism. Excesses, discontents and measures in travel and tourism* (pp. 152–167). CABI.

Dahles, H. (2002). The politics of tour guiding : Image management in Indonesia. *Annals of Tourism Research, 29*(4), 919–932. https://doi.org/10.1016/S0160-7383(01)00083-4

Dalembert, J., & Damoison, D. (2003). *Vodou: Un tambour pour les anges.* Autrement.

Davidas, L. (1997). *Chemin d'Identite. Leroi Jones/Amiri Barak et le fait culturel Africain-Americain.* Ibis Rouge.

Deneault, A. (2005). Esthetique coloniale, paradis fiscaux et vahines. In P. Blanchard & N. Bancel (Eds.), *Traces et Memoires coloniales en France* (pp. 286–296). Autrement.

Destin, Y. (2014). Haiti's prized presidential legacies. *The Journal of Haitian Studies, 20*(2), 191–203.

Dodds, R., & Butler, R. (Eds.). (2019). *Overtourism.* De Gruyter.

Ève di Chiara, C. (1988). *Le Dossier Haiti: Un pays en peril.* Talladier.

Foley, M., McGillivray, D., & McPherson, G. (2012). *Event Policy. From theory to strategy.* Routledge.

Fox, D., Gouthro, M. B., Morakabati, Y., & Brackstone, J. (2014). *Doing events research. From theory to practice.* Routledge.

Gilles, A. (2012). The social bond, conflict and violence in Haiti. https://www.prio.org/publications/7378. Accessed 13 February 2023.

Girault, C. (2010). Deux cents ans de malheur. *Le Nouvel Observateur*, No. 2359, 42–43.

Higate, P., & Henry, M. (2009). *Insecure spaces*. Zeb Books.

Jamal, T., & Budke, C. (2020). Tourism in a world with pandemics: Local-global responsibility and action. *Journal of Tourism Futures, 6*(2), 181–188. https://doi.org/10.1108/JTF-02-2020-0014

Kikuchi-Uehara, E., Nakatani, J. & Hirao, M. (2016). Analysis of factors influencing consumers' proenvironmental behavior based on life cycle thinking. Part II: Trust model of environmental information. *Journal of Cleaner Production*. https://doi.org/10.1016/j.jclepro.2016.03.011

Meliou, E., & Maroudas, L. (2010). Understanding tourism development: A representational approach. *Tourismos, 5*(2), 115–127.

Meylon-Reinette, S. (2010). De la dédiasporisation des jeunes Haïtiens à New-York. *Etudes Caribéennes, 16*. https://doi.org/10.4000/etudescaribeennes.4628

Minto-Coy, I., & Séraphin, H. (2017). Role of the diaspora in the emergence of Economic and Territorial Intelligence in Haiti. *International Journal of Business and Emerging Markets, 9*(1), 48–67. https://doi.org/10.1504/IJBEM.2017.10001090

Nolan, E., & Séraphin, H. (2019). Venice: Capacity and tourism. In R. Dodds & R. Butler (Eds.), *Overtourism* (pp. 139–151). De Gruyter.

Nunkoo, R. (2017). Governance and sustainable tourism: What is the role of trust, power and social capital? *Journal of Destination Marketing Management, 6*(4). https://doi.org/10.1016/j.jdmm.2017.10.003

Park, H. (2014). *Heritage tourism*. Routledge.

Paul, B., & Séraphin, H. (2014). L'Haitianité et la responsabilité sociale de la diaspora dans le développement d'Haïti. *Études Caribéennes, 29*. https://journals.openedition.org/etudescaribeennes/7161

Roc, N. (2008). *Haiti-Environment: from the 'Pearl of the Antilles' to desolation*. https://policycommons.net/artifacts/1485022/haiti-environment/2143963/. Accessed 13 April 2023.

Saint-Louis, F. (2000). *Le Vodou Haitien. Reflet d'une société bloquée*. Harmatan.

Sanchez, P., & Adams, K. (2008). The Janus-faced character of tourism in Cuba. *Annals of Tourism Research, 35*(1), 27–46. https://doi.org/10.1016/j.annals.2007.06.004

Séraphin, H. (2010). Quel avenir pour le tourisme en Haïti? *Revue Espaces, 281*, 4–6.

Séraphin, H. (2012). *Private and public sector initiative for the development of entrepreneurship in Haiti: The tourism industry, shouldn't it be the priority*, 2nd international conference in socially responsible and sustainable entrepreneurship and innovation, University of Southampton, 24–25 October 2012.

Séraphin, H. (2013). *The contribution of tour guides to destination understanding and image. The case of Haiti via an analysis of: 'Bonjour blanc, a journey through Haiti'*. International Research Forum on Guided Tours, Breda University of Applied Sciences (Netherlands), 4–6 April 2013.

Séraphin, H. (2014a). *Le Tourisme: L'Ouverture pour le Peuple de Toussaint?* Publibook.

Séraphin, H. (2014b). Bonjour blanc a journey through Haiti: An allegory of the tourism industry in Haiti. In S. Quinter & R. Baleiro, R. (Eds.), *Lit &Tour : ensaios sobre literatura e turismo* (pp. 355–381). Humus.

Séraphin, H. (2015). *Marketing and tourism. Research method for the segmentation of the Haitian diaspora*. International Journal of Arts & Sciences (IJAS) International Conference for Business and Economics, Florence (Italy), 16–19 June 2015.

Séraphin, H. (2020). COVID-19: An opportunity to review existing grounded theories in event studies. *Journal of Convention and Event Tourism, 22*(1), 3–35. https://doi.org/10.1080/15470148.2020.1776657

Séraphin, H., & Nolan, E. (2014). Voodoo in Haiti: A religious ceremony at the service of the 'Houngan' called tourism. In W. Frost & J. Laing (Eds.), *Rituals and traditional events in the modern world* (pp. 221–231). Routledge.

Séraphin, H., & Paul, B. (2015). La diaspora: Un levier pour le développement du tourisme en Haïti, *Mondes du Tourisme, 11*. http://tourisme.revues.org/990

Séraphin, H., & Yallop, A. (2020). *Overtourism and tourism education. A strategy for sustainable tourism futures*. Routledge.

Séraphin, H., Gladkikh, T., & Vothanh, T. (Eds.). (2020). *Overtourism: Causes, implications and solutions*. Palgrave Macmillan.

Séraphin, H., Smith, S., Scott, P., & Stokes, P. (2018a). Destination Management through organisational ambidexterity: A study of Haitian enclaves. *Journal of Destination Marketing and Management, 9*, 389–392. https://doi.org/10.1016/j.jdmm.2018.03.005

Séraphin, H., Yallop, A., Capatina, A., & Gowreesunkar, V. (2018b). Heritage in tourism organisations' branding strategy: The case of a post-colonial, post-conflict and post-disaster destination. *International Journal of Culture, Tourism and Hospitality Research, 12*(1), 89–105. https://doi.org/10.1108/IJCTHR-05-2017-0057

Smith, L. (2006). *Uses of heritage*. Routledge.

Tardieu, J. (2014). *Investor et S'investir en Haiti, un acte de foi*. Editions CIDIHCA.

Thomson, I. (2014). *Bonjour Blanc, a journey through Haiti*. Vintage.

Urry, J. (1990). *The Tourist Gaze: Leisure and travel in contemporary society.* Sage Publications.

UNESCO. (n.d.). *World heritage.* https://whc.unesco.org/en/about/. Accessed 12 April 2023.

Vo-Thanh, T., Séraphin, H, Okumus, F. & Koseoglu, M. (2020). Organizational ambidexterity in tourism research: A systematic review. *Tourism Analysis, 25*(1). https://doi.org/10.3727/108354220X15758301241701

World Bank. (2022). *GDP per capita Haiti.* https://data.worldbank.org/indica tor/NY.GDP.PCAP.CD?locations=HT. Accessed 12 April 2023.

World of Data. (2019). *Tourism in Haiti.* https://www.worlddata.info/america/ haiti/tourism.php. Accessed 12 April 2023.

CHAPTER 8

Sustainable Project Management of Green Spaces, Protected and Conserved Areas: Opportunities and Challenges

Malgorzata Radomska, Richard Clarke, and Denise Hewlett

8.1 Introduction

Many of our natural environments in rural areas and in urbanised contexts of public parks, green ecological corridors and wedges, survive in a dynamic, if not tumultuous, context, marked by change (Bretschger & Pittel, 2020; Head, 2022). This is a situation which has been exacerbated by the ongoing impacts of the pandemic (Crossley & Russo, 2022; Ding et al., 2022). Key challenges to their existence and consequently implications for their management, include an increasingly disconcerting and politically unstable world exacerbated for example by declining economies and financial systems, social unrest, wars, extensive migration of populations, all of which, has an impact on funding mechanisms required to

M. Radomska
PeopleScapes Research & Knowledge Exchange Centre, Department of Responsible Management and Leadership, Faculty of Business and Digital Technologies, University of Winchester, Winchester, UK
e-mail: Malgorzata.Radomska@winchester.ac.uk

N. Finneran et al. (eds.), *Managing Protected Areas*, https://doi.org/10.1007/978-3-031-40783-3_8

support the continued management of many of our urban greenspaces and protected areas. In addition, their managing agencies face additional demands on natural environments to help ameliorate environmental concerns including managing impacts of climate change, pollution (air, water, noise), natural disasters and biodiversity crises (Geldmann et al., 2015; Ranius et al., 2023; UNEP, 2022). And yet, often concurrently vast swathes of natural land rich in biodiversity are regularly being exploited for humanity's development purposes.

To work within this dynamic and highly challenging context demands that managing agencies of protected areas and greenspaces, frequently need to exercise their abilities in what is considered as best practice governance, planning and management (Pantaloni et al., 2022). To lead on such projects, requires an appropriately supportive organisation and project-oriented culture from within the managing agency. That agency can underpin the necessary processes for the effective management of multiple projects, with skilled project managers who can manage limited resources in an optimised way (Yan & Tang, 2021). There are many examples found in greenspace and protected area contexts of project-orientated organisations who can be characterised by their 'seemingly endless ability to carry out small and extensive management of internal and external projects, (of) projects that can be repetitive', i.e., related to land management regimes, but many 'projects can also be considered quite unique' (Gareis, 1991, 71). Additionally, the drive towards a project-orientated structure in these organisations, is driven by not only a need for change, but potentially could be interpreted as engineered through traditional funding mechanisms driven by national governments for which project teams need to competitively bid for funds with managing agencies and

R. Clarke (✉)
Environmental Consultant, Marlborough, UK
e-mail: Richard.Clarke069@btinternet.com

D. Hewlett
PeopleScapes Research & Knowledge Exchange Centre, Department of Responsible Management, University of Winchester, Winchester, UK

Bournemouth University, Dorset, UK

D. Hewlett
e-mail: denise.hewlett@winchester.ac.uk

from funding authorities. This is a common approach used in the UK and EU contexts.

Moreover, in this region, due to increasingly limited staff numbers, and yet a wide-ranging knowledge base required to effectively manage greenspaces and protected areas, it is quite usual to find a culture of sharing best practices, and knowledge. This is frequently supported through project work amongst a variety of agencies with responsibility for a given geographic area. Furthermore, similar threads of working amongst managing agencies' activities are evident. Following best practice governance guidelines, through stakeholder engagement, collaborative discussions and often multiple discourse of compromises, a range of well planned, often multiple projects are developed to enable conservation, recreation and/or development goals for an area. Such projects are frequently targeted to achieve planned objectives, and are designed so that they can, on completion, be evaluated in terms of their output, outcomes and benefits often for a wide variety of stakeholders. An important question regarding project management in green spaces is the extent to which such projects can be delivered successfully given the difficult context in which they operate. Another important point is whether such projects can be delivered in a sustainable way: a perspective frequently critiqued, that can be interpreted in multiple ways, but for the purpose of the focus of this chapter, sustainability is taken according to the IUCN as comprising of 4 key principles (Worboys et al., 2015):

- **Leadership**—considered essential for ensuring that sustainability is central to delivery and design.
- **Triple bottom line**—addressing environmental, social cultural and economic needs of an area.
- **Whole of life asset thinking**—programmes and projects should have due regard for the full extent of the life of programmes and projects.
- **Resource efficiency**—making maximum use of all-too-often, limited resources.

The aim of this chapter is to elaborate on a few selected aspects of project and policy management challenges in the subject of green spaces and protected area management. This has required drawing upon two bodies of knowledge: firstly, providing for context, from green spaces' initiatives, regulations and policy contexts in the UK and representing international contexts, on policies, standards and best practice guidelines in protected area contexts; and secondly, recognising project management as a discipline in its own right, we examine project management in greenspace and protected area contexts against principles, concepts and current key theories generated in the generic literature on project management. This has provided useful insights on how the management of projects in greenspace and protected area contexts might be enhanced. Examples of the development of effective policy concerned with green spaces and protected areas in the UK include: UK government's 30 by 30 targets; the IUCN's pilot project to progress 'Green Listing' certification across the UK: and the National Association for Areas of Outstanding Natural Beauty (NAAONB) along with the individual Areas of Outstanding Natural Beauty partnerships, joint Colchester Agreement. These policy initiatives, resemble projects in that they have needed to be delivered in a coordinated and comprehensive way. A selection of policy initiatives and projects will be developed in this chapter to help us better understand the challenges faced by project teams and provide us with potentially useful tips on how to improve project management practices in green spaces and protected area management.

8.2   Green Spaces, Protected and Conserved Areas: An Overview of International and UK Project Initiatives

Protected areas need to be well-governed, well-designed and well-managed in order to be able to conserve nature and provide social, economic, cultural and spiritual benefits (Jiricka-Pürrer et al., 2019; Puhakka et al., 2017; Verschuuren & Brown, 2019). Therefore, over decades, many international and national initiatives and guidelines have been developed to protect the nature found in these areas and support green space managers in their daily work. For example, in the UK, the Countryside Commission was established in 1968 and the Nature Conservancy Council in 1973. Both organisations aimed to provide protected area management with a sense of direction, outlined the aims and objectives to be met and equipped them with tools to assess their organisational performance (Fros et al., 2021; Jansson et al., 2019; Kelly, 2022; Young,

2011). At the international level the'30 by 30' initiative aims to designate 30% of the surface of our planet as protected areas by 2030 (see Chapter 1) and in addition, contribute to the UK's Net Zero Strategy (2021). This sets out policies and proposals for decarbonising all sectors of economy following the COVID-19 pandemic (Dinerstein et al., 2019; HM Government, 2021).

Looking at UK national legislation, it is important to refer to the 25-Year Environment Plan (HM Government, 2018). This laid the foundations of the current UK environmental policy as it has aimed to improve the environment within a generation. It focuses on activities that would contribute to: providing for clean air clean and plentiful water, thriving plants and wildlife; reducing the risks of harm from environmental hazards; using natural resources more sustainably and efficiently enhancing beauty, heritage, and engagement with the natural environment and mitigating and adapting to climate change. These objectives set out in the 25-Year Environment Plan, were reinforced in the UK Environment Act (2021) the adoption of which was to ensure that the UK's commitment towards green spaces and protected areas continued after Brexit (Lee, 2022; Reid, 2021).

Attention also needs to be paid to the Green Infrastructure Framework launched by Natural England in 2023 (Natural England, 2023). This action demonstrates the implementation of a commitment by the Government to Defra's 25-Year Environment Plan, which aims to increase the amount of green cover up to 40% in urban residential areas and creating a good quality greenspace in every local area across the country. The idea behind this action is not new. Yet it does reemphasise the importance of increasing access to natural green spaces for people's health and wellbeing and revives attention on setting a target for local government agencies to provide for public access to greenspaces within 15 minutes' walk from a residential area. In doing so, key aspirations are to contribute to improving people's health and wellbeing, enhancing air quality, contributing to support nature recovery, through which issues of social inequality and environmental decline might be addressed (see Navarrete-Hernandez & Laffan, 2019; Mosler & Hobson, 2021).

However, whereas national and international initiatives and legislation outline some ambitious strategic goals to be achieved in the near future, they provide little detailed information about how exactly their content will be achieved, who will be responsible for delivery and how to measure the progress, delivery or performance assessment that will be required. A common challenge for any strategic plan in an organisation, as in the case of the UK Government's 25-Year Environment Plan, is that it is often not followed by a realistic, well-designed and detailed operational plan: a plan that would be supported by developing specific output-oriented projects, and would be aligned with a strategic plan which could produce the desired benefits (Kjersem et al., 2017). Even more importantly is the ability to assess the extent to which the actions taken have produced a measurable outcome that has contributed to achieving the overarching goals. This is not a unique situation in the context of UK's Protected Areas, rather it is a situation which has been identified worldwide (Hockings et al., 2019). In seeking to address this issue, the International Union for Conservation of Nature or IUCN developed its Green List Certification that was launched in November 2017.

Green List Certification represents a global campaign for successful nature conservation, and has resulted in the production of the Green List of Protected and Conserved Areas Standard (IUCN, 2017). This provides a global benchmark for successful green spaces and protected area management. It provides a list of seventeen criteria under four components that are accompanied by fifty indicators which help to measure site performance and progress assessments as to the extent to which governance and management have been effective, equitable and sustainable, i.e., 'a standard that addresses the social, environmental or economic practices of a defined entity, or a combination of these' (ISEAL, 2023). Following these standards and using them to measure site performance is considered to help managers assess if their initiatives have contributed to the delivery of nature conservation results.

The three components of the IUCN Green List Standard, can be applied to management approaches of any green space or protected area. These are:

- Good governance,
- Sound design and planning,
- Effective management

and together they support the fourth component on achieving successful conservation outcomes (IUCN, 2017, 10–14). The good governance component unsurprisingly draws upon previously established good practice guidelines (c.f. Borrini-Feyerabend et al., 2013). As such at the minimum, is concerned with ensuring that governance arrangements and decision-making processes are transparent, that they are clearly defined and appropriately communicated and that they represent and address the interests of civil society and other key stakeholders. Also, any site management is expected to draw upon expert knowledge using adaptive and responsive management practices (Jaafar & Yusof, 2019). The sound planning component concerns enabling a good understanding of the social, economic and environmental context, knowing the key environmental values and attributes of a managed green space/protected area and designing a long-term sustainable management plan.

Effective management highlights the importance of developing a long-term management strategy for an area that emphasises economic conditions and has clear aims and objectives that are fundamentally supported by the adequate allocation of both financial and human resources.

However, as welcome as these criteria, components and indicators are, their construction is primarily related to process-oriented guidelines and these do not, we contend, place enough attention on key people skills needed for successful project delivery including leadership, team building and the abilities to communicate effectively and well. Multiple debates in generic project management literature demonstrates that this lack of attention on people's skills, is not uncommon in all forms of projects although the significance of such skills on successful project management and its delivery has been gaining increasing attention (c.f. Madsen, 2019; Whyte et al., 2022). The sections below discuss from a process and people aspect of project management, how principles of project and programme management have been identified in the framework set out in the IUCN's Green Listing process.

8.3 GOOD GOVERNANCE

In the generic area of project management, the term governance is directed specifically to project governance. In practice, it's a key concept that is frequently reported as an oversight by project management teams and/or by funding authorities (Müller, 2009; Turner, 2020; Turner & Müller, 2017). Good project governance aims to establish a set of policies, regulations and procedures that can enable a monitoring framework as to how a project is being delivered, how well or otherwise the project is achieving organisational objectives, meeting the aims and expectations of stakeholders and ensuring project viability (Ahola et al., 2014;

Too & Weaver, 2013). Also, in the current post-COVID-19 challenging, economic situation, improving the governance structure of a project enhances the probability of a project's success (Bednarz et al., 2021; Pinto, 2022). An important factor that will either enhance or impede project success, and is integral to governance and management practices, concerns stakeholder management (Aaltonen, 2011). Stakeholders (individuals, groups or organisations), whether primary (organisations who deliver the project, their suppliers, project owners), secondary (organisations or provides or services the project is depended on, unions, customers) or tertiary (local communities, interest groups, media, etc.) (Rolstadås & Schiefloe, 2017) all can enhance/pejoratively affect project outcome(s) (Littau et al., 2010). Therefore, it is crucial for a project manager to consider how to engage and develop strong relationships with a wide range of stakeholders (Pollack et al., 2017; Sefiani et al., 2018), how to establish communications with each of them, that is built on a relationship of trust, leadership and strong interpersonal skills (Aladpoosh et al., 2012; de Oliveira & Rabechini, 2019; Hartmann & Hietbrink, 2013; Mok et al., 2014). Moreover, sharing information about the project aims and objectives, listening to stakeholders' feedback and providing them with regular updates about project development should be managed strategically throughout the whole project life cycle (Bourne & Walker, 2005). For example, a lack of good understanding for internal stakeholders' views and skills might result in not using their full potential and consequently lead to the mismanagement of project resources (Beringer et al., 2012).

In cases of project management in protected areas, important stakeholders include the public, local communities and residents whose expectations and perceptions should be expressed and taken into consideration at least, during public consultations. The purpose of their engagement includes ensuring that their rights are recognised and that they are involved in management and decision-making processes. An important aspect of stakeholder communication management is choosing appropriate channels, tools and techniques (Turkulainen et al., 2015; Walley, 2013) and tailoring communication style to maintain stakeholders commitment, enthusiasm and prevent potential conflicts (Guo & Saxton, 2014; Helin et al., 2013). However, as practice often shows (see boxed case study below), activities in the contexts of protected area and greenspace management can attract controversy, often of political and economic design, and can become hugely complex. If not managed appropriately, competing objectives and visions might result in stakeholder conflicts that can at the very least result in project delay or even failure (Aalbers et al., 2019).

Notec Valley Landscape Park (Dolina Noteci Park Krajobrazowy)]
Notec Valley is an area located in Wielkopolska region, in the western part of Poland. It is characterised by a unique landscape and is inhabited by a range of endangered species that require conservation. At the beginning of 2023, intensive work started on the establishment of the Notec Valley Landscape Park. Based on the legislation drafted, the park is to be established on 42,000 ha., spreading over thirteen municipalities (BIP, 2023). It is to be the first landscape park in Wielkopolska region. The key aims of local government institutions, include the conservation of this unique green space and its inherent species, enhancing health and wellbeing of residents, and increasing tourism and recreation offers provided in the area, to support economic development. It was also hoped that the establishment of the park would enable local government to apply for national and European funding to enable improvements in green space infrastructure, and for economic development providing for a significant financial boost for the region.

However, the idea of the park was developed by government officials, in the Dolina Noteci Landscape Parks Team (*Zespół Parków Krajobrazowych Dolina Noteci*) without consulting with project stakeholders and they particularly omitted communications with local farmers who became strong opponents to the idea of establishing a landscape park in the region. Their main point of concern was that the park would have negative impacts on the management and development of their farms and, thus, would have an adverse impact on their income generation. As no public consultations took place when the idea of the park was being developed and there was no communication between the administrative officials and residents, misconceptions were widespread and opposition grew. A series of meetings was organised by Dolina Noteci Landscape Parks Team starting at the end of January 2023 to clarify concerns such as soil fertility, deforestation, hunting, farm closures and development of livestock buildings and to outline the potential benefits of the park establishment for the region and its residents (Danielewicz, 2023; Wolski, 2023). However, as these events were organised after the draft legislation for the park was shared with the public, the communication about this project was limited and the farmers did not feel that they were trusted partners whose views and interests would be taken into account. The farmers felt that when delivering this project, their economic activity would at best be limited, if not eliminated. They thought that their economic activities were perceived by decision-making authorities to have negative impacts on the natural environment.

8.4   Good Management: Sound Design and Planning and Leadership Qualities

Management is defined as 'the process of assembling and using sets of resources in a goal-directed manner to accomplish tasks in an organisation' (Hitt et al., 2011, 4): it involves establishing planning, organising, implementing and controlling processes, assembling and using allocated resources to achieve a desired outcome. However, delivering a successful project is not only about the management of processes and being adaptive to change (Salerno et al., 2015). It is also about individual behaviour in relation to other members of a project team (Crevani et al., 2010; Müller et al., 2018). It involves personal and professional development skills, a people perspective to project management, an ability to empower team members' good interpersonal skills (Gido & Clements, 1999, 85) and well-developed emotional intelligence to be able to deliver anticipated results, even when unexpected events happen and things go wrong (Goleman, 2004): all of which can be enhanced through engaging in mindful project management (Kutsch & Hall, 2020).

The adoption of 'balanced leadership' is important (Müller et al., 2017), by 'giving 'power to' others, empowering team members and using their power to allow for distributed forms of leadership to accomplish project tasks (Whyte et al., 2022, 2). Additionally, trust is a key and important principle, which is influenced by and affects the level of confidence in the team that a delegated task will be delivered. Providing team members with opportunities to learn new skills and improve their competences, helps develop the team's morale, promotes open communication, team cohesion, positive professional work relations based on mutual respect, support and understanding: all of which are key to contributing to the successful delivery of project outputs (Al-Ghazali, 2020; Raziq et al., 2018). Inspiring, encouraging and motivating the project team, co-creating and sharing a vision of a bright and prosperous future of an organisation (Keegan & Den Hartog, 2004; Raziq et al., 2018) can boost employees' work engagement. Similarly, performance, creativity, as well as innovative and entrepreneurial behaviour can also be boosted (Afsar et al., 2017; Ding et al., 2017; Jansen et al., 2009; Vaccaro et al., 2012). It is about what Barber and Warn (2005) have named as being able to be 'a firelighter', applying a proactive leadership and as the project manager, evoking 'both passion as well as reason' amongst stakeholders, to harness their emotional and cognitive commitment by promoting understanding of the potential benefits of the project, instead of being a reflective 'firefighter' focusing mainly on crises management and reactive problem-solving (Barber & Warn, 1033–35). This is exemplified in the work undertaken by the National Association for AONBs as part of their 'Taking the Lead' project 2017–19 (box text below).

The National Association for AONBs and 'Taking the Lead' project (2017–2019)]

The National Association for Areas of Outstanding Natural Beauty (NAAONB) is a charity that promotes the conservation, enhancement and understanding and public appreciation for a protected area network comprised of 46 Areas of Outstanding Natural Beauty (AONB) across England, Wales and Northern Ireland. The charity is governed by a board of trustees who follow the charity's memorandum and articles of association which set out the objects for the charity. These are, to:

- promote the conservation and enhancement of natural beauty in and around Areas of Outstanding Natural Beauty;
- advance the education, understanding and appreciation of the public in relation to the conservation and enhancement of natural beauty; and
- promote the efficiency and effectiveness of those organisations.

In the Autumn of 2017, the trustees held a special board meeting to review the function of the charity and the effectiveness of the AONB network. They identified some fundamental flaws in the charity and the operation of the AONB network. One of the areas of concern was how the individual AONB teams and their partnerships worked together because projects in protected areas often require collaboration with public agencies and private sector organisations at national and international scales of management. The AONB teams are based within a host local government authority and are influenced by that authority's culture and approach to management. This situation has resulted in the disparate collection of teams who have not functioned well as a network. There was little collaboration amongst them and few joint initiatives were undertaken. There was a lack of confidence amongst local teams to take responsibility and leadership of projects delivered at the national level due to what was considered as the organisational complexity of the AONB network, difficulties in the management of the geographically dispersed teams and a general view that benefits of collaboration were considered few and far between.

To address these challenges, a professional and personal development programme was offered to a cohort of staff. This was aimed at supporting them develop greater levels of collaborative working and leadership at the national level. This included personality profiling using psychometric tests, coaching and to monitor and evaluate the project, methods of constructive inquiry. The result was the 'Taking the Lead' (TTL) project. This took place over two years (2017–2019), and enabled good communication mechanisms to be formed between individuals taking part in the project. There was regular engagement between the senior officers in the AONBs' teams and with the TTL project team. From the perspective of trustees, participants and other staff, the project was deemed to be successful. This was reinforced by an evaluation report that was prepared by independent consultants who undertook one-to-one interviews with participants and other stakeholders.

The overall achievements included: increased collaboration between individuals in different AONB teams and many examples of staff taking on national roles. Participants reported increased self-awareness, greater confidence to deliver activities at a national level and dramatically increased levels of collaboration and cooperation between individual teams.

8.5   International Teams: Online Management and Hybrid Working

Managing international teams, is a common practice in the management of green spaces and protected areas, especially in terms of research and consultancy work where information and expertise on topics related to social, environmental and political sciences are warranted. Internationalisation of project teams, for project leaders and team members, brings to the fore additional challenges created by managing the diversity of people from different ethnic, national, linguistic, and socio-economic backgrounds, potentially if working online, of different IT knowledge and abilities, and who are in different geographical locations and time zones (Lientz & Rea, 2003, 3). The team members may well have different work and organisational cultures, different perceptions of leadership and expectations towards the project lead and different values and social norms. They may also have a culturally distinctive perception of 'project success' and might face linguistic challenges during communication processes. Therefore, making an international project team work efficiently requires from the project manager well-developed cross-cultural management skills (Haghirian, 2011, 124; Müller & Turner, 2010, 58; Schneider & Barsoux, 2003, 217).

The twenty-first century has significantly changed the way we communicate and the pace of our information exchange. Phone/tele-conferences arranged well in advance have been replaced by ad hoc MS Teams and Zoom meetings scheduled via apps that are easily accessible on our PCs, laptops and mobile phones. However, scheduling a team meeting at a time that would work for all members located in different time zones remains a challenge. Research has shown, living in one time zone but adjusting the work schedule to a different one might negatively impact health and wellbeing of team members and become a source of stress (Nurmi, 2011). Another challenge is how to organise efficient teamwork, given the various patterns of work delivery modes. Remote and flexible working has become a new reality of a workplace culture since the COVID-19 pandemic. Lockdowns and restricted movement regulations required workers to switch to online and hybrid working (Al-Habaibeh et al., 2021; Merchant, 2021). Some people and teams welcomed this new organisation of work and encourage the maintenance of flexible working policies in the post-pandemic world. Consequently, many companies have responded positively considering hybrid working as of benefit for both their employees and organisations (Ipsen et al., 2021).

Remote working can give employees a better work-life balance, but it can also make employees feel burnout and isolated by having lack of face-to-face social contact with their colleagues (Moglia et al., 2021). Some individuals would like to have the possibility of a choice and switch between office work and working from home depending on their tasks and day schedule. An important task for a project manager responsible for day-to-day delivery of a project is to think about how to organise the teamwork given those various modes of work delivery. How to ensure high productivity of a project team, how to work efficiently combining work of those who are in and outside the office or how to organise the communication amongst the team members if the technology is interrupted or when there is no Internet access available (Lippert & Dulewicz, 2018). This hybrid project management context demands refined leadership skills particularly in terms of communication, abilities to adapt, and ultimately to build trust amongst project team members and maintain team cohesion. Soft skills are key. A case study that covers each of the principles laid out in this section and their implications for management and communication is provided in the box case study below.

Significant Spaces Research Team

The Significant Spaces research team was brought together through common interests to investigate how greenspaces influence people's health and wellbeing. A multidisciplinary international team representing social, environmental, health and microbiome sciences of ten people was brought together. The skills and knowledge found in this team, were squarely focused on one aim: the design of a valid, robust and conceptualisation study that could advance much-needed knowledge as to how greenspaces enhanced people's health and wellbeing. Various methodologies needed to be agreed, synthesised and validated amongst the project members and this was challenging in order to also adhere, or indeed challenge, members' respective theoretical frameworks. Extensive knowledge in quantitative and qualitative methods was also required amongst all members in the team. To collate and test the environmental and landscape data required, the project team needed to engage with 54 external organisations comprised of environmental managers, landowner representatives from protected area agencies and public health practitioners.

The working up of this project took two years to complete, the result of which has been a fully tested conceptualisation of how environmental characteristics of distinct landscapes could impact people's health and wellbeing. An agreed project management framework of timelines and documentation was developed and agreed very early on in discussions. This was fundamental to the working of the project as each of the research team, held full-time jobs and had additional personal responsibilities.

The aims of the project were defined very early on in team discussions, as were also project deliverables, including timescales. These were revised and updated regularly, and information was disseminated on a regular basis. All project documentation adopted strict version control protocols. Documentation was archived by all members in the team in their own records, albeit one central SharePoint archive was created and archived according to date order. All documents were read only, unless access to any one document was required by a team member.

Time zone differences fundamentally challenged the organisation of project meetings, with some of the team working within EU time zone, and others in the Oceanic region. Such time differences required compromises on the part of each of the team members, to meet in the evenings or early in the mornings. To keep meetings within time frames agreed, agendas and clear instructions were sent out by email to all in the team allowing for any preparation on the part of team members to deliver responses to key questions posed by others in the team. Minutes of all meetings were brief bullet points and sent out within seven to ten hours of each of the meetings. Hybrid working amongst those in the UK occasionally took place, but with those in Australia, New Zealand, and in Spain, all staff had to be proficient in the Microsoft Teams and 365 contexts. All meetings started and ended punctually.

Key skills of each team member were identified very early on. This helped to divide the team into mini working groups led by one representative, whose role was to manage activities to key deadlines agreed by the whole team. This also meant that not all team members had to attend all project team meetings instead the representative of each of the key stages attended on behalf of the mini working group. This additionally was useful in that one person amongst the ten in the team, was identified as the key representative working with external organisations, updating team members, and collating data from the organisations that in turn could be passed amongst the research team for testing. Fortunately, few cultural differences were experienced, all in the team spoke English fluently, including those whose first language was not English.

8.6 Effective Management

The effective management of protected areas is not possible without sufficient allocation of resources, especially of financial resources. Yet funding is always challenging in this area and invariably will reflect the political climate of the time. Increasingly we are experiencing in the protected area context decreasing funding streams, and in urban areas the demise of the public park has in the English context, for example, been regrettably reported upon for some time. This situation warrants project teams abilities and skills in terms of external income generation, needing to have a good command of funding opportunities, and in turn, when identified, the ability to orientate activities to meet strategic funding priorities of potential sponsors. Additionally, bid writing skills and knowledge of tendering and bidding processes are essential (Blume-Kohout & Adhikari, 2016; Lindgreen et al., 2019), all of which can take time to develop and will invariably need skills in project budgeting and costs estimations. A lack of these skills, especially the underestimation of project costs, is highlighted as a key challenge in delivering projects successfully, resulting in the delay of project outputs and frequently increasing project costs.

Exacerbating challenges in budget management, sharply rising inflation as currently being experienced worldwide, will additionally challenge accurate forecasts on costs, and management of project expenditure. In some cases, funders/sponsors might be able to extend their financial support to a project further. In other cases, projects might remain unfinished, bringing financial losses (Flyvbjerg et al., 2003, 12–21; Miller & Lessard, 2000, 14) and holds the potential to damage reputation, impeding successful future bids for funds. Therefore, it is important to demonstrate to a potential funder that not only a project constitutes good value for money, given its potential impact and the estimated total costs, but also that proper governance mechanisms and processes are developed and that they can be implemented by a proactive, effective project leader.

Another important issue in protected areas, which is directly linked with funding, is what happens once the project has been delivered and the funding has ended. Organisations that innovate through projects do so by delivering tailored interventions in each topic (Hornstein, 2015). Once the project is delivered, it and the team are closed, and their outputs become mainstream business as usual (BAU). In a business organisation, it is assumed that the costs of transition and integration into BAU have been considered when assessing the project affordability. However, that might not be the case in protected areas where organisations have limited resources and are heavily dependent on limited funds due to increased context of financial austerity in the public sector (Carmona et al., 2019; Mell, 2020; Whitten, 2019). Projects delivered in protected areas are projects delivered for the public good, to protect areas/locations because of their natural, economical or cultural values and not primarily for the income generation (Dobson, 2018, 75–75). Therefore, once a project has been delivered, it usually requires either obtaining further external funding to manage its outputs and benefits, requiring a team reorganisation, relying on the good will of staff to absorb additional work *gratis*, holding the potential to exceed or stretch their capacities, or potentially having to consider the abandonment of other initiatives due to zero funds, staff shortages, political redirection of local funds and/or a recruitment freeze, for example (cf. Khan & Munira, 2021).

Funding of the National Association for Areas of Outstanding Natural Beauty (NAAONB)

The NAAONB has a staff compliment equivalent to less than four full-time employees funded by the funds received from government and has reserves in the order of £70,000 for its charitable activities. However, in order to deliver any national activities, campaigns or programmes, additional funding is always needed. One of the potential funders that was willing to support the initiatives of the NAAONB was the Heritage Lottery Fund (HLF), now the National Heritage Lottery Fund. Under their Resilient Heritage Funding Programme, they provide grants to strengthen organisations, and build the capacity of staff and volunteers to better manage heritage in the long term. The NAAONB was successful in securing a grant from the HLF of £170,000 to cover the costs of the 'Taking the Lead' project development and delivery (2017–2019). This success was a result of an intensive collaboration between the charity's Development Manager and an experienced external bid writer who together wrote the funding application and shaped the project leading to successful award of the funds by demonstrating measurable benefits of the project. However, once the project finished, there was no further financial support, highlighting the challenges of project delivery by organisations who are heavily dependent on public funding.

8.7 CONCLUSION

The purpose of this chapter was to overview project management practices in green spaces and protected areas, examine challenges and consider opportunities for improvement. National and international project initiatives provide useful guidelines on the direction and scope of the work of twenty-first-century, post-pandemic project managers whose work focuses on conserving nature and managing public access to protected areas and greenspaces. There is a lot to be learnt, especially from the generic project management literature, on how to deliver such projects successfully and in a sustainable manner for which communication skills and stakeholder engagement are critically important especially if working in a hybrid mode using available IT technology.

This approach increasingly demands a new context of virtual leadership to engage and collaborate effectively in project teams that are invariably comprised of multidisciplinary and international team members. Soft skills are key. The development of these skills can be enhanced using psychometric testing and analysis, albeit we do recognise that some of these tools have been criticised for their inaccuracies. Yet in the case studies provided, the use of psychometric testing with individuals, then at the team level and then at the organisational level, showed the importance of developing greater understanding with individuals about their preferred working style and additionally demonstrated the importance to team-working of understanding the diversity of working styles amongst colleagues. Through appreciation for the value of complementary working styles, we can effectively co-create activities and project manage them. Only through this enhanced awareness by individuals and teams can we approach complex tasks and deliver them in an efficient way.

What is also overwhelmingly obvious worldwide is that many protected areas and other forms of green and blue spaces, of value to the public, exist in a state of financial insecurity. This situation warrants additional, if not enhanced abilities of managers and project teams to be able to identify opportunities for external income generation as a key operational priority. This means that bid writing will continue to be one of the most important management skills.

It also seems reasonable to rethink the idea of project management delivery in green spaces' management. This we contend, needs greater emphasis on project delivery that is both aligned with the organisation's strategic goals and priorities, and that is delivered in a sustainable manner, balancing costs and benefits of project delivery. We would also recommend that albeit Green Finance Initiatives are evident across the EU, further research into income generation strategies of green places and protected areas in public organisations is critically needed. The results of such a study could (a) inform best practices in project management from a financial perspective; (b) identify and evaluate potential collaborative and funding opportunities, to improve organisational efficiency; (c) there is even, we contend, a potential to establish communities of funding practice amongst staff working in the protected area/green space management contexts; and (d) through these points, there is a potential to address the age-old concern for how outcomes and outputs of projects can be continued long after a project has been defunded and the project team has been disbanded.

Finally, given the challenges of project management practices in green spaces and protected areas, an emphasis on leadership training and soft skills development would be useful in project management training. Such attention would address criticisms of the discipline of project management for being primarily a process-driven practice. Project management processes and procedures are essentially supportive to a manager's role and are relatively easy to review, manage and learn. However, working efficiently with people requires going beyond processes and the knowledge of approaches to team management. It requires the development and application of sound leadership, of soft skills and of a greater mindfulness for context and for the value of the people perspective to project management.

References

Aalbers, C., Kamphorst, D., & Langers, F. (2019). Fourteen local governance initiatives in greenspace in urban areas in the Netherlands. Discourses, success and failure factors, and the perspectives of local authorities. *Urban Forestry and Urban Greening, 42*, 82–99. https://edepot.wur.nl/478342

Aaltonen, K. (2011). Project stakeholder analysis as an environmental interpretation process. *International Journal of Project Management, 29*(2), 165–183. https://doi.org/10.1016/j.ijproman.2010.02.001

Afsar, B., Badir, Y., Saeed, B., & Shakeer, H. (2017). Transformational and transactional leadership and employee's entrepreneurial behavior in knowledge-intensive industries. *The International Journal of Human Resource Management., 28*, 307–332. https://doi.org/10.1080/09585192.2016.124 4893

Ahola, I., Ruuska, K. A., & Kujala, J. (2014). What Is Project Governance and What Are Its Origins? *International Journal of Project Management, 32*(8), 1321–1332. https://doi.org/10.1016/j.ijproman.2013.09.005

Aladpoosh, H., Shaharoun, A., & Saman, M. (2012). Critical features for project stakeholder management: A systematic literature review. *International Journal of Applied Systemic Studies., 4*(3), 150–167. https://doi.org/10.1504/IJASS. 2012.051130

Al-Ghazali, B. (2020). Transformational leadership, career adaptability, job embeddedness and perceived career success: A serial mediation model. *Leadership and Organization Journal, 41*, 993–1013. https://doi.org/10.1108/LODJ-10-2019-0455

Al-Habaibeh, A., Watkins, M., Waried, K., & Javareshk, M. (2021). Challenges and opportunities of remotely working from home during Covid-19 pandemic. *Global Transitions, 3*, 99–108. https://doi.org/10.1016/j.glt.2021.11.001

Barber, E., & Warn, J. (2005). Leadership in project management: From firefighter to firelighter. *Management Decision, 43*(7/8), 1032–1039. https://doi.org/10.1108/00251740510610026

Bednarz, A., Borkowska-Bierć, M., & Matejun, M. (2021). Managerial responses to the onset of the COVID-19 pandemic in healthcare organizations project management. *International Journal of Environmental Research and Public Health, 18*(22), 12082. https://doi.org/10.3390/ijerph182212082

Beringer, C., Jonas, D., & Gemünden, H. (2012). Establishing project portfolio management: An exploratory analysis of the influence of internal stakeholders' interactions. *Project Management Journal, 43*(6), 16–32. https://doi.org/10.1002/pmj.21307

BIP (Biuletyn Informacji Publicznej). (2023). Projekt Uchwały Sejmiku Województwa Wielkopolskiego w sprawie Parku Krajobrazowego Dolina Noteci. Available at: Uchwała (umww.pl). Accessed 4 May 2023.

Blume-Kohout, M., & Adhikari, D. (2016). Training the scientific workforce: Does funding mechanism matter? *Research Policy, 45*(6), 1291–1303. https://doi.org/10.1016/j.respol.2016.03.011

Borrini-Feyerabend, G., Dudley, N., Jaeger, T., Lassen, B., Pathak Broome, N., Phillips, A., & Sandwith, T. (2013). Governance of protected areas: From understanding to action. Best Practice Protected Area Guidelines Series No. 20, Gland, Switzerland: IUCN. xvi + 124pp.

Bourne, L., & Walker, D. (2005). The paradox of project control. *Team Performance Management, 11*, 5/6, 157–178. https://doi.org/10.1108/13527590510617747

Bretschger, L., & Pittel, K. (2020). Twenty key challenges in environmental and resource economics. *Environnemental and Resource Economics, 77*, 725–750. https://doi.org/10.1007/s10640-020-00516-y

Carmona, M., Hanssen, G., Lamm, B. et al. (2019). Public space in an age of austerity. *Urban Design International, 24*, 241–259. https://link.springer.com/article/10.1057/s41289-019-00082-w

Countryside Act (1983). Available at: Countryside Act 1968 (legislation.gov.uk). Accessed 22 April 2023.

Crevani, L., Lindgren, M., & Packendorff, J. (2010). Leadership, not leaders: On the study of leadership as practices and interactions. *Scandinavian Journal of Management, 26*, 77–86. https://doi.org/10.1016/j.scaman.2009.12.003

Crossley, A., & Russo, A. (2022). Has the pandemic altered public perception of how local green spaces affect quality of life in the United Kingdom? *Sustainability, 14*, 7946. https://doi.org/10.3390/su14137946

Danielewicz, M. (2023). *Nowy park krajobrazowy w Dolinie Noteci. Pojechałam sprawdzić, dlaczego rolnicy go nie chcą.* Gazeta Wyborcza Poznań. Available at: Nowy park krajobrazowy w Dolinie Noteci. Pojechałam sprawdzić, dlaczego rolnicy go nie chcą (wyborcza.pl). Accessed 23 March 2023.

De Oliveira, G., & Rabechini (Jr), R. (2019). Stakeholder management influence on trust in a project: A quantitative study. *International Journal of Project Management, 37*(1), 131–144. https://doi.org/10.1016/j.ijproman.2018.11.001

Dinerstein, E.,Vynne, C., Sala, E., Joshi, A., Fernando, S. et al. (2019). A global deal for nature: Guiding principles, milestones, and targets. *Science Advances, 5*(4), eaaw2869. https://doi.org/10.1126/sciadv.aaw2869

Ding, A., Cenci, J., & Zhang, J. (2022). Links between the pandemic and urban green spaces, a perspective on spatial indices of landscape garden cities in China. *Sustainable Cities and Society, 85*, 104046. https://doi.org/10.1016/j.scs.2022.104046

Ding, X., Li, Q., Zhang, H., Sheng, Z., & Wang, Z. (2017). Linking transformational leadership and work outcomes in temporary organizations: A social identity approach. *International Journal Project Management, 35*, 543–556. https://doi.org/10.1016/j.ijproman.2017.02.005

Dobson, J. (2018). From contest to context: Urban green space and public policy. *People, Place and Policy, 12*(2), 72–83. https://doi.org/10.3351/ppp.2018.3824435278

Environment Act (2021). Available at: Environment Act 2021 (legislation.gov.uk). Accessed March 23 2023.

Flyvbjerg, B., Bruzelius, N., & Rothengatter, W. (2003). *Megaprojects and risk: An anatomy of ambition.* Cambridge University Press.

Fros, H., Hagemann, F., Sang, Å., & Randrup, T. (2021). Striving for inclusion—A systematic review of long-term participation in strategic management of urban green spaces. *Frontiers in Sustainable Cities, 3.* https://doi.org/10.3389/frsc.2021.572423

Gareis, R. (1991). Management by projects: The management strategy of the 'new' project-oriented company. *International Journal of Project Management, 9*(2), 71–76. https://doi.org/10.1016/0263-7863(91)90062-Z

Geldmann, J., Coad, L., & Barnes, M. (2015). Changes in protected area management effectiveness over time: A global analysis. *Biological Conservation, 191*, 692–699. https://doi.org/10.1016/j.biocon.2015.08.029

Gido, J., & Clements, J. (1999). *Successful project management*. South-Western Collage Publishing.

Goleman, D. (2004). What makes a leader? *Harvard Business Review, 82*(1), 82–91. https://hbr.org/2004/01/what-makes-a-leader. Accessed 22 April 2023.

Guo, C., & Saxton, G. (2014). Online stakeholder targeting and the acquisition of social media Capital. *International Journal of Non-Profit and Voluntary Sector Marketing, 19*(4), 286–300. https://doi.org/10.1002/nvsm.1504

Haghirian, P. (2011). *Successful cross-cultural management: A guide for international management*. Business Expert Press.

Hartmann, A., & Hietbrink, M. (2013). An exploratory study on the relationship between stakeholder expectations, experiences, and satisfaction in road maintenance. *Construction Management Economics, 31*(4), 345–358. https://doi.org/10.1080/01446193.2013.768772

Head, B. (2022). *Wicked problems in public policy*. Springer/Palgrave Macmillan.

Helin, S., Jensen, T., & Sandstrom, J. (2013). Like a battalion of tanks: A critical analysis of stakeholder management. *Scandinavian Journal of Management, 29*(3), 209–218. https://doi.org/10.1016/j.scaman.2012.11.010

Hitt, M., Black, S., & Porter, L. (2011). *Management*. Pearson.

HM Government. (2018). A green future: Our 25 year plan to improve the environment. 25-year-environment-plan.pdf (publishing.service.gov.uk). Accessed 24 March 2023.

HM Government (2021). Net Zero Strategy. Build Back Greener. net-zero-strategy-beis.pdf (publishing.service.gov.uk). Accessed 24 March 2023.

Hockings, M., Hardcastle, J., Woodley, S. et al. (2019). The IUCN green list of protected and conserved areas: setting the standard for effective area-based conservation. *Parks, 25*(2), https://doi.org/10.2305/IUCN.CH.2019.PARKS-25-2MH.en

Hornstein, H. (2015). The integration of project management and organizational change management is now a necessity. *International Journal of Project Management, 33*(2), 291–298. https://doi.org/10.1016/j.ijproman.2014.08.005

Ipsen, C., van Veldhoven, M., Kirchner, K., & Hansen, J. P. (2021). Six key advantages and disadvantages of working from home in Europe during COVID-19. *International Journal of Environmental Research and Public Health, 18*(4), 1826. https://doi.org/10.3390/ijerph18041826

ISEAL (International Social and Environmental Accreditation and Labelling Alliance). (2023). ISEAL Credibility Principles. ISEAL Credibility Principles. Accessed 23 March 2023.

IUCN. (2017). *IUCN green list of protected and conserved areas: Standard, Version 1.1. The global standard for protected areas in the 21st Century*. IUCN.

Jaafar, K., & Yusof, S. (2019). Project management evolution: From traditional to responsive project management. In S. Yusof & K. Jaafar (Eds.), *The digital project management evolution: Essential case studies from organisations in the Middle East* (pp. 1–5). Productivity Press. https://doi.org/10.4324/978042 9266508

Jansen, J., Vera, D., & Crossan, M. (2009). Strategic leadership for exploration and exploitation: The moderating role of environmental dynamism. *The Leadership. Quarterly, 20,* 5–18. https://doi.org/10.1016/j.leaqua.2008. 11.008

Jansson, M., Vogel, N., Fors, H., & Randrup, T. (2019). The governance of landscape management: New approaches to urban space development. *Landscape Research, 44,* 952–965. https://doi.org/10.1080/01426397.2018.153 6199

Jiricka-Pürrer, A., Tadini, V., Salak, B., Taczanowska, K., Tucki, A., & Senes, G. (2019). Do protected areas contribute to health and well-being? A cross-cultural comparison. *International Journal of Research and Public Health, 16*(7), 1172. https://doi.org/10.3390/ijerph16071172

Keegan, A., & Den Hartog, D. (2004). Transformational leadership in a project-based environment: A comparative study of the leadership styles of project managers and line managers. *International Journal Project Management, 22,* 609–617. https://doi.org/10.1016/j.ijproman.2004.05.005

Kelly, M. (2022). Habitat protection, ideology and the British nature state: The politics of the Wildlife and Countryside Act 1981. *The English Historical Review, 137*(586), 847–883. https://doi.org/10.1093/ehr/ceac112

Khan, M., & Munira, S. (2021). Climate change adaptation as a global public good: Implications for financing. *Climatic Change, 167,* 50. https://doi.org/ 10.1007/s10584-021-03195-w

Kjersem, K., Jünge, G., & Emblemsvåg, J. (2017). Project execution strategy and planning challenges. In H. Lödding, R. Riedel, K-D. Thoben, G. von Cieminski, & D. Kiritsis (Eds.), *Advances in production management systems. The path to intelligent, Collaborative and sustainable manufacturing* (APMS 2017. IFIP Advances in Information and Communication Technology, vol. 514). https://doi.org/10.1007/978-3-319-66926-7_28

Kutsch, E., & Hall, M. (2020). *Mindful project management* (2nd ed.). Routledge. https://doi.org/10.4324/9780429259579

Lee, M. (2022). Brexit and the environment bill: The future of environmental accountability. *Global Policy, 13*(Suppl. 2), 119–127. https://doi.org/10. 1111/1758-5899.13061

Lientz, B., & Rea, K. (2003). *International project management.* Routledge.

Lindgreen, A., Di Benedetto, C., Verdich, C., Vanhamme, J., Venkatraman, V., et al. (2019). How to write really good research funding applications. *Industrial Marketing Management, 77,* 232–239. https://doi.org/10.1016/j.ind marman.2019.02.015

Lippert, H., & Dulewicz, V. (2018). A profile of high-performing global virtual teams. *Team Performance Management, 24*(3/4), 169–185. https://doi.org/10.1108/TPM-09-2016-0040

Littau, P., Jujagirl, N., & Adlbrecht, G. (2010). 25 years of stakeholder theory in project management literature (1984–2009). *Project Management Journal., 41*(4), 17–29. https://doi.org/10.1002/pmj.20195

Madsen, S. (2019). *The power of project leadership. 7 keys to help you transform from project manager to project leader.* Kogan Page.

Mell, I. (2020). The impact of austerity on funding green infrastructure: A DPSIR evaluation of the Liverpool Green & Open Space Review (LG&OSR), UK. *Land Use Policy, 91*, 104284. https://doi.org/10.1016/j.landusepol.2019.104284

Merchant, J. (2021). Working online due to the Covid-19 Pandemic: A research and literature review. *Journal of Analytical Psychology, 66*(3), 484–505. https://doi.org/10.1111/1468-5922.12683

Miller, R., & Lessard, D. (Eds.). (2000). *The strategic management of large engineering projects: Shaping institutions, risks, and governance.* Massachusetts Institute of Technology.

Moglia, M., Hopkins, J., & Bardoel, A. (2021). Telework, hybrid work and the United Nation's sustainable development goals: Towards policy coherence. *Sustainability, 13*(16), 1–28. https://doi.org/10.3390/su13169222

Mok, K., Shen, G., & Yang, J. (2014). Stakeholder management studies in mega construction projects: A review and future directions. *International Journal of Project Management, 33*(2), 446–457. https://doi.org/10.1016/j.ijproman.2014.08.007

Mosler, S., & Hobson, P. (2021). Close-to-nature heuristic design principles for future urban green infrastructure. *Urban Planning, 6*(4), 67–79. https://doi.org/10.17645/up.v6i4.4451

Müller, R. (2009). *Project governance.* Gower Publishing.

Müller, R., Packendorff, J., & Sankaran, S. (2017). Balanced leadership: A new perspective for leadership in organizational project management. In S. Sankaran, R. Muller, & N. Druin (Eds.), *Cambridge handbook of organizational project management* (pp. 186–199). https://doi.org/10.1017/9781316662243.018

Müller, R., Sankaran, S., Drouin, N., Vaagaasar, A., & Bekker, M. (2018). A theory framework for balancing vertical and horizontal leadership in projects. *International Journal of Project Management, 36*(1), 83–94. https://doi.org/10.1016/j.ijproman.2017.07.003

Müller, R., & Turner, R. (2010). *Project-oriented leadership.* Gower Publishing Limited.

Natural England (2023). Green infrastructure framework. Green Infrastructure Home (naturalengland.org.uk). Accessed March 24 2023.

Nature Conservancy Council Act (1973). https://www.legislation.gov.uk/ukpga/1973/54/contents. Accessed 29 March 2023.

Navarrete-Hernandez, P., & Laffan, K. (2019). A greener urban environment: Designing green infrastructure interventions to promote citizens' subjective wellbeing. Landscape and Urban Planning, 191, 103618. https://doi.org/10.1016/j.landurbplan.2019.103618

Nurmi, N. (2011). Coping with coping strategies: How distributed teams and their members deal with the stress of distance, time zones and culture. Stress and Health, 27, 123–143. https://doi.org/10.1002/smi.1327

Pantaloni, M., Marinelli, G., Santilocchi, R., Minelli, A., & Neri, D. (2022). Sustainable management practices for urban green spaces to support green infrastructure: An Italian case study. Sustainability, 14, 4243. https://doi.org/10.3390/su14074243

Pinto, J. (2022). Reassessing project practices, research, and theory in a post-Covid reality. International Journal of Information Systems and Project Management, 10, 4. https://doi.org/10.12821/ijispm100401

Pollack, J. M., Barr, S., & Hanson, S. (2017). New venture creation as establishing stakeholder relationships: A trust-based perspective. Journal of Business Venturing Insights, 7, 15–20. https://doi.org/10.1016/j.jbvi.2016.12.003

Puhakka, R., Pitkänen, K., & Siikamäki, P. (2017). The health and well-being impacts of protected areas in Finland. Journal of Sustainable Tourism, 25, 1830–1847. https://doi.org/10.1080/09669582.2016.1243696

Ranius, T., Widenfalk, L., Seedre, M., et al. (2023). Protected area designation and management in a world of climate change: A review of recommendations. Ambio A Journal of the Human Environment, 52, 68–80. https://doi.org/10.1007/s13280-022-01779-z

Raziq, M., Borini, F., Malik, O., Ahmad, M., & Shabaz, M. (2018). Leadership styles, goal clarity, and project success: Evidence from project-based organizations in Pakistan. Leadership and Organization Development Journal, 39, 309–323. https://doi.org/10.1108/LODJ-07-2017-0212

Reid, C. (2021). Mapping post-Brexit environmental law. ERA Forum, 21(4), 655–665. https://doi.org/10.1007/s12027-020-00627-5

Rolstadås, A., & Schiefloe, P. (2017). Modelling project complexity. International Journal of Managing Projects in Business, 10(2), 295–314. https://doi.org/10.1108/IJMPB-02-2016-0015

Salerno, M., Gomes, L., Da Silva, D., Bagno, R., & Freitas, S. (2015). Innovation processes: Which process for which project? Technovation, 35, 59–70. https://doi.org/10.1016/j.technovation.2014.07.012

Schneider, S., & Barsoux, J. (2003). Managing across cultures. Financial Times, Prentice Hall.

Sefiani, Y., Davies, B., Bown, R., & Kite, N. (2018). Performance of SMEs in Tangier: The interface of networking and wasta. *EuroMed Journal of Business, 13*(1), 20–43. https://doi.org/10.1108/emjb-06-2016-0016

Too, E., & Weaver, P. (2013). The management of project management: A conceptual framework for project governance. *International Journal of Project Management, 32*(8), 1382–1394. https://doi.org/10.1016/j.ijproman.2013.07.006

Turkulainen, V., Aaltonen, K., & Lohikoski, P. (2015). Managing project stakeholder communication: The Qstock Festival case. *Project Management Journal, 46*(6), 74–91. https://doi.org/10.1002/pmj.21547

Turner, R. (2020). How does Governance influence decision making on projects and in project-based organizations? *Project Management Journal, 51*(6), 670–684. https://doi.org/10.1177/8756972820939769

Turner, R., & Müller, R. (2017). The governance of organizational project management. In S. Sankaran, R. Müller, & N. Drouin (Eds.), *Cambridge handbook of organizational project management* (pp. 75–91). Cambridge University Press. https://doi.org/10.1108/IJMPB-10-2017-0113

UNEP (United Nations Environment Programme). (2022). *5 key drivers of the nature crisis.* https://www.unep.org/news-and-stories/story/5-key-drivers-nature-crisis#:~:text=The%20biggest%20driver%20of%20biodiversity,conversion%20to%20other%20land%20uses. Accessed 22 April 2023.

Vaccaro, I., Jansen, J., Van den Bosch, F., & Volberda, H. (2012). Management innovation and leadership: The moderating role of organizational size. *Journal of Management Studies, 49*, 28–51. https://doi.org/10.1111/j.1467-6486.2010.00976.x

Verschuuren, B., & Brown, S. (2019). *Cultural and spiritual significance of nature in protected areas. Government, management and policy.* Routledge.

Walley, P. (2013). Stakeholder management: The sociodynamic approach. *International Journal of Managing Projects in Business, 6*(3), 485–504. https://doi.org/10.1108/IJMPB-10-2011-0066

Whitten, M. (2019). Blame it on austerity? Examining the impetus behind London's changing green space governance. *People, Place and Policy, 12*(3), 204–224. https://doi.org/10.3351/ppp.2019.8633493848

Whyte, J., Naderpajouh, N., Clegg, S., Matous, P., Pollack, J., & Crawford, L. (2022). Project leadership: A research agenda for a changing world. *Project Leadership and Society, 3*, 1–9. https://doi.org/10.1016/j.plas.2022.100044

Wolski, M. (2023). Park Krajobrazowy Dolina Noteci—spotkanie informacyjne w Urzędzie Gminy w Chodzieży. Park Krajobrazowy Dolina Noteci—spotkanie informacyjne w Urzędzie Gminy w Chodzieży | Chodzież Nasze Miasto. Accessed 5 April 2023.

Worboys, G., Lockwood, M., Kothari, A., et al. (Eds.). (2015). *Protected area governance and management.* ANU Press.

Yan, S., & Tang, J. (2021). Optimization of green space planning to improve ecosystem services efficiency: The case of Chongqing Urban Areas. *International Journal of Research and Public Health, 18*(16), 8441. https://doi.org/10.3390/ijerph18168441

Young, R. (2011). Planting the living city: Best practices in planning green infrastructure—Results from major U.S. cities. *Journal of the American Planning Association, 77*, 368–381. https://doi.org/10.1080/01944363.2011.616996

(Re)Connecting with Nature: Exploring Nature-Based Interventions for Psychological Health and Wellbeing

Debra Gray, Denise Hewlett, Julie Hammon,
and Stephanie Aburrow

9.1 INTRODUCTION

There is now much evidence, from across the world, that access to nature and outdoor spaces, including 'green' and 'blue' spaces such as woodlands, parks, allotments, gardens and riverside or seaside settings,

D. Gray (✉)
PeopleScapes Research & Knowledge Exchange, University of Winchester, Winchester, UK
e-mail: Debra.Gray@winchester.ac.uk

D. Hewlett
PeopleScapes Research & Knowledge Exchange Centre, Department of Responsible Management, University of Winchester, Winchester, UK

Bournemouth University, Dorset, UK

D. Hewlett
e-mail: denise.hewlett@winchester.ac.uk

© The Author(s) 2024
N. Finneran et al. (eds.), *Managing Protected Areas*,
https://doi.org/10.1007/978-3-031-40783-3_9

produces a range of positive psychological benefits (Alcock et al., 2014; Annerstedt et al., 2012; Bratman et al., 2012; de Vries et al., 2003; Grahn & Stigsdotter, 2003; Groenewegen et al., 2006; Hartig et al., 2003; Lovell et al., 2020; Maas et al., 2009a, 2009b; Thompson et al., 2012; Ulrich et al., 1991; van den Berg et al., 2010; White et al., 2013; White et al., 2017). Such benefits include reductions in stress, anxiety and depression (Beyer et al., 2014; Roe et al., 2013; Sarkar et al., 2018; van den Berg, et al., 2016; Vujcic et al., 2017), as well as improvements in attention and relaxation, self-esteem, mood and confidence (Barton & Pretty, 2010; Hartig et al., 2014; Houlden et al., 2018; Houlden et al., 2019). These benefits accrue at all ages. For example, exposure to nature is consistently associated with benefits for child and adolescent mental health, including improvements in emotional wellbeing and resilience (e.g. see Chawla, 2015; Chawla et al., 2014) and lower rates of depression (e.g. see Maas et al., 2009a). Likewise, studies of older adults show good evidence of reduced anxiety and improved cognitive functioning (e.g. see Sia et al., 2020).

On this basis, it has been argued that simply being in—or near to—nature has a salutogenic effect, acting as a buffer between our stressful daily lives and our mental health. However, it is also the case that such natural environments can shape our behaviour in ways that promote positive psychological experiences, and in turn our wellbeing. For example, accessing nature can lead adults and children to be physically more active, which can have important benefits for common mental illnesses such as anxiety and depression (Barton et al., 2016; Barton & Pretty, 2010). People who use the natural environment for physical activity at least once per week have about half the risk of poor mental health compared with those who do not do so; and each extra weekly use of the natural environment for physical activity reduces the risk of poor mental health by a further six per cent (Mitchell, 2013). In addition to this, natural spaces also offer chances to have positive interactions with others, thereby

J. Hammon · S. Aburrow
Dorset AONB, Dorset, UK
e-mail: julie.hammon@dorsetcouncil.gov.uk

S. Aburrow
e-mail: steph.aburrow@dorsetcouncil.gov.uk

combating loneliness and promoting a sense of collective identity and resilience that is central to our sense of psychological wellbeing (Bowe et al., 2020; Gray & Stevenson, 2020; Jetten et al., 2012). Evidence shows that living in an environment with more parks, along with the presence of trees and grass in neighbourhoods, encourages greater use of outdoor spaces, and increases social contact with others (Brennan et al., 2017; Maas et al., 2009a; Sugiyama et al., 2008). In terms of these behaviours, blue spaces (rivers, lakes and coasts) are as important as green environments: it is not the colour that matters but the opportunity to behave and respond in a particular way (White et al., 2017).

These findings have led to urgent calls for ways to reprioritise the connection of people to the environment, as a key public health strategy (e.g. see Charles et al., 2018). One key example of this, is the Healthy Parks, Health People (HPHP) initiative (see Maller et al., 2009), which was first started by Parks Victoria, but is now a global movement including countries across Europe as well as South Korea, Singapore, Canada and the USA to name but a few. The focus of HPHP is to provide a framework for promoting the many health and wellbeing benefits of connecting with nature, with a strong focus on mental health. Such an approach is argued to have a series of wider 'co-benefits', beyond the health and wellbeing of any one person (Robinson & Breed, 2019). Mental health conditions are one of the main causes of the overall disease burden worldwide, at an estimated global cost of £1.6 trillion per year (Mental Health Taskforce, 2016). Health systems across the world have not yet responded adequately to the current burden of mental disorders, and the gap between the need for treatment and its provision is wide (WHO, 2001). Within this context, it is argued that nature can provide cost-effective and low-risk solutions (Allen et al., 2014; Townsend & Weerasuriya, 2010).

It has also been argued that improving nature-people connections is vital for promoting broader pro-environmental behaviours, and thereby achieving climate change goals, by overcoming what has been called the 'the extinction of experience with nature' (Soga & Gaston, 2016). A lack of personal experiences with nature limits our understanding of environmental challenges and appropriate actions to address those challenges (Cajete & Williams, 2020). Therefore, it is theorised that expanding people's emotional connections to nature, and promoting deeper conceptualisations of the natural world through nature-based interventions, will improve public support for a move towards more sustainable cities (Ives et al., 2018). This is seen to be particularly vital for children and adolescents, as a number of studies, from across numerous countries, provide support for the important role of meaningful childhood experiences in nature as a predictor of behaviours to protect the environment (Chawla, 2020). For this reason, there are now many programmes, operating at a global scale, (e.g. Nature Clubs for Families by the Children & Nature Network), which are focused on trying to find ways to encourage young people, and their families, to be closer to nature (D'Amore & Chawla, 2017).

9.2   Moving Towards Nature-Based Interventions

Given these arguments, many have pushed for the development of evidenced-based interventions of natural, blue green settings, which are aimed at being truly health-creating, moving people towards a state of complete physical, mental and social wellbeing. Such interventions span a wide spectrum of activities and levels of engagement (see Fig. 9.1), some of which are focused on 'bringing nature to people' by greening of places in people's nearby environment, particularly where greenspace access is currently limited, such as in urban public spaces, hospitals, classrooms and offices. Others, however, are aimed at 'bringing people to nature' by encouraging and facilitating adults and children to actively participate in nature-based activities (Bowler et al., 2010; Bragg & Atkins, 2016). More recently, there has been a move towards providing 'nature on prescription', in an effort to integrate such activities into routine medical practice (Garside et al., 2020; Husk et al., 2019; Husk et al., 2020; Robinson et al., 2020). Such programmes bring together concepts of 'non-medical' prescriptions (e.g. for exercise or diet) pioneered by general practitioners in New Zealand in the 1990s, with the more recent social prescribing movement (Robinson et al., 2020). Examples include Park Rx in the USA and Canada and 'green social prescribing' in the UK, and generally involve health care providers 'prescribing' nature-based interventions in the community.

Nature-based interventions can vary considerably in their aims (Shanahan et al., 2019). Some are targeted at the general population as part of health promotion, especially those without the ability or opportunity to engage with greenspace as part of their usual lifestyle, e.g. the Danish TEACHOUT project which focuses on providing outdoor education in primary schools (Nielsen et al., 2016). However, others are

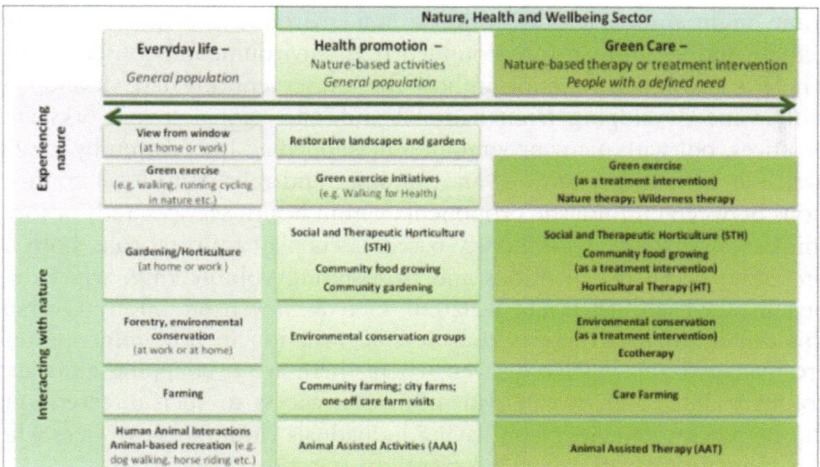

Fig. 9.1 Mapping nature-based interventions (*Source* Bragg and Atkins 2016, 22)

targeted as therapeutic interventions to address the specific needs of a specific group; sometimes called 'green care' (Sempik, 2010; Sempik & Bragg, 2016), 'ecotherapy' (Hinds & Jordan, 2016) or simply 'nature-based interventions'. Examples include programmes developed for at-risk youth, those living with a specific mental health condition, adults, and children with learning disabilities, those with a drug or alcohol addiction history, adults on probation and individuals living with dementia and their carers (Bragg & Atkins, 2016; see box text below for a case study of the development of such an intervention in Dorset, UK). These categories are, of course, not mutually exclusive, as people can move between a treatment intervention, health promotion programme and everyday activity within nature as their health and wellbeing needs dictate.

The types of activities that make up nature-based interventions can also vary considerably, ranging from wilderness therapy, social and therapeutic horticulture, facilitated environmental conservation, care farming, ecotherapy, nature-based arts and crafts, and animal-assisted interventions (see Bragg & Leck, 2017; Jepson et al., 2010 for an overview). Likewise, in some cases, the activity takes place in nature, but nature is not the focus (i.e. nature as 'background' to green exercise), whereas in others (and some might argue most cases) the natural environment is the focus of the activity and very much foregrounded (e.g. environmental conservation). There is also much variation in how these interventions have developed, with some developing from national and/or regional frameworks and practices, but with many programmes developed ad hoc, driven by locally available knowledge, settings and funding, and reliant on local connections between enthusiastic proponents within health services and/or local third sector groups. Community-based assets play a pivotal role, both in providing social interventions and in fostering volunteering, which has benefits for health (Bowe et al., 2020; Gray & Stevenson, 2020). Overall, this can mean that there is considerable variation in how interventions are managed and delivered, even where there are overlapping aims and contexts. Moreover, it can also mean that access to such interventions can be patchy, dependent on having individuals within local services who can champion nature-based interventions (Garside et al., 2020).

Stepping into Nature (SiN)

Stepping into Nature is a programme of work that aims to help people be happier and healthier by connecting people with nature, using Dorset's natural and cultural landscape to provide activities and sensory rich places for older adults, including those with living with dementia, and their care partners. Dorset has a higher proportion of people over the age of 65 than the national UK average, with a predicted increase of around six per cent by 2031. It is estimated that there are currently 13,000 people 65 or over living with dementia, with over 50,000 unpaid carers. We know that living with dementia, or providing care for a loved one, can affect daily lives. People can become isolated from society, their confidence, skills and physical fitness wane and their health and wellbeing deteriorates. However, it is important to remember that continued ill health in old age is not inevitable, improving social and emotional wellbeing, and healthy behaviours, can increase the time people can be independent and active in later life. Maintaining physical and cognitive function and increasing resilience are more likely to continue if built into everyday life.

SiN's core belief is that being in nature and sharing an activity is a crucial part of living well. Regardless of where we live, nature is free and on our doorstep for our entire lives. However, many people experience barriers that stop them fully benefitting from this multisensory environment. Working with older people, and a network of organisations, SiN provided opportunities that promoted these healthy changes. From 2017 to 2023, two rounds of funding totalling £710,000 were secured from the National Lottery Community Fund. Through this funding, the team worked in collaboration with the health, environment and community sectors to:

- **Upskill** staff and volunteers in the environment sector to become dementia-friendly through dementia awareness sessions linked to the natural environment. By increasing understanding of dementia and appreciation of needs, providers could adapt their activities and settings to become more inclusive and were better equipped to cater for varying abilities or unexpected situations.
- **Support** community groups and organisation, via grants and advice, to deliver activities or improve inclusivity of greenspaces.
- **Evaluate** whether engagement with the natural environment led to improved physical and emotional wellbeing, reduced social isolation and feelings of loneliness, increased motivation, independence, confidence and life skills with support of Public Health Dorset.
- **Promote** a trusted, high-quality, inclusive brand that was created to helped promote various activities by reducing fragmentation of information, creating a consistent message and increasing the capacity of activity providers.
- **Provide networking** opportunities for activity providers to connect to the target audience and health sector organisations through working groups, national conferences and Picnic in the Park events.
- **Pilot projects** as new ideas come up and test different approaches, for example the development of nature buddies' network, a one-to-one support service focused on linking people together to enjoy nature together on a more personalised level.

Over the lifetime of the project, more people have been engaged with nature by providing a variety of activities in different locations and other opportunities for people to connect with and increase the amount of time they spend in nature. This includes a range of different events including Sing and Stroll and Wellbeing Walks, see below. Through working in partnership with other NGOs, and local authorities, SiN has delivered over 400 activities which have been attended by over 3000 people.

Over 6,000 physical resources have been shared (such as history walks guides, seasonal books, a seasonal art and writing box, wild writing packs) and many more downloaded from online. Post activity, participants reported immediate feelings of fun, happiness and pleasure and found the activities interesting, enjoyable to do something different or liked to learn a new skill. From observations it was also noted that people looked visibly relaxed as an activity progressed—shoulders rested, open body language and sounding more confident. The inclusive approach of the Stepping into Nature activities enabled participants to meet new people within a 'no pressure environment'. During activities people chatted, sharing stories, experiences, knowledge with each other and in some cases providing informal support through meeting others with similar issues or situations. Organisations reported that their capacity and confidence to deliver meaningful activities for this audience had increased and by being part of the project, staff and volunteers felt better equipped to be able to provide the support needed.

While there were many benefits, there were also several challenges that SiN faced on the journey. Overall, activity take-up was low at the start of the project as it took time to build a brand that gave some quality assurance and trust. There were common barriers related to the physical environment, often where environments were not suitable (e.g. because of stiles, uneven ground, mud or poor signage), or because of a lack of information that enabled people to make an informed decision about whether a particular environment was suitable for them (e.g. in relation to parking, facilities, directions). Some service users didn't want to explore the area alone, because it felt unsafe, or they didn't know the route. Moreover, a continuous stigma around dementia was conveyed, where providers and public weren't aware of the symptoms and how to support people. In addition, COVID-19 impacted on their activities in many ways. The project supported was ongoing throughout the pandemic, but all group activities were put on hold. Staff in SiN had to learn and adapt their approach from direct delivery to the development of self-led activities, such as accessible introductory history walk guides and seasonal books along with resources that encouraged connection to others and nature through stories, poetry and art. These, along with other resources in the community, were promoted via our virtual festival Picnic in the Parks website, developed with Active Dorset and Dorset Local Nature Partnership.

Building the team's learning and experience, the team is now designing the next steps, the development of a wider Health and Nature Dorset collaboration which aims to help connect organisations that are working towards similar goals together. Through this, the aim is to identify ways in which collaboration can be improved to enhance the nature-based wellbeing offers in Dorset. For more information, refer to: https://www.dorsetaonb.org.uk/project/stepping-into-nature/.

9.3 Mapping the Evidence Base

The diversity of nature-based interventions, in terms of the range of different interventions and therapeutic opportunities available to people, make it difficult to understand what kinds of nature-based interventions (or elements of these) work best for whom, where and when. The evidence base is diluted by the considerable degree of variation in participants, mechanisms and outcomes (Bowler et al., 2010). There is some evidence that such interventions show great promise, in terms of improvements in psychological wellbeing, cognitive functioning, coping ability and reductions in social isolation, across diverse diagnoses, spanning from obesity to schizophrenia (e.g. Annerstedt & Wahrborg, 2011; Bragg & Atkins, 2016). Qualitative work in this area also highlights a broad and wide-reaching perceived impacts on wellbeing, mood and functioning from participants (see Garside et al., 2020). However, despite a large amount of research, the evidence for the outcomes and benefits of nature-based interventions is limited, relying on unvalidated measures, small sample sizes, often do not include a comparator group, have limited follow-up evidence and there are very few randomised control studies (Bragg & Atkins, 2016; Charles et al., 2018; Garside et al., 2020; Husk et al., 2019; Husk et al., 2020; Lovell et al., 2015). There are also very few longitudinal studies, and these are needed. Overall, there is a lack of robust evidence that such interventions are effective, for whom and what is needed to make them work (or work better). Key questions remain concerning mechanisms of change, context and the replicability of interventions in different environments, and access and the potential for such interventions to address (or worsen) health inequalities. Each of these is covered in more depth below.

Mechanisms of Change

Some have questioned why such evidence is needed, given that there is such consensus in the published literature that nature contributes to enhanced wellbeing, mental development and personal fulfilment. However, it is also the case that much of evidence about the health and wellbeing benefits of nature are based at the population level—focused on the quantity or proximity of greenspace within a specific place, and its relationship to levels of wellbeing (e.g. Alcock et al., 2014). It is not always clear how these relationships at a population level will—or should—translate down into interventions in ways that produce specific wellbeing benefits. Indeed, many interventions are not theoretically driven, or the rationale for the intervention that informed its design, delivery and intervention components are not always entirely explicit. Broader theories from environmental or evolutionary psychology, are sometimes used, including Biophilia (Wilson, 1986), Attention Restoration Theory (ART; Kaplan & Kaplan, 1989), Psycho-evolutionary Theory (PET; Ulrich et al., 1991), Supportive Environment Theory (Grahn et al., 2017) and Contemplative Landscape Theory (Olszewska et al., 2018). However, it is not yet explicit how these theories relate to a clear understanding and integration of active elements (the mechanism and mediators of change) that can produce hypothesised results.

Added to this, is the fact that health and wellbeing is itself a complex construct with multiple determinants, and there are multiple hypothesised (but not always agreed) pathways through which nature is meant to relate to health (see Hartig et al., 2014). Often this is not well recognised, amid a growing tendency to present 'health as much simpler than it actually is' (Wolf, 2010, 84). Potential mechanisms are likely to include sensory-perceptual and immunological processes, air quality, learning new skills, the restorative qualities of nature and reduced social isolation (Kuo, 2015; Shanahan et al., 2015). However, what is not clear is whether *interventions* that incorporate these mechanisms are more successful than those that do not, or indeed what the 'optimal dose' of these factors would be (Olszewska et al., 2018). Some have argued that we need to ask if such an 'optimal dose' of nature can even exist, given the diverse ways in which individuals experience and interpret nature (e.g. Bell et al., 2019). There is also the risk that inappropriate nature-based activity components, or group dynamics exacerbates or worsens existing mental health conditions. This is a concern for health care professionals who are asked to prescribe such interventions to their patients, and as a result, referral rates are currently very low (Van den Berg, 2017). To begin to address some of these questions, a more convincing explanatory framework is needed, that specifies the main pathways and mechanisms of these interventions, and which considers the ways these may be impacted by an intervention (Van den Berg, 2017).

Intervention Context

As with the mechanisms of change, the context within which nature-based interventions occur is currently under-theorised—or indeed often unspecified, beyond being a local natural environment of some sort. However, key questions remain about the unique qualities of a setting, or how varying configurations of greenspace could offer different opportunities for physical exercise, psychological restoration, relaxation or social connections, leading to different mental health outcomes (Alcock et al., 2015). The limited work in this area has produced mixed results. For example, well-maintained and tended natural spaces have been found to produce more significant improvements in affect and mood (e.g. Martens et al., 2011). However, such findings do not necessarily hold at a population level (e.g. Alcock et al., 2015). Importantly, there is some debate about whether all natural environments are beneficial to mental health. For example, there is some evidence that low in prospect and high in refuge (e.g. dense wooded areas) can increase stress levels and fatigue and thereby, negatively impact on mental health (Gatersleben & Andrews, 2013; see also Milligan & Bingley, 2007; Van den Berg & Ter Heijne, 2005).

Landscape perceptions are important to consider here, as empirical work has consistently demonstrated that such perceptions are key to how people experience and use space (Hägerhäll et al., 2018; Knez & Eliasson, 2017; Korpela et al., 2001; Shanahan et al., 2015). Indeed, there is already much research that demonstrates that access to urban greenspaces (such as parks) does not translate directly into actual use. Rather, it is people's perception of spaces that is an influential factor for or against using a park (Byrne & Wolch, 2009). Perceived quality and maintenance of a space is important, as empirical studies show that such spaces tend to be seen as safer and therefore used more widely (Jansson et al., 2013; Milligan & Bingley, 2007). It is also well established that this matters more for some groups than others.

Research from the UK, the USA, Sweden, China, Mexico and South Africa highlights that perceptions of quality and maintenance in urban greenspaces are vitally important to women, as it is linked to perceptions of safety, which is crucial for their decisions to use such spaces (e.g. see Mayen Huerta & Utomo, 2022). The perceptions of others are also important. Public greenspaces, such as parks, can be contested spaces, where ideologies about 'who belongs where clash' (Manzo,

2003, 55). For example, we know that young people's use of public spaces—including urban greenspaces such as parks—is often controlled by (typically) adult others in ways that limit their use of such spaces (Gray & Manning, 2014, 2022; Gray et al., 2021). Likewise, such places can invoke a strong sense of place identity, which in turn can foster a sense of 'togetherness', self-esteem, self-worth and self-pride (Korpela et al., 2001). Places can act as symbolic repositories of national and cultural values (Twigger-Ross & Uzzell, 1996) in ways that are meaningful to psychological health and wellbeing.

These findings would indicate that *where* interventions take place is an important consideration and more research is needed to fully understand how such landscapes and other forms of greenspaces in urban and rural areas can—and should—be used as a resource for psychological wellbeing. Moreover, it also raises questions about how transferable evidence is between different contexts and whether, for example, the outcomes of nature-based interventions in highly differing environments, such as densely forested mountains, large urban parks and coastal landscapes, reasonably ought to be compared. This requires more research on how different types and characteristics of greenspaces are perceived and/or experienced in a multisensory manner, through sight, hearing, touching and smell, explicitly linking such perceptions to specific mental health outcomes (see also Alcock et al., 2015).

Impact on Health Inequalities

There is evidence that inequalities in health are lower in greener communities, meaning that providing effective nature-based interventions that (re)connect people to nature can be an important way to help to reduce health inequalities (e.g. Mitchell & Popham, 2008; Roe et al., 2013). However, there is a need for further research to ensure that benefits are maximised while not increasing the health inequalities it is trying to eliminate. In part, this relates to already existing, and well-documented concerns about barriers to accessing natural environments (Boyd et al., 2018; Cole et al., 2017; Richardson & Mitchell, 2010). What is clear from this research is that barriers faced by different groups (and in different countries) are multiple and complex and include a range of physical (e.g. relating to topography and facilities) and sociocultural (e.g. feeling 'out of place') factors, which will feed into people's willingness to engage with, and benefit from, nature-based interventions in different ways (see Wolff et al. 2022 for an overview).

Such features have the possibility of exacerbating inequalities through processes of enrolment, engagement and adherence (Husk et al., 2019; Husk et al., 2020). For example, as documented in the case study in the box text, a lack of facilities such as toilets, sitting spots or challenging topography can make it harder to recruit older adults or those living with a disability to interventions in those spaces. Likewise, many health care professionals are reluctant to prescribe nature if they do not think that their patients will be able to adhere to the prescription, because they believe that their patients lack resources in terms of time, money, not having access to nearby outdoor spaces due to transportation and availability, and not having motivation (Christiana et al., 2017). This means that those who are at greatest risk of poor mental health may be the least likely to be offered an intervention based in nature. There is also the danger that nature-based interventions can reproduce already existing spatial inequalities through unequal provision and the availability of resources (e.g. sites).

We know that good quality greenspaces are unequally distributed, with those living in economically deprived areas having the least available good quality public greenspace (Schüle et al., 2019). However, it is also the case that having access to a good range of good quality local greenspaces is a vital element for the development and provision of nature-based interventions (Robinson et al., 2020). Meaning that there can be significant differences in access to such interventions across different geographical areas in ways that can further entrench existing spatialised health inequalities. Indeed, in their study, Robinson et al. (2020) found an association between the abundance of greenspace near to GP surgeries and the likelihood of a GP providing green prescriptions. As they note, this raises several important questions about whether (and how) the lack of available services/infrastructure equates to more limited provision. This would suggest that a dual approach is needed that includes both improvements to the provision of, and access to, good quality and inclusive natural environment, that speak to a diverse range of needs and preferences, alongside interventions that work to improve people's engagement with those environments. However, this clearly has an impact on the management of those environments, which could become overwhelmed with demand.

Overall, what this points to, is the need to critically examine the assumption that the health benefits of green or blue spaces will be the same across all population groups. Caution needs to be taken about trans-ferring the findings from one population group to another, if they have different sociocultural perspectives on the natural environment, as well as differing values on how the natural environment relates to health. Currently, there is little evidence about how nature-based interventions vary in experience across different populations, e.g. which delivery modes or activities are most valued by participants, whether it is better for activity groups to comprise only those with particular needs, conditions or ages, or whether these should be mixed or how best to harness the group effects for positive interaction (Garside et al., 2020). There are also some key groups which are currently under-represented in the body of evidence. For example, despite the fact that many of the policies about improving nature connections are specifically aimed at children, and the fact that many countries already have historically developed nature programmes for children (e.g. The Children and Nature Initiative in the USA; Louv, 2011), most green (and social) prescriptions currently are targeted at adults, and there is relatively little research with young people. This means that we know much less about how such interventions are experienced or effective for children and young people (see Garside et al., 2020; Kondo et al., 2020). However, given the rising mental health issues within this age group—particularly the post-COVID-19 pandemic (Ma et al., 2021)—and the fact that 75% of all mental health issues that start in childhood continue to adulthood, this is an important group on which to focus attention (Kondo et al., 2020).

9.4   CONCLUSIONS

Given the prevalence and burden of mental ill health worldwide, there is a pressing need for interventions and solutions that provide effec-tive, equitable and cost-effective ways of promoting better mental health and wellbeing. Nature-based interventions, that seek to prioritise the (re)connection of people to nature, are an integral part of this solu-tion, with great promise to address the social determinants of mental health. However, this field as a whole is complex and diverse and, while nature-based interventions are not new, models and processes for inte-grating these into public health systems (e.g. through green prescriptions) are in their infancy. Currently, little is understood about how green (or social) prescribing interacts with the health service or the capacity of the community sector to offer activities or manage the demands of the health service moving forward. In addition, without engaging effectively with practitioners those areas will not be available and not be managed appropriately to allow access to nature. As the field matures, there is a great need for robust study designs, and a greater focus on what works, understanding mechanisms of change, and clarifying the health associations for different contexts and population groups. Such interventions are not without risks to patients and finances, and there are clear consequences of developing interventions that offer poor outcomes (and hence poor value for money), because they are developed without evidence about what should be offered or the processes that are required to support them, resulting in patients not getting a green prescription that is appropriate to their needs. All stakeholders in the pathway from commissioning and promotion, to referral, and the development and delivery of sustainable, fundable interventions, require a better understanding of nature-based interventions.

REFERENCES

Alcock, I., White, M., Lovell, R., Higgins, S., Osborne, N., Husk, K., & Wheeler, B. (2015). What accounts for 'England's green and pleasant land'? A panel data analysis of mental health and land cover types in rural England. *Landscape and Urban Planning, 142*, 38–46. https://doi.org/10.1016/j.landurbplan.2015.05.008

Alcock, I., White, M., Wheeler, B., Fleming, L., & Depledge, M. (2014). Longitudinal effects on mental health of moving to greener and less green urban areas. *Environmental Science & Technology, 48*(2), 1247–1255. https://doi.org/10.1021/es403688w

Allen, J., Balfour, R., Bell, R., & Marmot, M. (2014). Social determinants of mental health. *International Review of Psychiatry, 26*(4), 392–407. https://doi.org/10.3109/09540261.2014.928270

Annerstedt, M., Östergren, P.-O., Björk, J., Grahn, P., Skärbäck, E., & Währborg, P. (2012). Green qualities in the neighbourhood and mental health–results from a longitudinal cohort study in Southern Sweden. *BMC Public Health, 12*(1), 337. https://doi.org/10.1186/1471-2458-12-337

Annerstedt, M., & Wahrborg, P. (2011). Nature assisted therapy: Systematic review of controlled and observational studies. *Scandinavian Journal of Public Health, 39*, 371388. https://doi.org/10.1177/1403494810396400

Barton, J., Bragg, R., Wood, C., & Pretty, J. (2016). *Green exercise.* Routledge.

Barton, J., & Pretty, J. (2010). What is the best dose of nature and green exercise for improving mental health? A multi-study analysis. *Environmental Science & Technology, 44*(10), 3947–3955. https://doi.org/10.1021/es903183r

Bell, S., Leyshon, C., Foley, R., & Kearns, R. (2019). The "healthy dose" of nature: A cautionary tale. *Geography Compass, 13*(1), e12415. https://doi.org/10.1111/gec3.12415

Beyer, K., Kaltenbach, A., Szabo, A., Bogar, S., Nieto, F., & Malecki, K. (2014). Exposure to neighborhood green space and mental health: Evidence from the survey of the health of Wisconsin. *International Journal of Environmental Research and Public Health, 11*(3), 3453–3472. https://doi.org/10.3390/ijerph110303453

Bragg, R., & Atkins, G. (2016). *A review of nature based interventions for mental health care* (Natural England Commissioned Reports, 204). http://publications.naturalengland.org.uk/publication/4513819616346112. Accessed 14 April 2023.

Bragg, R., & Leck, C. (2017). *Good practice in social prescribing for mental health: The role of nature based interventions* (Natural England Commissioned Reports, 228). http://publications.naturalengland.org.uk/publication/5134438692814848. Accessed 14 April 2023.

Bratman, G., Hamilton, J., & Daily, G. (2012). The impacts of nature experience on human cognitive function and mental health. *Annals of the New York Academy of Sciences, 1249*(1), 118–136. https://doi.org/10.1111/j.1749-6632.2011.06400.x

Brennan, M., Mills, G., & Ningal, T. (2017). *Dublin tree canopy study*. https://mappinggreendublin.com/dublin-canopy-project/. Accessed 14 April 2020.

Bowe, M., Gray, D., Stevenson, C., McNamara, N., Wakefield, J., Kellezi, B., et al. (2020). A social cure in the community: A mixed-method exploration of the role of social identity in the experiences and well-being of community volunteers. *European Journal of Social Psychology, 50*(7), 1523–1539. https://doi.org/10.1002/ejsp.2706

Bowler, D., Buyung-Ali, L., Knight, T., & Pullin, A. (2010). *The importance of nature for health: Is there a specific benefit of contact with green space?* https://environmentalevidence.org/project/the-importance-of-nature-for-health-is-there-a-specific-benefit-of-contact-with-green-space-systematic-review/. Accessed 14 April 2023.

Boyd, F., White, M., Bell, S., & Burt, J. (2018). Who doesn't visit natural environments for recreation and why: A population representative analysis of spatial, individual and temporal factors among adults in England. *Landscape and Urban Planning, 175*, 102–113. https://doi.org/10.1016/j.landurbplan.2018.03.016

Byrne, J., & Wolch, J. (2009). Nature, race, and parks: Past research and future directions for geographic research. *Progress in Human Geography, 33*(6), 743–765. https://doi.org/10.1177/0309132509103156

Cajete, G., & Williams, D. (2020). Eco-aesthetics, metaphor, story, and symbolism: An indigenous perspective. A conversation. In A. Cutter-MacKenzie-Knowles, K. Malone, & E. Barrat Hacking (Eds.), *Research handbook on childhood nature: Assemblages of childhood and nature research* (pp. 1707–1733). Springer.

Charles, C., Keenleyside, K., Chapple, R., Kilburn, B., Salah van der Leest, P., Allen, D., et al. (2018). *Home to us all: How connecting with nature helps us care for ourselves and the Earth.* https://www.cbd.int/doc/strategic-plan/Post2020/postsbi/C&nn2.pdf. Accessed 14 April 2023.

Chawla, L. (2015). Benefits of nature contact for children. *Journal of Planning Literature, 30*(4), 433–452. https://doi.org/10.1177/0885412215595441

Chawla, L. (2020). Childhood nature connection and constructive hope: A review of research on connecting with nature and coping with environmental loss. *People and Nature, 2*(3), 619–642. https://doi.org/10.1002/pan3.10128

Chawla, L., Keena, K., Pevec, I., & Stanley, E. (2014). Green schoolyards as havens from stress and resources for resilience in childhood and adolescence. *Health and Place, 28,* 1–13. https://doi.org/10.1016/j.healthplace.2014.03.001

Christiana, R., Battista, R., James, J., & Bergman, S. (2017). Pediatrician prescriptions for outdoor physical activity among children: A pilot study. *Preventive Medicine Reports, 5,* 100–105. https://doi.org/10.1016/j.pmedr.2016.12.005

Cole, H., Garcia, L. M., Connolly, J., & Anguelovski, I. (2017). Are green cities healthy and equitable? Unpacking the relationship between health, green space and gentrification. *Journal of Epidemiology and Community Health, 71*(11), 1118–1121. https://doi.org/10.1136/jech-2017-209201

D'Amore, C., & Chawla, L. (2017). Many children in the woods: Applying principles of community-based social marketing to a family nature club. *Ecopsychology, 9*(4), 232–240. https://doi.org/10.1089/eco.2017.0019

De Vries, S., Verheij, R., Groenewegen, P., & Spreeuwenberg, P. (2003). Natural environments—Healthy environments? An exploratory analysis of the relationship between greenspace and health. *Environment and Planning A, 35*(10), 1717–1731. https://doi.org/10.1068/a35111

Garside, R., Orr, N., Short, R., Lovell, B., Husk, K., McEachan, R., et al. (2020). *Therapeutic nature: Nature-based social prescribing for diagnosed mental health conditions in the UK* (Final Report for DEFRA). https://arc-swp.nihr.ac.uk/wp/wp-content/uploads/2021/06/15138_TherapeuticNature-Finalreport.pdf. Accessed 14 April 2023.

Gatersleben, B., & Andrews, M. (2013). When walking in nature is not restorative—The role of prospect and refuge. *Health and Place, 20,* 91–101. https://doi.org/10.1016/j.healthplace.2013.01.001

Grahn, P., Pálsdóttir, A., Ottosson, J., & Jonsdottir, I. (2017). Longer nature based rehabilitation may contribute to a faster return to work in patients with reactions to severe stress and/or depression. *International Journal of Environmental Research and Public Health, 14*(11), 1310. https://doi.org/10.3390/ijerph14111310

Grahn, P., & Stigsdotter, U. (2003). Landscape planning and stress. *Urban Forestry and Urban Greening, 2,* 1–18. https://doi.org/10.1078/1618-8667-00019

Gray, D., & Manning, R. (2014). 'Oh my god, we're not doing nothing': Young people's experiences of spatial regulation. *British Journal of Social Psychology, 53*(4), 640–655. https://doi.org/10.1111/bjso.12055

Gray, D., & Manning, R. (2022). Constructing the places of young people in public space: Conflict, belonging and identity. *British Journal of Social Psychology, 61*(4), 1400–1417. https://doi.org/10.1111/bjso.12542

Gray, D., Manning, R., & Oftadeh-Moghadam, S. (2021). Using mixed and multi-modal methods in psychological research with young people. In R. Locke & A. Lees (Eds.), *Mixed-methods research in wellbeing and health* (pp. 156–173). Routledge.

Gray, D., & Stevenson, C. (2020). How can 'we' help? Exploring the role of shared social identity in the experiences and benefits of volunteering. *Journal of Community & Applied Social Psychology, 30*(4), 341–353. https://doi.org/10.1002/casp.2448

Groenewegen, P., Van den Berg, A., De Vries, S., & Verheij, R. (2006). Vitamin G: Effects of green space on health, well-being, and social safety. *BMC Public Health, 6*(1), 1–9. https://doi.org/10.1186/1471-2458-6-149

Hägerhäll, C., Ode Sang, Å., Englund, J., Ahlner, F., Rybka, K., Huber, J., & Burenhult, N. (2018). Do humans really prefer semi-open natural landscapes? A cross-cultural reappraisal. *Frontiers in Psychology, 9,* 822. https://doi.org/10.3389/fpsyg.2018.00822

Hartig, T., Evans, G., Jamner, L., Davis, D., & Gärling, T. (2003). Tracking restoration in natural and urban field settings. *Journal of Environmental Psychology, 23*(2), 109–123. https://doi.org/10.1016/S0272-4944(02)00109-3

Hartig, T., Mitchell, R., De Vries, S., & Frumkin, H. (2014). Nature and health. *Annual Review of Public Health, 35*(1), 207–228. https://doi.org/10.1146/annurev-publhealth-032013-182443

Hinds, J., & Jordan, M. (2016). *Ecotherapy. Theory, research and practice.* Palgrave.

Houlden, V., de Albuquerque, J. P., Weich, S., & Jarvis, S. (2019). A spatial analysis of proximate greenspace and mental wellbeing in London. *Applied Geography, 109,* 102036. https://doi.org/10.1016/j.apgeog.2019.102036

Houlden, V., Weich, S., Porto de Albuquerque, J., Jarvis, S., & Rees, K. (2018). The relationship between greenspace and the mental wellbeing of adults: A systematic review. *PLoS ONE, 13*(9), e0203000. https://doi.org/10.1371/journal.pone.0203000

Husk, K., Blockley, K., Lovell, R., Bethel, A., Lang, I., Byng, R., & Garside, R. (2020). What approaches to social prescribing work, for whom, and in what circumstances? A realist review. *Health and Social Care in the Community, 28*(2), 309–324. https://doi.org/10.1111/hsc.12839

Husk, K., Elston, J., Gradinger, F., Callaghan, L., & Asthana, S. (2019). Social prescribing: Where is the evidence? *British Journal of General Practice, 69*(678), 6–7. https://doi.org/10.3399/bjgp19X700325

Ives, C., Abson, D., Von Wehrden, H., Dorninger, C., Klaniecki, K., & Fischer, J. (2018). Reconnecting with nature for sustainability. *Sustainability Science, 13*(5), 1389–1397. https://doi.org/10.1007/s11625-018-0542-9

Jansson, M., Fors, H., Lindgren, T., & Wiström, B. (2013). Perceived personal safety in relation to urban woodland vegetation—A review. *Urban Forestry and Urban Greening, 12*(2), 127–133. https://doi.org/10.1016/j.ufug.2013.01.005

Jepson, R., Robertson, R., & Cameron, H. (2010). *Green prescription schemes: mapping and current practice.* NHS Scotland. https://dspace.stir.ac.uk/bitstream/1893/12871/1/Jepson_2010_Green_Prescription_Schemes.pdf. Accessed 14 April 2023.

Jetten, J., Haslam, S., & Haslam, C. (2012). The case for a social identity analysis of health and well-being. In J. Jetten (Ed.), *The social cure: Identity, health and wellbeing* (pp. 3–19). Routledge/Psychology Press.

Kaplan, R., & Kaplan, S. (1989). *The experience of nature: A psychological perspective.* Cambridge University Press.

Knez, I., & Eliasson, I. (2017). Relationships between personal and collective place identity and well-being in mountain communities. *Frontiers in Psychology, Sec Environmental Psychology, 8.* https://doi.org/10.3389/fpsyg.2017.00079

Kondo, M., Oyekanmi, K., Gibson, A., South, E., Bocarro, J., & Hipp, J. (2020). Nature prescriptions for health: A review of evidence and research opportunities. *International Journal of Environmental Research and Public Health, 17*(12), 4213. https://doi.org/10.3390/ijerph17124213

Korpela, K., Hartig, T., Kaiser, F., & Fuhrer, U. (2001). Restorative experience and self-regulation in favorite places. *Environment and Behavior, 33*(4), 572–589. https://doi.org/10.1177/00139160121973133

Kuo, M. (2015). How might contact with nature promote human health? Exploring promising mechanisms and a possible central pathway. *Frontiers in Psychology, 6,* 1093. https://doi.org/10.3389/fpsyg.2015.01093

Louv, R. (2011). Reconnecting to nature in the age of technology. *The Futurist*, *45*(6), 41–45. http://www.wfs.org/content/futurist/november-december-2011-vol-45-no-

Lovell, R., Husk, K., Cooper, C., Stahl-Timmins, W., & Garside, R. (2015). Understanding how environmental enhancement and conservation activities may benefit health and wellbeing: A systematic review. *BMC Public Health*, *15*(1), 1–18. https://doi.org/10.1186/s12889-015-2214-3

Lovell, R., White, M., Wheeler, B., Taylor, T., & Elliott, L. (2020). *A rapid scoping review of health and wellbeing evidence for the green infrastructure standards*. European Centre for Environment and Human Health, University of Exeter Medical School. For: Natural England, Department for the Environment, Food and Rural Affairs, Public Health England, and Ministry for Housing, Communities and Local Government. https://beyondgreenspace.files.wordpress.com/2020/10/neer015-a-rapid-scoping-review-of-health-and-wellbeing-evidence-for-the-framework-of-green-infrastructure-standards-final-draft-sept-2020-1.pdf. Accessed 14 April 2023.

Maas, J., Van Dillen, S., Verheij, R., & Groenewegen, P. (2009a). Social contacts as a possible mechanism behind the relation between green space and health. *Health and Place*, *15*(2), 586–595. https://doi.org/10.1016/j.healthplace.2008.09.006

Maas, R., Verheij, S., DeVries, P., Spreeuwenberg, F., Schellevis, P., & Groenewegen, P. (2009b). Morbidity is related to a green living environment. *Journal of Epidemiology and Community Health*, *63*, 967–973. https://doi.org/10.1136/jech.2008.079038

Maller, C., Townsend, M., St Leger, L., Henderson-Wilson, C., Pryor, A., Prosser, L., & Moore, M. (2009). Healthy parks, healthy people: The health benefits of contact with nature in a park context. *The George Wright Forum*, *26*(2), 51–83.

Ma, L., Mazidi, M., Li, K., Li, Y., Chen, S., Kirwan, R., et al. (2021). Prevalence of mental health problems among children and adolescents during the COVID-19 pandemic: A systematic review and meta-analysis. *Journal of Affective Disorders*, *293*, 78–89. https://doi.org/10.1016/j.jad.2021.06.021

Manzo, L. (2003). Beyond house and haven: Toward a revisioning of emotional relationships with places. *Journal of Environmental Psychology*, *23*(1), 47–61. https://doi.org/10.1016/S0272-4944(02)00074-9

Martens, D., Gutscher, H., & Bauer, N. (2011). Walking in 'wild' and 'tended' urban forests: The impact on psychological well-being. *Journal of Environmental Psychology*, *31*(1), 36–44. https://doi.org/10.1016/j.jenvp.2010.11.001

Mayen Huerta, C., & Utomo, A. (2022). Barriers affecting women's access to urban greenspaces during the COVID-19 pandemic. *Land*, *11*(4), 560. https://doi.org/10.3390/land11040560

Mental Health Taskforce. (2016). *The five year forward view for mental health.* A report from the independent Mental Health Taskforce to the NHS in England. NHS England. www.england.nhs.uk/mentalhealth/taskforce/. Accessed 14 April 2023.

Milligan, C., & Bingley, A. (2007). Restorative places or scary spaces? The impact of woodland on the mental well-being of young adults. *Health and Place, 13*(4), 799–811. https://doi.org/10.1016/j.healthplace.2007.01.005

Mitchell, R. (2013). Is physical activity in natural environments better for mental health than physical activity in other environments? *Social Science and Medicine, 91,* 130–134. https://doi.org/10.1016/j.socscimed.2012.04.012

Mitchell, R., & Popham, F. (2008). Effect of exposure to natural environment on health inequalities: An observational population study. *The Lancet, 372*(9650), 1655–1660. https://doi.org/10.1016/S0140-6736(08)61689-X

Nielsen, G., Mygind, E., Bølling, M., Otte, C., Schneller, M., Schipperijn, J., & Bentsen, P. (2016). A quasi-experimental cross-disciplinary evaluation of the impacts of education outside the classroom on pupils' physical activity, well-being and learning: The TEACHOUT study protocol. *BMC Public Health, 16*(1), 1–15. https://doi.org/10.1186/s12889-016-3780-8

Olszewska, A., Marques, P., Ryan, R., & Barbosa, F. (2018). What makes a landscape contemplative? *Environment and Planning B: Urban Analytics and City Science, 45*(1), 7–25. https://doi.org/10.1177/0265813516660716

Richardson, E., & Mitchell, R. (2010). Gender differences in relationships between urban green space and health in the United Kingdom. *Social Science and Medicine, 71*(3), 568–575. https://doi.org/10.1016/j.socscimed.2010.04.015

Robinson, J., & Breed, M. (2019). Green prescriptions and their co-benefits: Integrative strategies for public and environmental health. *Challenges, 10*(1), 9. https://doi.org/10.3390/challe10010009

Robinson, J., Jorgensen, A., Cameron, R., & Brindley, P. (2020). Let nature be thy medicine: A socioecological exploration of green prescribing in the UK. *International Journal of Environmental Research and Public Health, 17*(10), 3460. https://doi.org/10.3390/ijerph17103460

Roe, J., Thompson, C., Aspinall, P., Brewer, M., Duff, E., Miller, D., et al. (2013). Green space and stress: Evidence from cortisol measures in deprived urban communities. *International Journal of Environmental Research and Public Health, 10*(9), 4086–4103. https://doi.org/10.3390/ijerph10094086

Sarkar, C., Webster, C., & Gallacher, J. (2018). Residential greenness and prevalence of major depressive disorders: A cross-sectional, observational, associational study of 94 879 adult UK Biobank participants. *The Lancet Planetary Health, 2*(4), e162–e173. https://doi.org/10.1016/S2542-5196(18)30051-2

Schüle, S., Hilz, L., Dreger, S., & Bolte, G. (2019). Social inequalities in environmental resources of green and blue spaces: A review of evidence in the WHO European region. *International Journal of Environmental Research and Public Health, 16*(7), 1216. https://doi.org/10.3390/ijerph16071216

Sempik, J. (2010). Green care and mental health: Gardening and farming as health and social care. *Mental Health and Social Inclusion, 14*(3), 15–22. https://doi.org/10.5042/mhsi.2010.0440

Sempik, J., & Bragg, R. (2016). Green care: Nature based interventions for vulnerable people. In J. Barton, R. Bragg, & C. Wood (Eds.), *Green exercise* (pp. 116–129). Routledge.

Shanahan, D., Astell-Burt, T., Barber, E., Brymer, E., Cox, D., Dean, J., et al. (2019). Nature-based interventions for improving health and wellbeing: The purpose, the people and the outcomes. *Sports, 7*(6), 141. https://doi.org/10.3390/sports7060141

Shanahan, D., Fuller, R., Bush, R., Lin, B., & Gaston, K. (2015). The health benefits of urban nature: How much do we need? *BioScience, 65*(5), 476–485. https://doi.org/10.1093/biosci/biv032

Sia, A., Tam, W., Fogel, A., Kua, E., Khoo, K., & Ho, R. (2020). Nature-based activities improve the well-being of older adults. *Scientific Reports, 10*(1), 1–8. https://doi.org/10.1038/s41598-020-74828-w

Soga, M., & Gaston, K. J. (2016). Extinction of experience: The loss of human–nature interactions. *Frontiers in Ecology and the Environment, 14*(2), 94–101. https://doi.org/10.1002/fee.1225

Sugiyama, T., Leslie, E., Giles-Corti, B., & Owen, N. (2008). Associations of neighbourhood greenness with physical and mental health: Do walking, social coherence and local social interaction explain the relationships? *Journal of Epidemiology and Community Health, 62*(5), e9–e9. https://doi.org/10.1136/jech.2007.064287

Thompson, C., Roe, J., Aspinall, P., Mitchell, R., Clow, A., & Miller, D. (2012). More green space is linked to less stress in deprived communities: Evidence from salivary cortisol patterns. *Landscape and Urban Planning, 105*(3), 221–229. https://doi.org/10.1016/j.landurbplan.2011.12.015

Townsend, M., & Weerasuriya, R. (2010). *Beyond blue to green: The benefits of contact with nature for mental health and well-being.* Deakin University. http://www.hphpcentral.com/wp-content/uploads/2010/09/beyond blue_togreen.pdf. Accessed 14 April 2023.

Twigger-Ross, C., & Uzzell, D. (1996). Place and identity processes. *Journal of Environmental Psychology, 16*(3), 205–220. https://doi.org/10.1006/jevp.1996.0017

Ulrich, R., Simons, R., Losito, B., Fiorito, E., Miles, M., & Zelson, M. (1991). Stress recovery during exposure to natural and urban environments. *Journal of*

Environmental Psychology, 11(3), 201–230. https://doi.org/10.1016/S0272-4944(05)80184-7

Van den Berg, A. (2017). From green space to green prescriptions: Challenges and opportunities for research and practice. *Frontiers in Psychology, 8,* 268. https://doi.org/10.3389/fpsyg.2017.00268

Van den Berg, A., Maas, J., Verheij, R., & Groenewegen, P. (2010). Green space as a buffer between stressful life events and health. *Social Science and Medicine, 70*(8), 1203–1210. https://doi.org/10.1016/j.socscimed.2010.01.002

Van den Berg, A., & Ter Heijne, M. (2005). Fear versus fascination: An exploration of emotional responses to natural threats. *Journal of Environmental Psychology, 25*(3), 261–272. https://doi.org/10.1016/j.jenvp.2005.08.004

Van den Berg, M., Van Poppel, M., Van Kamp, I., Andrusaityte, S., Balseviciene, B., Cirach, M., et al. (2016). Visiting green space is associated with mental health and vitality: A cross-sectional study in four European cities. *Health and Place, 38,* 8–15. https://doi.org/10.1016/j.healthplace.2016.01.003

Vujcic, M., Tomicevic-Dubljevic, J., Grbic, M., Lecic-Tosevski, D., Vukovic, O., & Toskovic, O. (2017). Nature based solution for improving mental health and well-being in urban areas. *Environmental Research, 158,* 385–392. https://doi.org/10.1016/j.envres.2017.06.030

White, M., Alcock, I., Wheeler, B., & Depledge, M. (2013). Would you be happier living in a greener urban area? A fixed-effects analysis of panel data. *Psychological Science, 24*(6), 920–928. https://doi.org/10.1177/0956797612464659

White, M., Pahl, S., Wheeler, B., Depledge, M., & Fleming, L. (2017). Natural environments and subjective wellbeing: Different types of exposure are associated with different aspects of wellbeing. *Health and Place, 45,* 77–84. https://doi.org/10.1016/j.healthplace.2017.03.008

Wilson, E. (1986). *Biophilia.* Harvard University Press.

Wolf, J. (2010). Against breastfeeding (sometimes). In J. Metzel & A. Kirkland (Eds.), *Against health: How health became the new morality* (pp. 83–92). New York University Press.

Wolff, E., Rauf, H. A., Diep, L., Natakun, B., Kelly, K., & Hamel, P. (2022). Implementing participatory nature-based solutions in the Global South. *Frontiers in Sustainable Cities, 4,* 956534.

World Health Organization. (2001). *The World Health Report 2001. Mental health: New understanding, new hope.* World Health Organization. http://www.who.int/whr2001/. Accessed 14 April 2023.

CHAPTER 10

Significant Spaces: Exploring the Health and Wellbeing Impacts of Natural Environments

Denise Hewlett, Debra Gray, Richard Gunton, Tom Munro, Sheela Agarwal, Martin Breed, Chris Skelly, Philip Weinstein, Ainara Terradillos, Natalia Lavrushkina, and Danny Byrne

10.1 INTRODUCTION

For decades we have seen much concern about a global epidemic of chronic, non-communicable diseases, with increasing recognition of the potential significant impacts on population health and wellbeing, quality of life, health systems and the economy (Egorov et al., 2016; WHO,

D. Hewlett (✉)
PeopleScapes Research & Knowledge Exchange Centre, Department of Responsible Management, University of Winchester, Winchester, UK
e-mail: denise.hewlett@winchester.ac.uk

Bournemouth University, Dorset, UK

D. Gray
PeopleScapes Research & Knowledge Exchange, University of Winchester, Winchester, UK
e-mail: Debra.Gray@winchester.ac.uk

© The Author(s) 2024
N. Finneran et al. (eds.), *Managing Protected Areas*,
https://doi.org/10.1007/978-3-031-40783-3_10

167

2018, 2022a). There is a particular concern about our urban environments in which 55% of the world's population live: by 2050, this proportion is expected to increase to 68% (WHO, 2022b). This situation presents significant challenges for the 4.2 billion people living in cities, including: the increase in sedentary lifestyles in areas dominated by private car use and with inadequate public transportation, poor housing, inadequate sanitation, impacts of climate change and heat island effects (WHO, 2022b), environmental pollution, soil and water contamination, noise pollution and air pollution.

R. Gunton
Winchester Business School, University of Winchester, Winchester, UK
e-mail: Richard.Gunton@winchester.ac.uk

T. Munro
Dorset AONB, Dorset, UK
e-mail: tom.munro@dorsetcouncil.gov.uk

S. Agarwal
University of Plymouth Business School, Plymouth, UK
e-mail: s.agarwal@plymouth.ac.uk

M. Breed
Flinders University, Bedford Park, SA, Australia
e-mail: martin.breed@flinders.edu.au

C. Skelly
Department of Health and Social Care, London, UK

P. Weinstein
School of Public Health, Faculty of Health and Medical Sciences, University of Adelaide, Adelaide, SA, Australia
e-mail: philip.weinstein@adelaide.edu.au

A. Terradillos
Universidad de Sevilla, Andalusia, Spain

N. Lavrushkina
Faculty of Management, Bournemouth University, Bournemouth, UK
e-mail: nlavrushkina@bournemouth.ac.uk

D. Byrne
Independent Consultant, Hampshire, UK
e-mail: byrne729@btinternet.com

Indeed, as but one example, the World Health Organisation (WHO) has determined that 99% of the global population are breathing air that exceeds WHO guideline limits (WHO, 2022c). Within this context, nature—and more broadly, green space—are increasingly being considered as a key public health intervention, based on considerable evidence that such environments can be good for our health, and can play a substantial part in enhancing our quality of life. Since the 1960s, researchers have attempted to determine what has generally been assumed for millennia: that nature and greenness or *viriditas* provides benefits to public health and wellbeing, and that good, spatial planning designs can enhance these benefits (Dahlgren & Whitehead, 1991, 2021; Hancock & Perkins, 1985; Marcus & Sachs, 2014).

Access to green spaces has been associated with a range of improvements to self-reported general health, including lower prevalence of diagnosed morbidities, increased longevity, less premature mortality and more rapid recovery from illness (Mitchell & Popham, 2008; Rojas-Rueda et al., 2019; Van den Berg et al., 2015). Greener neighbourhoods are associated with improvements in outcomes across a range of common health conditions, including anxiety, depression, cardiovascular disease (CVD), stroke and diabetes, and are also correlated with better heart health (Aitken, 2021; Grazuleviciene et al., 2015) and generally better psychological wellbeing (Gascon et al., 2016; Gray et al., 2021; Houlden et al., 2018; Van den Berg et al., 2016). Moreover, there is some evidence that these can be particularly important for dealing with health inequalities. In a landmark study conducted by Mitchell and Popham (2008, 1655), they found that 'populations that are exposed to the greenest environments also have lowest levels of health inequality related to income deprivation'. Blue space research, albeit of more recent attention, presents an additionally strong body of evidence on the benefits of blue spaces to our health, with people living near, or being able to view coastlines being generally healthier, having fewer symptoms of stress, and generally being more satisfied with their lives than those living inland (Grellier et al., 2017).

The COVID-19 pandemic has emphasised how interconnected people and places are, and how much people value green spaces, especially for those in urbanised locations. Much data is still being collated on the impact of the pandemic on our relationship with nature. Yet what has been published already demonstrates clearly how much people value nature and natural environments be this related to experiencing nature in private gardens (Pouso et al., 2020), public parks, National Parks and National Forests (ONS, 2021) or simply through viewing nature in green or blue spaces (Corley et al., 2021; Powers Tomasso et al., 2021). For example, one such study, GreenCOVID conducted across Ireland, Spain and England, demonstrated not only the value people placed on nature in the pandemic, but also suggested how people consider nature affects their health, their sense of individual and collective wellbeing (Garrido-Cumbrera et al. 2021, 2022; Guzman 2020, 2021).

10.2   Nature-Health Pathways

While there is now a robust body of evidence—and some general understanding and agreement—that access to nature can produce a range of psychological, physical and social benefits to people's health and wellbeing, there has been less agreement on the pathways through which these impacts on health are meant to accrue (see Hartig et al., 2014; Kuo, 2015 for an overview). Indeed, over the last 50 years, many theoretical and conceptual frameworks have been developed to explain the link between nature and health (see Table 10.1).

There are equally a large number of possible causal mechanisms posited. Indeed, in their review of the plausible pathways by which nature might promote health, Kuo (2015) identified no fewer than 21 plausible causal pathways, with some pathways understood better than others. Typically, however, most research has focused on four: environmental conditions, physical activity, relaxation and stress and social integration. Those working on environmental conditions, e.g., increased environmental microbial diversity, have highlighted the role that this can play in improving immunity and the risk of infectious diseases (Flies et al., 2017, 2018; Robinson et al., 2022). Likewise, those working on physical activity have demonstrated that better access to green spaces can increase physical activity levels, which is in turn linked to improvements in mental and physical health, e.g., improvements in sleep and obesity (Barton & Pretty, 2010; James et al., 2015). In terms of stress reduction, it is clear that both 'green' and 'blue' spaces can foster stress reduction, restoration and relaxation (Hartig et al., 2014; Roe et al., 2013; Thompson et al., 2012). Finally, in terms of social interaction, there is some evidence that nature can increase chances to have positive interactions with others, thereby reducing loneliness and promoting a sense of collective identity that is central to our sense of psychological wellbeing, though it must

Table 10.1 Theoretical and philosophical foundations

Theory	Development
Biophilia	Social psychologist, Erich Fromm (Gunderson, 2014) Biologist Edward O. Wilson (1993) Biophilia Hypothesis, 'the innately emotional affiliation of human beings to other living organisms' (Kellert & Wilson, 1993, 31)
Prospect-refuge theory	Geographer Jay Appleton, links to environmental aesthetics (1975): public preferences/perceptions of landscapes relate to what is considered to be needed for survival. Greatest preference for being able to see clear views from what might be considered as a safe space (prospect) and not being able to be seen (refuge). Clear implications for visitor management in leisure/tourism contexts: for examples, people who are unwell or tired, women, prefer more refuge whereas teenagers' preference would be to be seen Heerwagen and Orians (1993): implications for landscape design, i.e., availability of shelters and waymarking
Stress reduction theory	Introduced by Ulrich (1981): access to natural environments can have stress reducing properties and urban characteristics/ settings impede the process
Attention restoration theory	Theory developed by Kaplan and Kaplan (1989): People's ability to concentrate improves after spending time in nature or looking at nature scenes
Mandala of Health	Conceptualised by Hancock and Perkins (1985), the Mandala of Health presents a bio-psycho-social-environmental thematic framework depicting multiple determinants of public health. Similarly in 1991, Dahlgren and Whitehead developed the Rainbow Model, and reviewed its influence on policy, research and practice in 2021 (Dahlgren & Whitehead, 2021)

be said that social pathways remain understudied relative to the others (Gray & Manning, 2014, 2022; Maas et al., 2009; Sugiyama et al., 2008).

As pointed out by Kuo (2015), the multiplicity of pathways by which nature might impact on health lends much credibility to the fact that nature promotes health. Moreover, given the large number of potential pathways, the cumulative effect of these pathways on health at a population level might be quite large. However, it remains that a central conceptual framework that explains the nature/health link has yet to be agreed upon, although some have tried to put forward what such a framework might look like, e.g., by proposing a central pathway such as immunity (e.g., Kuo, 2015) or specifying domains of pathways (see Marselle et al., 2021). While it could be argued that the wealth of evidence, and the challenges of establishing causality in the area, make such an endeavour unnecessary, it is also the case that the lack of a central framework that links nature to health is challenging because it limits its use in both public health strategy and in policy, and because it holds back the development of nature-based interventions in healthcare settings (Marselle et al., 2021; Chapter 8). A better understanding of the central mechanisms by which nature can impact on health is needed to guide health research and policy. We will pick up this point later in the chapter when we look at gaps in the evidence base.

10.3   Implementing Nature in Policy, Planning and Design

Given evidence of the health benefits of natural environments, the protection and enhancement of green (and more recently of blue) spaces as a public health strategy has been promoted by planning authorities, public health institutions, protected area agencies and other government bodies worldwide. Much of this is done in recognition of the fact that many people live without access to quality natural spaces that provide for rest, leisure, walking and other opportunities to increase our activities and enhance our lifestyles (PHE, 2020). This is also a context which demonstrates spatial and health inequalities, with the poorest and most disadvantaged presenting with the poorest health outcomes, and some of the lowest access to quality natural environments: all of which points to the necessity for improvements in our natural environment in planning processes and designs (Honey-Rosés et al., 2021; PHE, 2020).

In urban areas, this has meant the promotion of nature, the enhanced management of public parks, and increasing use of pocket parks, green wedges, and corridors of green infrastructure (de Oliveira, 2020). Local authorities aim to include not only enhancing public health and well-being by directly engaging the public in naturalistic spaces, but also the enhanced management of such spaces works towards maintaining our food, water and energy security against global challenges of climate change and natural disasters. An ecosystem approach to managing our greenspaces, "seeks to optimise the synergies between nature, society and the economy" through providing for nature-based solutions (Faivre et al., 2017, 509) that can result in both social **and** biodiversity benefits (IUCN, 2020). Such is the interest in blue spaces in Europe in helping to tackle public health challenges (particularly following the pan-European project BlueHealth), that blue infrastructure strategies are becoming increasingly evident alongside green infrastructure approaches to enabling healthier active communities in urban areas (Grellier et al., 2018), even contributing to enhancing the notion of a sense of place in communities (see British Academy, 2016).

In rural locations, national parks and other forms of designated protected areas are promoted by tourism and leisure providers, turning many of these locations into tourism destinations in their own right. Popularity for these spaces is evident through the relatively constant public demand for accessing these areas. This is particularly the case since public restrictions on movement, driven by governments to contain the COVID-19 pandemic, were lifted. The result was a phenomenal increase in numbers of visitors to coastlines and rural spaces across the EU and in North America two examples (McLanahan, 2020; Rose, 2021). The situation for management agencies was overwhelming and has called into question how these areas can continue to be managed for tourism use, while maintaining their environmental qualities (see Chapter 14). Our natural environments are therefore providing opportunities for engagement with nature, while also facing significant challenges from the continuing degradation of our 'natural capital'. The impacts of climate change and the exceedances of the natural environment's capacity to sustain the multiple challenges are increasing the pressure on ecosystem functions. The cumulative impacts on our natural capital assets are particularly disconcerting as soil, water and biodiversity underpin healthy ecosystems that in themselves provide a wide range of essential and fundamental services that sustain our livelihoods and wellbeing.

The political will to encourage people's engagement with greenspaces and support the management of these areas, is invariably expressed through policy formulation, and informed and implemented by urban and landscape management and planning services. Policy and strategic direction for the interconnection of protected area agencies with public health services is evident through activities driven by global institutions such as the International Union for the Conservation of Nature (IUCN) (see Box). These also link to the United Nations Sustainable Development Goals, especially related to nature-based solutions, sustainable land management and planning and health and wellbeing. Additional strategic direction is found worldwide at regional levels, including as key examples, in parts of Europe, North America, Latin America, Canada, Australia and New Zealand.

IUCN World Commission on Protected Areas (WCPA) Health and Well-Being Specialist Group

The IUCN WCPA Health and Well-being Specialist Group promotes the health and wellbeing benefits of nature across conservation, public health and other sectors. Its activities build upon previous work progressed through the 'Healthy Parks Healthy People' programme of work. Key aims include: facilitating partnerships and collaborations among organisations to influence policies and plans across the sectors; building and communicating the body of evidence on benefits of nature for human health and well-being; and encouraging the development of standard metrics to measure co-benefits.

There are three key objectives of the Health and Well-being Specialist Group.

- Contribute to further building the evidence and knowledge base on health and wellbeing interdependencies between natural planetary ecosystems and human populations.
- Mainstream the knowledge of health and wellbeing implications of nature across the conservation, health and other sectors through the development of interdisciplinary materials, case studies, tools and programmes.
- Facilitate partnerships at a global, regional, national and sub-national scale between entities working on environmental health and human health to influence policies and plans across sectors that support programmes in parks and protected areas.

The Health and Well-being Specialist group works with other teams within the IUCN whose remit aims to connect people with nature. These include #NatureForAll, Urban Conservation Strategies and Ecosystem Services groups.

Taking the Australian case, initiated by Parks Victoria, 'Healthy Parks Healthy People' (HPHP) was created to promote the value of the environment to people's health. What has become a hugely successful programme that includes greenspace agencies and government departments working with public health practitioners, a number of initiatives including park prescriptions, free access to what are branded as greenspaces for health, has encouraged additional health and environment alliances at various scales in the USA, Canada, South Korea, Scotland and in Europe. These apply the principles of HPHP in partnership with sectors including tourism, leisure, complementary medicine working with environmental managers, and many have national strategies. Such are discussions across Europe, that initiatives are validated even further by activities and specialist working groups led by the EUROPARC Federation, encouraging collaborations between protected area and greenspace managers working with public health officers.

Drawing on the UK context specifically, such activities have been promoted by national governing bodies, including Public Health England (PHE, 2020), and are being endorsed within 'Department of Environment Food and Rural Affairs' (Defra) 25-Year Environment Plan, resulting in activities being implemented by protected area agencies, local authorities and greenspace managers frequently engaging with local communities (box text below). Moreover, there is increasing policy emphasis on designing and developing interventions that support people's interactions with natural environments, with some clinicians actively prescribing patients' engagement in greenspaces for health reasons as part of preventative and reactive treatments drawing upon social and green prescribing (e.g., PHE, 2020; van den Berg, 2017).

Charitable (NGO) Foundations for Action

The Parks Foundation is an independent charity, established in 2015, devoted to enhancing Bournemouth, Christchurch and Poole's parks and green spaces. It was born out of a desire to create inspirational parks that improve people's health and wellbeing, reconnect people with nature and bring diverse communities together. Although a young organisation, The Foundation has achieved a lot in a short space of time by using the therapeutic nature of being out of doors to improve people's wellbeing while also increasing biodiversity in urban greenspaces. Its Parks in Mind project, which started in 2017, is a green social prescribing project designed to improve people's physical and mental wellbeing through the delivery of inclusive activities and volunteering opportunities. Delivered in parks that are frequently found in areas characterised by deprivation, one such case includes Boscombe. This is one of the five per cent most deprived areas of the UK. Boscombe residents have the lowest life expectancy of all Bournemouth wards and the highest level of hospital admissions for self-harm. Unemployment rates are more than three times the national average. Up to 35% of residents in the area do not have access to a garden or outdoor space, making their park-based activities a crucial part of their participants' lives.

Our Parks in Mind programme is a combination of nature conservation, arts-based therapy and wellbeing activities. In any month, participants may enjoy gardening, tree planting, mindfulness walks, tai chi or star gazing—all for free. Participants are either referred to the project (perhaps through their GP, a community mental health team or alcohol/drug rehabilitation scheme) or they can self-refer. The Foundation can evidence the difference it is making to people's lives too with 92% of participants stating that their mental health and overall wellbeing has improved and 98% feeling the quality of their lives has improved. One participant told them:

> "Thank goodness for these sessions. They have greatly helped me to adjust from the physical restrictions and mindset of the pandemic, to venturing outside again, to enjoy nature, a wide range of well-being and crafting activities, and socialising with different people".

As well as Parks in Mind, the Foundation delivers another project called Green Heart Parks, which currently works across 15 community parks in the area (see Fig. 10.1). This work reimagines parks from being green deserts into wildlife rich havens which increase biodiversity and engage local residents through inspiring and educational activities. During 2022, The Foundation delivered 550 activities to 5,880 people. With a focus on community growing and breaking down barriers to participation for non-park users, this community ground-up project helps both wildlife and neighbourhoods thrive.

Fig. 10.1 Green Heart Parks

Vibrant community cafes within some of the area's Green Heart Parks provide volunteering opportunities for people with special education needs and learning disabilities. The cafes increase the amount of time people spend in nature by providing a 'loo, brew and something to do', they also reduce antisocial behaviour by having a presence in the park. Police call outs to one park reduced by 44% since the community park café was opened. What's more, customers know that when they're buying a cuppa at the café, the money's reinvested back into the area's parks too. The Foundation works alongside the local authority (BCP Council) who have overall responsibility to maintain and develop the area's parks; the charity works alongside them with the aim to raise much needed funds for improvement and enhancements above what the council provides. The charity is publicly funded through donations from grants, individuals, major donors and their trading activities. You can find out more about their work at www.parksfoundation.org.uk.

10.4   What Are the Evidence Gaps?

A substantial body of evidence has already been collated on natural health services and the subject area is considered to be fast-moving towards maturity (Van den Berg, 2017). It has been an area of work that has attracted much academic attention, yet important gaps remain. Here we consider some of those gaps and make recommendations for future research. To date, research attention has primarily focused on urban settings. Communities living in rural and coastal areas are typically under-studied, despite the fact that in the case of the UK for example, nearly 10 million people (17% of the population; Defra, 2021) live in these areas, that are characterised by complicated health patterns (LGA, 2017). While there is some evidence that psychological and physical health may be better in rural areas, partly because of exposure to natural environments, it is also the case that issues of deprivation, isolation, inaccessibility to local health and recreational services, can lead to poorer health and wellbeing outcomes.

It is also difficult to compare inequality and health differences between urban and rural areas, as causes for these may vary across these two contexts. Further research at a local level is essential to better understand the complex health experiences of residents in rural and coastal areas. For example, there is increasing evidence in urban areas that soil and/or airborne microbial diversity plays an important role in improving human health through enhancing immune status and helping to ameliorate disease risk (Liddicoat et al., 2019, 2020; Mhuireach et al., 2016) (see Fig. 10.2). Yet to date, the application of human–environment microbiome interactions in rural settings remains underexplored.

Linked to this urban focus, limited attention has been paid, in detail, to the range of green spaces, their environmental components and the importance of ecological conditions to health and wellbeing. Research has generally employed rather simplistic conceptualisations of 'green space' (Frumkin et al., 2017; Lovell et al., 2020; Wheeler et al., 2015). This has resulted in limited regard for geographic complexities, including greenspace types, their environmental characteristics and ecological condition (Frumkin et al., 2017; Lovell et al., 2020; Wheeler et al., 2015). As such, we have limited knowledge of the importance of biodiversity in natural spaces, the range of complex landscapes, their characters, their configurations, their functionality (e.g., agricultural, pastoral), or how

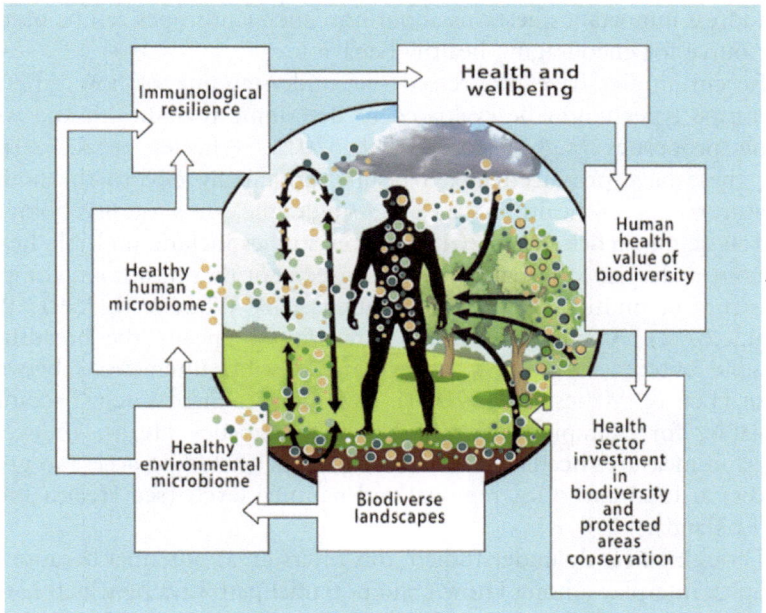

Fig. 10.2 Microbiome process, influences and impacts (*Source* Adapted from Mills et al. 2017)

varying landscape types, their microbial diversity, environmental condition, characteristics and even their publicly perceived 'special qualities' can improve health and wellbeing (see also Marselle et al., 2021).

There is also a need to better understand people's landscape preferences and their perceptions of different landscape types. Extensive research demonstrates that perceptions of landscape types, impact on how people use a space (Gatersleben & Andrews, 2013; Jansson et al., 2013; Knez & Eliasson, 2017; Stigsdotter et al., 2017) and is related to improvements in physical and psychological health (Fuller et al., 2007; Sandifer et al., 2015). Yet, much of this work also focuses on urban settings, with more limited research on 'wild' natural environments (Hägerhäll et al., 2018). Research is therefore needed that brings together population-level data with socio-cultural data collected at local spatial levels (Wheeler et al., 2015)—particularly in rural areas—in order to address important questions about how such landscapes can be used as a resource for encouraging healthy behaviour.

Recent studies have progressed our understanding of how different landscape types might be evaluated to determine their health and well-being properties (e.g., Alcock et al., 2015; Wheeler et al., 2015). Yet, these have produced mixed results, potentially due to the limited measurements of health used, e.g., single measures of psychological distress (e.g., Alcock et al., 2015). Fewer studies include multiple health outcomes, despite long-held universal agreement that health is a complex construct of multiple dimensions (Dahlgren & Whitehead, 1991; Patz et al., 2012). As a result, studies often fail to specify the breadth of possible outcomes and pathways, or the possible interactions between these (Lee & Maheswaran, 2010). This data is increasingly becoming available, for example Office of National Statistics Health Index for England measures health across three domains (including access to green spaces) at local authority, regional and national levels (see Health Index for England).

Though currently understudied, this offers great potential because the complex interplay among known and potential pathways by which nature impacts on could be key to developing conceptual frameworks that work in this setting (Marselle et al., 2021). For example, Annerstedt et al. (2012) found that certain green qualities (e.g., tranquillity and space) only impacted on the risk of poor mental health when physical activity was considered. There is a need for research that takes multifactorial and multiple-pathway approaches to understand complex relationships between a broad set of health and wellbeing outcomes with natural environment types, their ecological condition, their characteristics and their qualities.

Given these gaps in our knowledge, what is evident is that further research is needed to improve the evidence base for strategic and policy decisions about the role of natural environments in health and wellbeing. Addressing these gaps requires us to recognise the heterogeneity of landscapes (including those in rural areas), focusing on distinctive landscape types, characters, and ecological condition, and to examine a broad set of physical, psychological and social health and wellbeing outcomes. Moreover, it highlights the need for critical evaluations of the interactions between natural capital and associated non-use values, in order to provide a more nuanced understanding of the relationships between green spaces, and health and wellbeing. This calls for a multifactorial, and transdisciplinary enquiry that engages diverse landscapes, environmental characteristics, populations, applied sciences and stakeholders (see Box).

A Conceptual Framework for Examining Impacts of Multiple Environmental Factors on people's Health and Wellbeing

A fundamental aim of the 'Significant Spaces' project has been to address many of the gaps in the current body of research relating to the impact of green spaces on health. In particular, a key aim has been to redirect attention from urban green spaces to rural spaces, where many protected and designated landscapes, valued for the quality and extent of their biodiversity and socio-cultural assets, are found. Such spaces could be considered to be 'optimum' greenspaces, based on the extensive range of space and ecosystem services they provide. However, we know relatively little about the impact of such 'optimum spaces' on human health and wellbeing, nor about their economic value, with the result that the benefits that such spaces bring have been frequently overlooked or ignored in decision-making. Indeed, many of the original aims for conserving protected and designated landscapes in rural areas have been challenged and more generally, specifically in the UK example, many green open spaces are being degraded or lost due to a lack of economic incentive to justify their protection and/or conservation.

'Significant Spaces' calls for a new programme of research for which as exemplary case study areas, the 38 Areas of Outstanding Natural Beauty (AONBs) in England and Wales are identified. In this national case study framework, landscape planning practice and methodologies are brought together with those from social, economics, behavioural, biological and applied sciences, in order to examine the health and wellbeing properties of AONB designated landscapes. A three-stage approach is suggested in order to build a framework that acknowledges, and enables research to examine, the multiple determinants of physical health and mental wellbeing. Shown in see Fig. 10.3, are a number of layers of data collection required. Working from the bottom layer upwards, the first stage involves the analysis of existing landscape, topographical and ecological data in order to construct a national typology of landscapes by type, by their respective characteristics and by their condition, for each of the 38 AONBs in England and Wales.

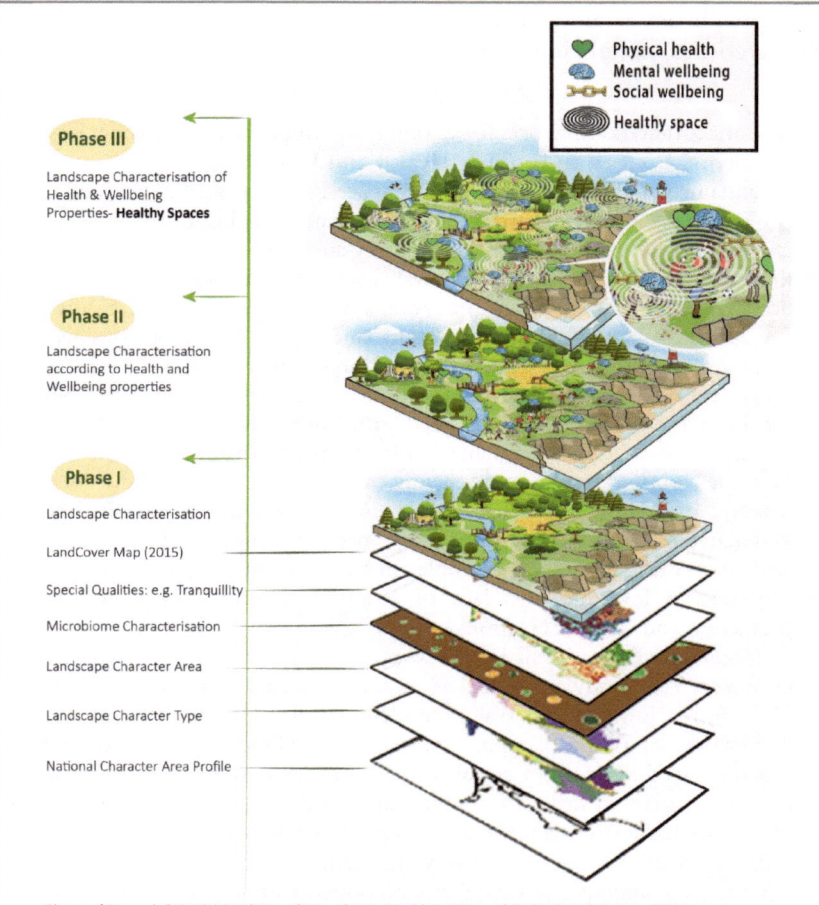

Phases of Research & Spatial Overlaying of Data. Three pictorial depictions of the landscape characterisations and evaluations created.Phase III represents the final **Landscape Characterisation of Health & Wellbeing** to be constructed for each AONB.

Fig. 10.3   Phases of research and spatial overlaying of data

This stage includes analysing the quality of microbiomes from soil extracted from a range of rural, semi-rural and urban environments in and around each of the AONBs. The subsequent, second stage requires an initial phase of collating data on a number of health and wellbeing factors of visitors to and residents in and around an AONB. Subsequently a comparative, quantifiable evaluation would be progressed to determine what impact might be determined on people's health and wellbeing outcomes of their access to and experience in a range of landscape types, of distinct landscape characteristics, and of varying ecosystem health/quality. For example, the quality of microbiomes collected at stage one, would be compared with physical health data provided by local residents in and visitors to the AONBs during stage two. The third and final stage, depicted as the top layer in Fig. 10.3, represents a model of a space according to its environmental characteristics, condition and effect it could have on people's health and wellbeing, should they access that area. This model constructed according to a ranking of health and wellbeing benefits gained from each landscape type would additionally inform the creation of landscape characterisations in each of the 38 AONBs resulting in an innovative model of healthy spaces that will enhance both the body of academic research and inform policy design and its implementation.

The results of this framework would feed into, and build on, existing environmental and cultural records of Landscape Character Assessments (LCAs) across the AONB network; resulting in the design and construction of an additional unique and innovative layer to the suite of existing LCAs that will subsequently feed into national, regional and local planning, development and productivity strategies that are aimed at enhancing public health, spatial planning, environmental management and economic performance. Consequently, 'Significant Spaces' would fill an important gap in our understanding of the health and environmental value of designated spaces and would inform how this importance should impact on how we plan and manage for these spaces at local and national levels. Additionally, it would address pressing concerns around the statutory purpose of designated landscapes in the twenty-first century, through an evaluation of these spaces in terms of their health and wellbeing benefits.

10.5   CONCLUSION

There is now a large body of work that evidences the beneficial relationship between natural environments and human health and wellbeing, and it has been noted by many that this field is fast-moving towards maturity. However, it is also the case that key gaps in our understanding remain. Most notably, there is a need to move beyond a research focus on urban environments, and a need for more research which interrogates the complexities of landscape characteristics and quality, as well as incorporating multidimensional understandings of health and wellbeing. What is needed, is robust and replicable evidence that recognises the inherent complexities of studying landscapes as a geographical area of interactions between people and place. Only this will enable a better understanding of the central mechanisms by which nature can impact on health, which is sorely needed in order to guide health research and policy. Indeed, in contrast to the reams of research in this area, albeit there is universal understanding that green blue spaces are good for our health, consequently nature-based interventions are emphasised, little of the evidence suggested on relationships between greenspaces and their influence on health has actually made its way into practice, raising questions as to how research and its findings can be applied.

There is a clear need to link, in the development and design of research programmes, to the practical tools that are used in the design and development of greenspaces on the ground. We have provided one example in the box text above of how this might work, through the Health and Wellbeing Landscape Character Assessment. We would argue that it is through the integration of research with planning tools such as these that research can effectively feed into national, regional and local spatial and environmental planning and development that is aimed at enhancing if not targeting specific populations and public health activities.

References

Aitken, W. (2021, August 27–30). *Green neighbourhoods linked with better heart health*. European Society of Cardiology Congress. https://www.escardio. org/The-ESC/Press-Office/Press-releases/Green-neighbourhoods-linked-with-better-heart-health. Accessed 17 February 2023.

Alcock, I., White, M., Lovell, R., Higgins, S., Osborne, N., et al. (2015). What accounts for 'England's green and pleasant land'? A panel data analysis of mental health and land cover types in rural England. *Landscape and Urban Planning, 142*, 38–46. https://doi.org/10.1016/j.landurbplan.2015.05.008

Annerstedt, M., Ostergren, P. O., Bjork, J., Grahn, P., Skarback, E., & Wahrborg, E. (2012). Green qualities in the neighbourhood and mental health—Results from a longitudinal cohort study in Southern Sweden. *BMC Public Health, 12*. https://doi.org/10.1186/1471-2458-12-337

Appleton, J. (1975). *The experience of landscape*. Wiley.

Barton, J., & Pretty, J. (2010). What is the best dose of nature and green exercise for improving mental health? A multi-study analysis. *Environmental Science and Technology, 44*(10), 3947–3955. https://doi.org/10.1021/es903183r

Breed, M., Cross, A., Wallace, K., Bradby, K., Flies, E., et al. (2018). Ecosystem restoration: A public health intervention. *EcoHealth, 18*, 269–271. https://doi.org/10.1007/s10393-020-01480-1

British Academy. (2016). *Where we live now: Making the case for place-based policy.* https://www.thebritishacademy.ac.uk/documents/277/Where-we-live-nowmaking-case-for-place-based-policy.pdf. Accessed 30 December 2022.

Corley, J., Okely, J., Taylor, A., Page, D., Welstead, M., et al. (2021). Home garden use during COVID-19: Associations with physical and mental well-being in older adults. *Journal of Environmental Psychology, 73*, 101545. https://doi.org/10.1016/j.jenvp.2020.101545

Dahlgren, G., & Whitehead, M. (1991). *Policies and strategies to promote social equity in health.* Institute for Futures Studies.

Dahlgren, G., & Whitehead, M. (2021). The Dahlgren-Whitehead model of health determinants: 30 years on and still chasing rainbows. *Public Health, 199*, 20–24. https://doi.org/10.1016/j.puhe.2021.08.009

Defra. (2021). Rural Population and Migration. Available from: www.gov.uk/government/statistics/rural-population-and-migration/ruralpopulation-and-migration

De Vries, S., Verheij, R., Groenewegen, P., & Spreeuwenberg, P. (2003). Natural environments—Healthy environments? An exploratory analysis of the relationship between greenspace and health. *Environment and Planning A, 35*(10), 1717–1731. https://doi.org/10.1068/a35111

Egorov, A., Mudu, P., Braubach, M., & Martuzzi, M. (Eds.). (2016). *Urban green spaces & health: A review of the evidence.* WHO Regional Office for Europe. https://www.euro.who.int/__data/assets/pdf_file/0005/321971/Urban-green-spaces-and-health-review-evidence.pdf. Accessed 13 April 2023.

Faivre, N., Fritz, M., Freitas, T., de Boissezon, B., & Vandewoestijne, S. (2017). Nature-based solutions in the EU: Innovating with nature to address social, economic and environmental challenges. *Environmental Research, 159*, 509–518. https://doi.org/10.1016/j.envres.2017.08.032

Flies, E., Skelly, C., Lovell, R., Breed, M., Philips, D., & Weinstein, P. (2018). Cities, biodiversity and health: We need healthy urban microbiome initiatives. *Cities and Health, 2*(2), 143–150. https://doi.org/10.1080/23748834.2018.1546641

Flies, E., Skelly, C., Singh Negi, S., Prabhakaran, P., Liu, Q., et al. (2017). Biodiverse green spaces: A prescription for global urban health. *Frontiers in Ecology and the Environment, 15*(9), 510–516. https://doi.org/10.1002/fee.1630

Frumkin, H., Bratman, G., Breslow, S., Cochran, B., & Kahn, P., Jr. (2017). Nature contact and human health: A research agenda. *Environmental Health Perspectives, 125*(7). https://doi.org/10.1289/EHP1663

Fuller, R. A., Irvine, K. N., Devine-Wright, P., Warren, P. H., & Gaston, K. J. (2007). Psychological benefits of greenspace increase with biodiversity. *Biology Letters, 3*(4), 390–394.

Garrido-Cumbrera, M., Foley, R., Braçe, O., Correa-Fernández, J., López-Lara, E., Guzman, V., Gonzalez Marin, A., & Hewlett, D. (2021). Perceptions of Change in the Natural Environment produced by the First Wave of the COVID-19 Pandemic across Three European countries. Results from the GreenCOVID study. *Urban Forestry & Urban Greening, 64,* 127260.

Garrido-Cumbrera, M., Foley, R., Correa-Fernández, J., Gonzalez-Marin, A., Brace, O., & Hewlett, D. (2022). The importance for wellbeing of having views of nature from and in the home during the COVID-19 pandemic. Results from the GreenCOVID study. *Journal of Enviornmental Psychology, 83,* 101864.

Gascon, M., Triguero-Mas, M., Martinez, D., Dadvand, P., Rojas-Rueda, D., et al. (2016). Residential green spaces and mortality: A systematic review. *Environment International, 86,* 60–67. https://doi.org/10.1016/j.envint.2015.10.013

Gatersleben, B., & Andrews, M. (2013). When walking in nature is not restorative—The role of prospect and refuge. *Health and Place, 20,* 91–101. https://doi.org/10.1016/j.healthplace.2013.01.001

Gray, D., & Manning, R. (2014). 'Oh my god, we're not doing nothing': Young people's experiences of spatial regulation. *British Journal of Social Psychology, 53*(4), 640–655. https://doi.org/10.1111/bjso.12055

Gray, D., & Manning, R. (2022). Constructing the places of young people in public space: Conflict, belonging and identity. *British Journal of Social Psychology, 61*(4), 1400–1417. https://doi.org/10.1111/bjso.12542

Gray, D., Manning, R., & Oftadeh-Moghadam, S. (2021). Using mixed and multi-modal methods in psychological research with young people. In R. Locke & A. Less (Eds.), *Mixed-methods research in wellbeing and health* (pp. 156–173). Routledge.

Grazuleviciene, R., Danileviciute, A., Dedele, A., Vencloviene, J., Andrusaityte, S., et al. (2015). Surrounding greenness, proximity to city parks and pregnancy outcomes in Kaunas cohort study. *International Journal of Hygiene and Environmental Health, 218*(3), 358–365. https://doi.org/10.1016/j.ijheh.2015.02.004

Grellier, J., White, M., Albin, M., Bell, S., Elliott, L., et al. (2017). BlueHealth: A study programme protocol for mapping and quantifying the potential benefits to public health and well-being from Europe's blue spaces. *British*

Medical Journal Open, 7(6), e016188. https://doi.org/10.1136/bmjopen-2017-016188

Gunderson, R. (2014). Erich Fromm's ecological messianism: The first biophilia hypothesis as humanistic social theory. *Humanity and Society, 38*(2), 182–204. https://doi.org/10.1177/01605976145291

Guzman, V., Garrido-Cumbrera, M., Braçe, O., Hewlett, D., & Foley, R. (2020). Health and wellbeing under COVID-19: The GreenCovid Survey. *Irish Geography, 53*(2).

Guzman, V., Garrido-Cumbrera, M., Braçe, O., Hewlett, D., & Foley, R. (2021). Associations of the natural and built enviornment with mental health and wellbeing during COVID-19: Irish perspectives from the GreenCOVID study. The Lancet Online. Available from: https://www.thelancet.com/journals/langlo/article/PIIS2214-109X(21)00128-5/fulltext

Hägerhäll, C., Ode Sang, Å., Englund, J., Ahlner, F., Rybka, K., et al. (2018). Do humans really prefer semi-open natural landscapes? A cross-cultural reappraisal. *Frontiers in Psychology, 9*, 822. https://doi.org/10.3389/fpsyg.2018.00822

Hancock, T., & Perkins, F. (1985). The Mandala of health: A conceptual model and teaching tool. *Family and Community Health, 8*(3), 1–10. https://doi.org/10.1097/00003727-198511000-00002

Hartig, T., Mitchell, R., De Vries, S., & Frumkin, H. (2014). Nature and health. *Annual Review of Public Health, 35*, 207–228. https://doi.org/10.1146/annurev-publhealth-032013-182443

Heerwagen, J., & Orians, G. (1993). Affect and aesthetics: Humans, habitats and aesthetics. In S. Kellert & E. Wilson (Eds.), *The biophilia hypothesis* (pp. 138–172). Island Press.

Honey-Rosés, J., Anguelovski, I., Chireh, V. K., Daher, C., Konijnendijk van den Bosch, C., Litt, J. S., Mawani, V., McCall, M. K., Orellana, A., Oscilowicz, E., Sánchez, U., Senbel, M., Tan, X., Villagomez, E., Zapata, O., & Nieuwenhuijsen, M. J. (2021). The impact of COVID-19 on public space: An early review of the emerging questions–design, perceptions and inequities. *Cities & Health, 5*(sup1), S263–S279. https://doi.org/10.1080/23748834.2020.1780074

Houlden, V., Weich, S., de Albuquerque, J., Jarvis, S., & Rees, K. (2018). The relationship between greenspace and the mental wellbeing of adults: A systematic review. *PloS One, 13*(9). https://doi.org/10.1371/journal.pone.0203000

James, P., Banay, R., Hart, J., & Laden, F. (2015). A review of the health benefits of greenness. *Current Epidemiology Reports, 2*, 131–142. https://doi.org/10.1007/s40471-015-0043-7

Jansson, M., Fors, H., Lindgren, T., & Wiström, B. (2013). Perceived personal safety in relation to urban woodland vegetation—A review. *Urban Forestry and Urban Greening, 12*(2), 127–133. https://doi.org/10.1016/j.ufug. 2013.01.005

Kaplan, R., & Kaplan, S. (1989). *The experience of nature: A psychological perspective.* Cambridge University Press.

Kellert, S., & Wilson, E. (Eds.). (1993). *The biophilia hypothesis.* Island Press.

Knez, I., & Eliasson, I. (2017). Relationships between personal and collective place identity and well-being in mountain communities. *Frontiers in Psychology, 8.* https://doi.org/10.3389/fpsyg.2017.00079

Kuo, M. (2015). How might contact with nature promote human health? Promising mechanisms and a possible central pathway. *Frontiers in Psychology, 6.* https://doi.org/10.3389/fpsyg.2015.01093

Lee, A., & Maheswaran, R. (2010). The health benefits of urban green spaces: A review of the evidence. *Journal of Public Health, 33*(2), 212–222. https://doi.org/10.1093/pubmed/fdq068

Lemes de Oliveira, F. (2020). *Green wedge urbanism.* Bloomsbury.

Liddicoat, C., Sydnor, H., Cando-Dumancela, C., Dresken, R., Liu, J., et al. (2020). Naturally-diverse airborne environmental microbial exposures modulate the gut microbiome and may provide anxiolytic benefits in mice. *Science of the Total Environment, 701,* 1–11. https://doi.org/10.1016/j.scitotenv. 2019.134684

Liddicoat, C., Weinstein, P., Bissett, A., Gellie, N., Mills, J., et al. (2019). Can bacterial indicators of a grassy woodland restoration inform ecosystem assessment and microbiota-mediated human health? *Environment International, 129,* 105–117. https://doi.org/10.1016/j.envint.2019.05.011

Local Government Association. (2017). *Health and wellbeing in rural areas.* Report Produced in Conjunction with Public Health England. https://www.local.gov.uk/sites/default/files/documents/1.39_Health%20in%20rural%20areas_WEB.pdf. Accessed 13 April 2023.

Lovell, R., White, M., Wheeler, B., Taylor, T., & Elliott, L. (2020). *A rapid scoping review of health and wellbeing evidence for the green infrastructure standards.* European Centre for Environment and Human Health, University of Exeter Medical School. For: Natural England, Department for the Environment, Food and Rural Affairs, Public Health England, and Ministry for Housing, Communities and Local Government. https://beyondgreenspace.files.wordpress.com/2020/10/neer015-a-rapid-scoping-review-of-health-and-wellbeing-evidence-for-the-framework-of-green-infrastructure-standards-final-draft-sept-2020-1.pdf. Accessed 13 April 2023.

Maas, J., Verheij, R., de Vries, S., Spreeuwenberg, P., Schellevis, F., & Groenewegen, P. (2009). Morbidity is related to a green living environment. *Journal of Epidemiology & Community Health, 63*(12), 967–973. https://doi.org/10.1136/jech.2008.079038

Marcus, C., & Sachs, N. (2014). *Therapeutic landscapes.* Wiley.

Marselle, M., Hartig, T., Cox, D. D., Bell, S., Knapp, S., et al. (2021). Pathways linking biodiversity to human health: A conceptual framework. *Environment International, 150*, 106420. https://doi.org/10.1016/j.envint.2021.106420

McLanahan, P. (2020, December 10). The newest challenges for Europe's Parks: A surge of new nature lovers. *New York Times.* https://www.nytimes.com/2020/12/10/travel/european-parks-pandemic.html. Accessed 30 June 2022.

Mhuireach, G., Johnson, B., Altrichter, A., Ladau, J., Meadow, J., et al. (2016). Urban greenness influences airborne bacterial community composition. *Science of the Total Environment, 15*(571), 680–687. https://doi.org/10.1016/j.scitotenv.2016.07.037

Miller, D., Morris, S., Morrice, J., Roe, J., Brown, C., et al. (2012). *Blue health: Water, health and wellbeing.* https://www.hutton.ac.uk/sites/default/files/files/snc/CREW%20Blue%20health%20project%20FINAL.pdf. Accessed 13 April 2023.

Mills, J., Weinstein, P., Gellie, N., Weyrich, L., Lowe, A., & Breed, M. (2017). Urban habitat restoration provides a human health benefit through microbiome rewilding: The Microbiome Rewilding Hypothesis. *Restoration Ecology, 25*(6), 866–872. https://doi.org/10.1111/rec.12610

Mitchell, R., & Popham, F. (2008). Effect of exposure to natural environment on health inequalities: An observational population study. *The Lancet, 372*(9650), 1655–1660. https://doi.org/10.1016/S0140-6736(08)61689-X

ONS (Office of National Statistics). (2021, April 26). *How has lockdown changed our relationship with nature?* https://www.ons.gov.uk/economy/environmentalaccounts/articles/howhaslockdownchangedourrelationshipwithnature/2021-04-26. Accessed 28 December 2022.

ONS. (2022). *Health Index for England 2015–2022.* https://www.ons.gov.uk/releases/healthindexforengland2015to2020. Accessed 22 January 2023.

Patz, J., Corvalan, C., Horwitz, P., Campbell-Lendrum, D., Watts, N., et al. (2012). *Our planet, our health, our future.* World Health Organisation. http://www.iucnwhsg.org/sites/default/files/Human%20health%20and%20the%20Ri%20Conventions.pdf. Accessed 19 April 2023.

PHE (Public Health England). (2020). *Improving access to greenspace: A new review for 2020.* https://assets.publishing.service.gov.uk/government/uploads/system/uploads/attachment_data/file/904439/Improving_access_to_greenspace_2020_review.pdf. Accessed 29 August 2022.

Pouso, S., Borja, Á., Fleming, L., Gómez-Baggethun, E., White, M., & Uyarra, M. (2020). Contact with blue-green spaces during the COVID-19 pandemic lockdown beneficial for mental health. *Science of The Total Environment, 756.* https://doi.org/10.1016/j.scitotenv.2020.143984

Powers Tomasso, L., Yie, J., Cedeno Laurent, J., Chen, J., Catalano, P., & Spengler, J. (2021). The relationship between nature deprivation and individual wellbeing across urban gradients under COVID-19. *Environmental*

Research and Public Health, 2021(18), 1511. https://doi.org/10.3390/ije rph18041511

Robinson, J., Aronson, J., Daniels, C., Goodwin, N., Liddicoat, C., et al. (2022). Ecosystem restoration is integral to humanity's recovery from COVID-19. *The Lancet—Planetary Health, 6*(9). https://doi.org/10.1016/S2542-519 6(22)00171-1

Roe, J., Thompson, C., Aspinall, P., Brewer, M., Duff, E., et al. (2013). Green space and stress: Evidence from cortisol measures in deprived urban communities. *International Journal of Environmental Research and Public Health, 10*(9), 4086–4103. https://doi.org/10.3390/ijerph10094086

Rojas-Rueda, D., Nieuwenhuijsen, M., Gascon, M., Perez-Leon, D., & Mudu, P. (2019). Green spaces and mortality: A systematic review and meta-analysis of cohort studies. *The Lancet Planetary Health, 3*(11), e469–e477. https://doi.org/10.1016/S2542-5196(19)30215-3

Rose, A. (2021, June 13). *National park visitors surge as Covid-19 pandemic restrictions wane.* CNN. https://edition.cnn.com/travel/article/national-park-visitors-surge/index.html. Accessed 3 July 2022.

Sandifer, P., Sutton-Grier, A., & Ward, B. (2015). Exploring connections among nature, biodiversity, ecosystem services, and human health and well-being: Opportunities to enhance health and biodiversity conservation. *Ecosystem Services, 12*, 1–15. https://doi.org/10.1016/j.ecoser.2014.12.007

Stigsdotter, U., Corazon, S., Sidenius, U., Kristiansen, J., & Grahn, P. (2017). It is not all bad for the grey city—A crossover study on physiological and psychological restoration in a forest and an urban environment. *Health and Place, 46*, 145–154. https://doi.org/10.1016/j.healthplace.2017.05.007

Sugiyama, T., Leslie, E., Giles-Corti, B., & Owen, N. (2008). Associations of neighbourhood greenness with physical and mental health: Do walking, social coherence and local social interaction explain the relationships? *Journal of Epidemiology and Community Health, 62*(5), e9–e9. https://doi.org/10.1136/jech.2007.064287

Thompson, C., Roe, J., Aspinall, P., Mitchell, R., Clow, A., & Miller, D. (2012). More green space is linked to less stress in deprived communities: Evidence from salivary cortisol patterns. *Landscape and Urban Planning, 105*(3), 221–229. https://doi.org/10.1016/j.landurbplan.2011.12.015

Ulrich, R. S. (1981). Natural versus urban scenes: Some psychophysiological effects. *Environment and Behaviour, 13*, 523–556. https://doi.org/10.1177/0013916581135001

Van den Berg, A. (2017). From green space to green prescriptions: Challenges and opportunities for research and practice. *Frontiers in Psychology, 8*. https://doi.org/10.3389/fpsyg.2017.00268

Van den Berg, M., Van Poppel, M., Van Kamp, I., Andrusaityte, S., Balseviciene, B., et al. (2016). Visiting green space is associated with mental health and vitality: A cross-sectional study in four European cities. *Health and Place, 38*, 8–15. https://doi.org/10.1016/j.healthplace.2016.01.003

Van den Berg, M., Wendel-Vos, W., Van Poppel, M., Kemper, H., Van Mechelen, W., & Maas, J. (2015). Health benefits of green spaces in the living environment: A systematic review of epidemiological studies. *Urban Forestry and Urban Greening, 14*(4), 806–816. https://doi.org/10.1016/j.ufug.2015.07.008

Völker, S., & Kistemann, T. (2011). The impact of blue space on human health and well-being-Salutogenetic health effects of inland surface waters: A review. *International Journal of Hygiene and Environmental Health, 214*, 449–460. https://doi.org/10.1016/j.ijheh.2011.05.001

Wheeler, B., Lovell, R., Higgins, S., White, M., Alcock, I. et al., (2015). Beyond greenspace: An ecological study of population general health and indicators of natural environment type and quality. *International Journal of Health Geographics, 14*(17). https://doi.org/10.1186/s12942-015-0009-5

World Health Organisation. (2018). *Non-communicable diseases: Country profiles 2018*. World Health Organisation. https://apps.who.int/iris/handle/10665/274512. Accessed 13 April 2023.

World Health Organisation. (2022a). *Noncommunicable diseases*. https://www.who.int/news-room/fact-sheets/detail/noncommunicable-diseases#:~:text=Noncommunicable%20diseases%20(NCDs)%20kill%2041,%2D%20and%20middle%2Dincome%20countries. Accessed 30 December 2022.

World Health Organisation. (2022b). *Urban health*. https://www.who.int/health-topics/urban-health#tab=tab_1. Accessed 28 August 2022.

World Health Organisation. (2022c). *Air pollution*. https://www.who.int/health-topics/air-pollution#tab=tab_1. Accessed 28 August 2022.

CHAPTER 11

Judaism and Engagements with Nature: Theology and Practice

Christina Welch and Neil Amswych

11.1 Introduction

This chapter explores the nature-based praxis found with Judaism, with a focus upon the Reform Jewish tradition and notably the work of Neil Amswych, the former Rabbi of the Bournemouth Reform Synagogue, now Rabbi in Sante Fe, New Mexico, USA. Bournemouth is a seaside town on England's south coast in the county of Dorset. It is a town with a long and historic relationship with Judaism (Fox, 2021; JCR, 2005), although until this time, the relationship did not have an ecological focus. Rabbi Neil and Dr Christina Welch, a Religious Studies scholar, met through student visits which she organised as part of her undergraduate teaching at the University of Winchester, where she lectures. CW learnt about Rabbi Neil's environmental work during these discussions and became involved in the eco-action plan he developed. Rabbi Neil moved

C. Welch (✉)
University of Winchester, Winchester, UK
e-mail: Christina.Welch@winchester.ac.uk

N. Amswych
Temple Beth Shalom, Sante Fe, NM, USA

N. Finneran et al. (eds.), *Managing Protected Areas*,
https://doi.org/10.1007/978-3-031-40783-3_11

to Santa Fe in New Mexico in 2014 where he is Rabbi of the Temple Beth Shalom, and a self-confessed eco-Rabbi. This chapter does not explore his environmental work in the USA—although he is currently writing a book on Jewish Environmental Theology—but sets the wider Jewish theological context for Rabbi Neil's environmental and sustainability work when he was in Bournemouth (with Interfaith Dorset Education and Action (IDEA)) grounding it within global Jewish eco-action.

Judaism

Judaism is the oldest of the three monotheistic Abrahamic faiths, and as such Christianity and Islam both share many commonalities with Judaism. Judaism's history is written in the fabric of the Hebrew Bible which describes its close relation with God, as well as the main personalities and prophets who shaped the early history of the Jewish people from their exile in Egypt, their Exodus to the Promised Land and then diasporic dispersal into Babylon following the destruction of their Temple, and later across the lands of the Mediterranean. Small Jewish groups came to the UK at the time of the Norman Conquest. As Christianity forbade lending money, as non-Christians, Jewish communities in Europe found themselves close to the centres of mercantile power and were able to act as financiers. In spite of this anti-Judaic attacks were common in medieval Europe, and in 1290 King Edward I expelled all Jewish people from England and they only returned in the seventeenth century when Oliver Cromwell, the Lord Protector of England, sought finance to combat Catholic Spain's hold on the New World (sizeable numbers of Jewish people had been expelled from Spain and they rapidly formed a diaspora across the Atlantic worlds into North America, the Caribbean and Brazil). Diaspora and movement emphasise the leitmotif of the Jewish experience. Those Jewish communities with roots in the Iberian Peninsula and North Africa in particular became known as Sephardic (oriental) Jews. Communities of displaced Jewish communities from the medieval German Rhineland emigrated eastwards into what is now eastern Europe and western Russia where they formed a 'pale' of settlement and a cultural life celebrated in popular literature (e.g. the film *Fiddler on the Roof*). These Ashkenazi communities were again subject to anti-Judaic violence and many emigrated to the UK and the Americas from the nineteenth century. Some six million mainly Ashkenazi Jewish men, women and children were murdered by the Nazis during the Second World War. Today there are around 15 million people who identify as being Jewish, of these some eight million live in the USA, five million in Israel and the remainder dispersed across the globe.

11.2 FROM AN IDEA TO IDEA

In 2007, Rabbi Neil had an idea. He called together over 100 faith leaders from across Dorset, inviting them to consider climate change, and how the county's religions could help its almost 410,000 residents to become more climate-conscious. After a slow start, IDEA (Interfaith Dorset Education and Action) was formed with a remit to educate and empower faith communities across Dorset; it had an explicitly strong environmental ethos. Indeed, so eco-aware was IDEA that in 2008 Rabbi Neil presented *The Earth Charter* (EC, 2001–2021; Rockefeller, 2010) to Bournemouth Council, which subsequently became the first UK authority to adopt it. As a result of this, Rabbi Neil was chosen as the UK representative invited to speak at that year's Earth Charter's global online conference, and in 2012, he was invited to contribute to UNESCO's book on *The Earth Charter* (Amswych, 2012b).

Rabbi Neil didn't stop with IDEA though. In 2009, he was an instrumental part of what eventually became The Big Green Jewish Initiative (Lubetkin, 2013), a Jewish environmental and social action group, and by the end of 2010, having called together the 400 faith communities in Dorset, he invited their leaders or representatives to help create The Big Green Believer's Agreement, a predominantly Dorset-based interfaith call to action to help stop global climate change. It took a year of long and complex negotiations to agree on a final text of the Agreement, by its launch in November 2011, around 10,000 people were involved in its aims.

Having raised the consciousness of local people of various faiths, Rabbi Neil set about raising the ecological consciousness of Reform Jews across Britain. In 2010/2011 he helped the Movement for Reform Judaism, a movement covering 20% of all British Jews, create its own overarching sustainability plan and commit itself to work towards total carbon neutrality (JC, 2011; RJ:B, 2022; RJ:E&CC, 2022). Further, with Palm Oil cultivation a key driver of global warming through mass deforestation (UCS, 2013), in 2011 Rabbi Neil wrote an article in the national newspaper *The Jewish Chronicle*, which raised the issue of palm oil use in kosher food (Amswych, 2011). In brief, kosher is the name given to food (and beverages) that are permitted within Jewish dietary law, a law/lore rooted in history and religious traditions, and include preparation, processing and presentation, as well as what can and cannot be eaten and drunk (MJL:KF, 2002–2022).

Although Palm Oil is not listed anywhere as non-Kosher, the environmental ethos embedded in Judaism highlights the sustainability of resources, whilst its social focus calls for justice for all life, and as such using palm oil, which Amswych highlights as an unsustainable cash crop, can be perceived as non-halachic (against Jewish religious laws/lores). Not only does the slash-and-burn of forest to clear land in order to grow palm oil cash crops cause detestation to the local areas affecting air quality and reducing habitat for the creatures that live there, but it releases carbon into the atmosphere, which has a negative global impact. It was Rabbi Neil's dedication to environmentalism and mitigating the effects of climate change, and his work with and through IDEA, that led to him being shortlisted for the UK National Climate Week's 'Most Inspiring Leader' award in 2012. But why did Amswych decide to lead the fight on environmental and sustainability issues? Because he knew it was a moral issue for the planet and all that lived on/in it, and as a Rabbi (a Jewish teacher) and a man of faith, he needed to lead by example, and so to the Jewish theology that grounds his faith-based praxis.

11.3   Eco-Theology in Doctrine and Practice

Judaism is an ancient religion with rituals that still resonate with its pre-monotheistic roots (JVL:AJH, n.d.), many of which are based around the seasons and the needs of agricultural people such as *Shavuot*. This festival today celebrates the giving of the Torah on Mount Sinai as well as the summer grain harvest. It is cited as the Festival of Weeks in Exodus 34:22 and Deuteronomy 16:10, and the Day of First Fruits in Numbers 28:26, and was one of three pilgrimages when Jewish men would bring first fruits as offerings to God in the Temple. A later agriculturally based festival is *Tu-Bishvat*. Celebrated late January to mid-February depending on the year, with this festival the very almanac of Judaism echoes seasonality in its primarily lunar-solar calendar, with each new month starting with the new moon (JVL:JHF, n.d.). *Tu-Bishvat* is known as the new year of the trees and honours the Jewish law/lore of *Orlah*, the prohibition against eating fruit during the first three years of a tree's life, and *Ma'aser oni*, the obligation to set aside a portion of crops for the poor; an implicit act of social justice. Today, many Jews hold a *Seder*, an ordered ritual meal, on *Tu-Bishvat* to honour the seasons and God's created world, with some being more explicitly environmental in terms of symbolism and readings from the *Torah*, the Hebrew Bible and its commentaries (see Hazon,

2019a). *Tu Bish'vat* also fosters Jewish environmental action with calls for tree planting, and suggesting Jews decrease their use of materials that are made from trees by, for example, moving to using only recycled paper and/or using toilet paper made from more sustainable products such as bamboo.

Although brief and somewhat simplistic, *Tu Bish'vat* as a New Year festival has its roots in the *Mishnah*, the oral traditions of Judaism. It was one of four annual tithes where the Israelites would bring one-tenth of their fruits to the Ancient Temple in Jerusalem as an offering to God. Trees are symbolic in Judaism with the Torah often referred to as the Tree of Life (*Etz Chaim*) and there is a Biblical prohibition against destroying fruit trees in times of war (Deuteronomy 20:20) that ensures people can eat even during military sieges. By the Middle Ages, long after the destruction of the Second Temple in 70CE, *Tu Bish'vat* was no longer a ritual for the legal dating of trees, and by the sixteenth century, it had been reimagined into a holiday. The development of *Tu Bish'vat* was accomplished by 'Isaac Luria Ashkenazi, the father of contemporary Kabbalah' (Gilad, 2014); Kabbalah is a form of Jewish mysticism (Dennis, 2022). Central to Kabbalah is the concept of *Tikkun Olam*; *Tikkun* meaning healing or correction, and *Olam*, the world. According to Rabbi Luria, known as Holy Ari, *Tikkun* is the first commandment of Kabbalah as when God created the universe, God's:

> infinite power had to restrain itself to bring about the creation of a finite world..[and] the Divine essence chose to focus on...our earth [but]...the self-imposed contraction of power...proved too powerful for the world [and] sparks of divine light fell from the heavens into the lower realms, where the remain concealed in the material world of rocks, animals, and people...to restore the world to its original condition [humanity] must search for an gather the fragments. (Fenyvesi, 1998, 75)

This means that within Jewish understanding, every action for the good of nature is an action that helps repair the world (*Tikkun Olam*). Social action, practice, is, therefore, embedded in Jewish theology, doctrine.

The concept of *Tikkun Olam* is also embedded in numerous Jewish eco-educational programmes. Teva, for instance, aims to 'fundamentally transform Jewish education through experiential learning that fosters Jewish, ecological, and food sustainability' (Hazon, 2019b) in the USA.

Founded in 1994, Teva has educational programmes for children and young people aged from 2 years of age to 17 years of age which include summer camps, day schools and resources that can be used in the classroom. The Heschel School in Toronto, Canada, founded in 1996, incorporates as part of its curriculum a teaching and learning garden. Environmentalism and ecological literacy are in effect centred at this educational establishment through the Torah prescription to choose life (Deuteronomy 30:19) combined with the Jewish principle of *Tikkun Olam* which recreates the world, repairs damage and brings goodness to the world. By foregrounding gardening, the children learn about seasonality, the need to nurture to bring about and sustain life, and the requirement to share in the bounties of life; being stewards of their own garden brings about not only a deep understanding of Judaism as a spiritual way of life but of the world and all that is in it as Created and worthy of respect (Ijaz & Mawson, 2020, 118; 122).

Jewish theology, however, centres on respect for the Earth in more ways than one. The first of the Abrahamic religions, both Christianity and Islam, share with it the notion of divine creation, with God forming the Earth and also humanity. And in all three religions, humanity is mythically formed from the very substance of the Earth (the dust of the Earth, Genesis 2:7) and is charged with tilling and tending the Earth (Genesis 2:15). But not only is humanity divinely formed, humanity is told by God to be careful stewards of the Earth and all the creatures and plants with whom humanity shares the earth. Included in this stewardship is that humanity must take only what is necessary from the earth's bounty and store only what would be needed (*Bal Taschit*), including for *Shabbat*; the day of rest (Exodus 16:4–5). Here the link between environmentalism and social justice is clear; excess hoarding is problematic for all creation with greed impacting not only on all those unable to enjoy the produce of their labours through some taking more than their fair share, but by hoarding what is naturally provided, others (humans and non-human animals) are denied sustenance, and this includes the creatures who by feasting on fallers, help keep the soil productive.

Although *Shabbat* is a day off working for humanity (in remembrance that *Shabbat* was a day off from Creation for God), animals must be continued to be cared for on this day (Exodus 20:10), even if that is one's daily job (Neustadt, 2022). Animals must not be worked on *Shabbat*; it is their day of rest too. The land too has a *Shabbat*; every seven years it must be rested (Exodus 23:10–11). For Jews this is a commandment from God but on a practical level this ensures the soil remains productive (Belovski, 2022). As such, care of place, animals and people (so all of creation) is at the heart of Judaism and its ethics, with environmentalism, sustainability and social justice deeply embedded (Amswych, 2012a, 2012b; Tirosh-Samuelson, 2002; Troster, 2010, 255).

Given Judaism's inherent eco-ethos, it is unsurprising that as well as Rabbi Amswych and his environmental work in England (both within his own faith community and in wider interfaith settings), there are many grassroots and more organised Jewish initiatives that have responded to help repair the world in our time of catastrophic global climate change. In America, *The Adamah (Earth) Fellowship* in North-western Connecticut, USA, is a multi-denominational Jewish foundation dedicated to sustainable and organic farming, and 'constructing a meaningful connection with nature' (Immergut, 2008). *Urban Adamah* in California, which grew out of the Connecticut initiative, operates on similar principles but has incorporated permaculture into its educative practice enhancing the food for all social justice remit, as well as the need to keep land healthy (Gazek, 2012).

These are just two of the many American grassroots Jewish farming projects that foreground social and ecological sustainability (Berndtson & Geores, 2014), but they can be found from Australia (SA, 2022) to Israel (RAC, 2022), and covering many types of responses to climate change (Shalev, 2022). Indeed, Dov Maimon of the *Jewish People Policy Institute* has suggested that responses to climate change could, and indeed should, be a unifying cause for Jews regardless of their denomination (2021); there are multiple denominations within Judaism from varying Hasidic groups, often known as Ultra-Orthodox, through Modern Orthodox, Conservative, Reform and Liberal, to secular Jews whose identity is ethnic rather than religious (Kunin, 2009). As people of Israel, Jews are bound as much by blood as they are by belief (by being Jewish, even non-practising Jews share a bond based around Jewish myth), and central to Judaism in all its denominations, and for Jews as individuals seeking to live a spiritual way of life, is care for people and all that live in or on the planet.

The beginning of the world as understood in Judaism (as noted above with the concept of *Tikkun Olam*) centres on this concept of care. Thus whilst humanity may have been expelled from the original paradise of Eden (Genesis 3), the story is one that does not find the Earth denigrated. Unlike Christianity with its concept of the Fall and Original Sin following the disobedience of Adam and Eve to God's prohibition on eating from the Tree of the Knowledge of Good and Evil (Madueme & Reeves, 2014), sin in Judaism is accrued personally in two ways; through disobedience against God's laws/lores (*Bein Adam Lamakom*) which can be atoned for annually at Yom Kippur, and by not behaving correctly to (sinning against) other people (*Bein Adam L'chaveiro*) (Tauber, 2022). Further, in a Jewish feminist reading of Genesis, rather than Eve being (mis)understood as the first sinner (in Christianity it is generally understood that Eve was tempted by an evil serpent to take the fruit and then pursued Adam to taste [sin] also, Genesis 3:1–24), Eve can be understood as the first scientist, testing the information given to her by the serpent in the Garden of Eden about the Tree (Kadari, 1999; Meyers, 2021).

As well as the importance of trees, as explored above, the Earth too is extremely important in Judaism. With humanity being made from dust, and at death returning to dust (Genesis 3:19), the intimate connections between the very soil humans live on (and grow produce in), and the stuff they are Biblically made of is clear; damage to the earth is damage to oneself (Watling, 2009, 127). And whilst the places Jews historically resided (notably Israel and the Sinai desert) are ones where it is more obvious that a utilitarian view of environmentalism is needed, regardless of any religious concerns, in Judaism it is clear that God resides in the wild spaces too (Watling, 2009, 125); in effect, whilst keeping one's own land in good condition to ensure good harvests is sensible, looking after all the land everywhere is important to God, and thus should be important to God's stewards on earth—humanity. In the Bible, it is clear that with the rights that humanity has been given as stewards of the planet (such as being able to work the land to make it productive), comes the responsibility to avoid selfishness and bring wanton harm or suffering to all else that lives on/in planet earth. So, although the Bible informs people that humanity must be fruitful and multiply (must populate the earth, Genesis 1:28), the commandment to steward the planet (Genesis 1:26) is not one of species depletion and habitat destruction through human over-population; celibacy, however, is not encouraged in Judaism (Diamond, 2007).

According to the Torah, the world as created was good (Genesis 1:31) and full of God's presence, and thus it is only God who can bring about plagues, floods and droughts to destroy some of the world, and Biblically these are brought about to punish human wrongdoings (Genesis 9; Exodus 1; Isaiah 19). Regardless of human views on the justice of such acts, the mindset of God is beyond the scope of this chapter, but that these are written in the Torah as acts of justice, the topic of justice (*Tzedek*) is important to raise. Benstein (2014) in his essay about justice and land use is useful here. Humans are to act justly, and that means justice to both humans and other-than-humans. Rabbinic Judaism, for instance, states that water should not be an ownable resource and should be available to everyone and anyone including non-human creatures and plants (although humans collecting enough for their needs in cisterns is permitted).

Further, although land can be owned, the 'landless are guaranteed a free food supply though the laws of gleaning' (Benstein, 2014, 314; Leviticus 19:9–10, Deuteronomy 24:19–22). As such, all creation should have access to the resources they need to live, and the Noah's Ark story (see Genesis 5) stressed this, when it comes to created living things, all creatures regardless of their usefulness to humanity were worthy of life according to this Biblical tale. The Song of Creation, properly known as *Perek Shirah*, stresses this concept (Sefaria.org. n.d.). Composed c.1000 CE, it is an ancient text whose author is lost to history. It is not part of Jewish liturgy but contains many Biblical verses and at least one from the Talmud; the primary source of Jewish religious law (*Halakha*) and a guide for daily observant life comprising the *Mishnah* (written collection of Oral traditions), and the *Gemara* (Rabbinical analysis of the *Mishnah*). *Perek Shirah* talks of the heavens as the handiwork of God, and that all that dwell on and in the Earth are created. It speaks of the land being founded with wisdom and that all things that grew in it should not be needlessly destroyed (taking growing things for food is of course permissible).

Although as with all ancient religions, there is a utilitarian thrust in ensuring that land and working animals are properly cared for (including being given rest to recuperate) as this guaranteed continued productivity, Judaism affirms the right to a good/appropriate life for all creation. This arguably utilitarian background has stood Judaism well; their ancient ways of doing things are earthed in generation upon generation of practice, and rituals continue to echo the religion's agricultural religious roots. This is perhaps most evident in the triumvirate relationship between God, the people and the land (specifically the land of Israel) where the people are judged through the land. If they are good, God rewards them with rains and bountiful crops, but if not, then their crops fail (see Deuteronomy 11:13–17; Jeremiah 11–20). As such, it is perhaps of no surprise that one Rabbi has proposed a New Year for the Animals, calling for a day to match *Tu Bish'vat* where Jews can rethink who they are in relation to non-human animals, and realign how they all treat non-human animals, which is often far too frequently not as it should be (Wittenberg, 2020).

11.4   ISRAEL: PEOPLE AND PLACE

Although all creation is important in Judaism, arguably the most important of human places is Israel. This chapter will not enter into the politics of the State of Israel, but it is important to note that Israel is the name not only of a physical place, but importantly for the people (past, present and future), that are now known as Jews. Israel is the Jewish national homeland, and its landscape is embedded in Jewish history, culture, religion and identity. Israel was the home to the First and Second Temples, and was the Promised Land to the Jewish people when they were in exile (Neusner, 2000, 8). However, the Jewish Covenant with God specifies the people of Israel (people who are descended from Jacob) rather than Israel the place, and the *Kaddish*, the mourner's prayer, closes by asking for peace upon all the people of Israel (MJL:K, 2002–2022). But as Schweid has argued 'the idealization of [Israel the people has also] culminated in [the] absolute spiritualization' of Israel the place (2009: 538). As such, although there is a long history to Jewish environmentalism, especially in America (Jacobs, 2002), exploring Jewish environmentalism in Israel will cement the inherent link between people and place when exploring Judaism, as it is a country where, although officially secular, government policy and grassroots action frequently has a theological bent.

There are many charities and grassroots organisations in Israel dedicated to environmental issues, including the topic of tackling climate change (RAC, 2022), and there are also centres of education that incorporate environmentalism in their curriculum (Gerstenfeld, 2001, 10). At the 2021 United Nations Climate Change Conference (COP26), 120 Israelis including scientists, educators, activists and clerics attended; many of these were Jews although the country includes peoples from a variety of faiths. The Israeli delegation brought their inter-disciplinary approach to understanding and addressing global climate change and its consequences (Leichman, 2021). They noted that, for example, the Israeli government's Environment Minister is working with the Organisation for Co-operation and Development (OECD) on climate change issues, and that leading Jewish environmentalists have called for legal frameworks in the country to co-ordinate cabinet ministries in combating climate change, notably by reducing emissions targets combined with an increase in the use of renewable energy.

There is also a move in Israel to establish joint projects on waste treatments and cross-border water pollution, and work internationally in the region to advance water solutions for decreasing desertification (Bassist, 2022). A recent article in *The Times of Israel* noted that by the turn of this century, temperature rises combined with land degradation and sea level rises could severely affect food production, water availability and even make swathes of the African continent un-liveable. The potential outcomes of such severe environmental problems are catastrophic with both political turmoil, and climate migration strong likelihoods (Surkes, 2021). And whilst it may seem, in the wake of impending doom, that individuals can do little, in Judaism it is clear that theologically every little action helps. This can be seen in that traditionally, people living outside Israel have donated money for trees to be planted in Israel at *Tu Bish'vat*; although noticeably, recently this practice has shown to be environmentally damaging and detrimental to the country's natural eco-system and Jews outside Israel and now encouraged to plant trees (or pay for trees to be planted) in areas where they will do the most environmental good (Shofet, 2020). As people learn more about the planet and the impact of human actions as stewards of it, human actions can and do change, and as with Amswych's idea, small actions can develop into bigger plans and bring about change on a broader scale. Thus, with Judaism's inherent eco-theology, doctrine can support praxis and allow the Jewish people and work with nature and the planet as a whole.

11.5 Conclusion

This chapter has explored the role that Judaism as a spiritual way of life can play in environmentalism and sustainability. It has highlighted the Biblical emphasis on social justice in regard to nature and the environment, historically the utilitarian approach to the land and to animals that ensured productivity which plays out in the Jewish festivals such as *Shabbat* and *Tu Bish'vat*. It has noted the many grassroots initiatives that draw on the Kabbalistic concept of *Tikkun Olam* (healing the world) that bring this connection with the natural world into action, through educational programmes for children and young adults, to permaculture courses that are helping to make the desert green, and highlighted the role that Israel as a country has in governmental approaches to climate change.

Primarily though this chapter has focussed on the IDEA developed by Reform eco-Rabbi Neil Amswych who led an interfaith initiative in Dorset, England, to encourage people in the county and beyond of faith or no-faith to be more eco-conscious, he even managed to get organisations to sign the pledge to think more about nature, the climate and sustainable living and working, when developing policies. Amswych developed IDEA because Judaism is inherently eco-focussed, and as a Rabbi, his faith informs his actions. In Judaism, eco-doctrine supports eco-practice, and that for Jews globally being Jewish means taking their role as stewards of the Earth seriously. It also means they can support others in this goal, for *Tikkun Olam* means healing the whole world and all that live in it or on it.

REFERENCES

Amswych, N. (2011, October 6). The damaging oil in your kosher food. *Jewish Chronicle*. https://www.thejc.com/lets-talk/all/the-damaging-oil-in-your-kosher-food-1.28126. Accessed 14 January 2023.

Amswych, N. (2012a, June 24). *An exploration of spirituality from Santa Fe's Eco Rabbi*. http://rabbineilamswych.blogspot.com. Accessed 14 January 2023.

Amswych, N. (2012b). Beyond the sabbatical: A contemporary Jewish response to sustainable development. In R. Clugston & S. Holt (Eds.), *Exploring synergies between faith values and education for sustainable development* (pp. 16–20). https://earthcharter.org/library/exploring-synergies-between-faith-values-and-esd/. Accessed 14 January 2023.

Bassist, R. (2022, March 31). *Israel's environment mister strives for regional coalition on climate change*. Al-Monitor. https://www.al-monitor.com/origin als/2022/03/israels-environment-minister-strives-regional-coalition-climate-change. Accessed 14 January 2023.

Belovski, Z. (2022). *Shemittah and Shabbos*. Aish.com. https://aish.com/926 19919/. Accessed 14 January 2023.

Benstein, J. (2014/2006). *Tzedek* and the city: Justice, land use and urban life. In R. Bohannon (Ed.), *Religions and environments: A reader in religions, nature and ecology* (pp. 313–317). Bloomsbury.

Berndtson, R., & Geores, M. (2014). 'Let my people grow'. The Jewish farming movement: A bottom-up approach to ecological and social sustainability. In S. Brinn (Ed.), *The changing world religion map: Sacred places, identities, practices and politics* (pp. 297–322). Springer.

Dennis, G. (2022). *What is Kabbalah?* https://reformjudaism.org/beliefs-practi ces/spirituality/what-kabbalah. Accessed 14 January 2023.

Diamond, E. (2007). "And Jacob remained alone": The Jewish struggle with celibacy. In C. Olson (Ed.), *Celibacy and religious traditions* (pp. 41–64). Oxford University Press.

EC (Earth Charter). (2001–2021). *The Earth Charter.* https://earthcharter. org/. Accessed 14 January 2022.

Fenyvesi, C. (1998). Practical Kabbalah: A family history. In E. Bernstein (Ed.), *Ecology & the Jewish spirit: Where nature and the sacred meet* (pp. 69–79). Jewish Lights.

Fox, P. (2021). *Jews by the seaside: The Jewish hotels and guesthouses of Bournemouth.* Valentine Mitchell & Co Ltd.

Gazek, A. (2012, January 10). *Wake up and smell the kale! Farming for justice.* Upstart.org. https://upstartlab.org/na-52/. Accessed 14 January 2023.

Gerstenfeld, M. (2001). Jewish environmental studies: A new field. *Jewish Political Studies Review, 13*(1/2), 3–62.

Gilad, E. (2014, January 16). The history of Tu Bishvat: From legalistic debate to fruit-eating Bonaxa. *Haaretz.* https://www.haaretz.com/jewish/2014-01-16/ty-article/.premium/tu-bishvat-is-more-than-fruit-eating/0000017f-e6e8-dea7-adff-f7fb7dd20000. Accessed 14 January 2023.

Hazon. (2019a). *Resources.* https://hazon.org/commit-to-change/holidays/. Accessed 14 January 2023.

Hazon. (2019b). *Teva.* https://hazon.org/teva/. Accessed 14 January 2023.

Ijaz, N., & Mawson, P. (2020). Cultivating a Jewish eco-education framework: The Toronto Heschel School's teaching and learning garden. *Canadian Jewish Studies, 29,* 113–139.

Immergut, M. (2008). Adamah (Earth): Searching for and constructing a Jewish relationship to nature. *Worldviews: Global Religions, Culture, and Ecology, 12,* 1–24. https://doi.org/10.1163/156853508X276815

Jacobs, M. (2002). Jewish environmentalism: Past accomplishments & future challenges. In H. Tirosh-Samuelson (Ed.), *Judaism and ecology: Created world & revealed word* (pp. 449–480). Harvard University Press.

JC (*Jewish Chronicle*). (2011, February 10). Reform aims for Carbon Neutral Judaism. *Jewish Chronicle.* https://www.thejc.com/news/community/reform-aims-for-carbon-neutral-judaism-1.21161. Accessed 14 January 2023.

JCR (Jewish Communities and Records). (2005, August 21). *Bournemouth & Poole Jewish Community.* https://www.jewishgen.org/jcruk/community/Bournemouth.htm. Accessed 14 January 2023.

JVL:AJH (Jewish Virtual Library). (n.d.). *Ancient Jewish history: The birth and evolution of Judaism.* https://www.jewishvirtuallibrary.org/the-birth-and-evolution-of-judaism. Accessed 14 January 2023.

JVL:JHF. (n.d.). *Jewish holidays & festivals: The Jewish calendar: An overview.* Jewish Virtual Library. https://www.jewishvirtuallibrary.org/the-jewish-calendar. Accessed 14 January 2023.

Kadari, T. (1999). *Eve: Midrash and Aggadah*. The Shalvi/Hyman Encyclopedia of Jewish Women. https://jwa.org/encyclopedia/article/eve-midrash-and-aggadah#pid-15075. Accessed 14 January 2023.

Kunin, S. (2009). Judaism. In L. Woodhead, P. Fletcher, H. Kawanami, & D. Smith (Eds.), *Religions in the modern world: Traditions & transformations* (pp. 173–203). Routledge.

Leichman, A. (2021, October 31). 20 Israelis leading the way out of the climate crisis. *Israel21c*. https://www.israel21c.org/20-israelis-leading-the-way-out-of-the-climate-crisis/. Accessed 14 January 2023.

Lubetkin, S. (2013, February 6). *'Big Green Jewish Website' links to worldwide environmental resources*. Jewish Voices. https://www.jewishvoicesnj.org/articles/big-green-jewish-website-links-to-worldwide-environmental-resources/. Accessed 14 January 2023.

Madueme, H., & Reeves, M. (Eds.). (2014). *Adam, the fall, and original sin: Theological, biblical and scientific perspectives*. Baker Academic.

Maimon, D. (2021, November 1). Climate action in Israel as a bridge for Jewish peoplehood. *The Times of Israel*. https://blogs.timesofisrael.com/climate-action-in-israel-as-a-bridge-for-jewish-peoplehood/. Accessed 14 January 2023.

Meyers, C. (2021). *Eve: Bible*. The Shalvi/Hyman Encyclopedia of Jewish Women. https://jwa.org/encyclopedia/article/eve-bible. Accessed 14 January 2023.

MJL:K (My Jewish Learning). (2002–2022). *Text of the Mourner's Kaddish*. https://www.myjewishlearning.com/article/text-of-the-mourners-kaddish/ Accessed 14 January 2023.

MJL:KF (My Jewish Learning). (2002–2022). *Kosher food: What makes food kosher or not*. https://www.myjewishlearning.com/article/kosher-food/. Accessed 14 January 2023.

Neusner, J. (2000). Defining Judaism. In J. Neusner & A. Avery-Peck (Eds.), *The Blackwell companion to Judaism* (pp. 3–19). Blackwell.

Neustadt, D. (2022). *Feeding animals on Shabbos*. https://outorah.org/p/53825/. Accessed 14 January 2023.

RAC (Religion Action Center). (2022). *Environmentalism in Israel*. https://rac.org/environmentalism-israel. Accessed 14 January 2023.

RJ:B (ReformJudaism.org). (2022). *Blue prints for green living*. https://reformjudaism.org/blueprints-green-living. Accessed 14 January 2023.

RJ:E&CC (ReformJudaism.org). (2022). *Environment and climate change*. https://www.reformjudaism.org/issues/environment-and-climate-change. Accessed 14 January 2023.

Rockefeller, S. (2010). Earth Charter. In W. Jenkins (Ed.), *The Berkshire encyclopedia of sustainability: The spirit of sustainability* (pp. 114–117). Berkshire Press.

SA (Sydney Alliance). (2022). *Jewish sustainability*. https://www.sydneyalliance. org.au/jewish-sustainability-initiative. Accessed 14 January 2023.

Schweid, E. (2009 [1987]). Land of Israel. In A. Cohen & P. Mendes-Flohr (Eds.), *20th century Jewish religious thought: Original essays on critical concepts, movement and beliefs* (pp. 535–541). JPS.

Sefaria.org. (n.d.). *Perek Shirah* (A. N. Varady & N. Slifkin, Trans.) https:// www.sefaria.org/Perek_Shirah?tab=contents. Accessed 14 January 2023.

Shalev, A. (2022, July 26). How Jewish philanthropies deal with climate change. *The Jerusalem Post*. https://www.jpost.com/diaspora/article-713057. Accessed 14 January 2023.

Shofet, J. (2020, February 10). This Tu Bishvat, don't plant a tree in Israel. *The Times of Israel*. https://blogs.timesofisrael.com/this-tu-bishvat-dont-plant-a-tree-in-israel/. Accessed 14 January 2023.

Surkes, S. (2021, March 11). Is Israel burying its head in sand as climate change makes Mideast a hot mess? *Times of Israel*. https://www.timesofis rael.com/is-israel-burying-its-head-in-sand-as-climate-change-makes-mideast-a-hot-mess/. Accessed 14 January 2023.

Tauber, Y. (2022). *The two-way mirror*. https://www.chabad.org/library/ article_cdo/aid/110333/jewish/The-Two-Way-Mirror.htm. Accessed 14 January 2023.

Tirosh-Samuelson, H. (Ed.). (2002). *Judaism and ecology: Created world and revealed word*. Harvard University Press.

Troster, L. (2010). Judaism. In W. Jenkins (Ed.), *The Berkshire encyclopedia of sustainability: The spirit of sustainability* (pp. 254–257). Berkshire Press.

UCS (Union of Concerned Scientists). (2013). *Palm oil and global warming*. https://www.ucsusa.org/sites/default/files/legacy/assets/documents/glo bal_warming/palm-oil-and-global-warming.pdf. Accessed 14 January 2023.

Watling, T. (2009). *Ecological imaginations in the world religions: An ethnographic analysis*. Continuum.

Wittenberg, J. (2020, August 21). *The Jewish new year for animals—Why this is so important*. Heart and Mind. http://jonathanwittenberg.org/environment/ thejewish-new-year-for-animals/. Accessed 14 January 2023.

CHAPTER 12

Islam and Engagements with Nature; Theology and Practice

Christina Welch and Fahima B. Rahman

12.1 Introduction

This chapter explores not theory into practice, but theology into practice. Perhaps more accurately, it reflects on engagements of Muslims with the natural world with a central case study based around Fahima (F), an English Muslim. F is women of colour with a Bangladeshi background and is observant in her faith. Her contributions to this chapter should be self-evident; she provides the Muslim aspect of this co-authored chapter. Christina (C) lectures in Religious Studies, and Islam is one of the religious traditions she teaches her undergraduate students about, including Islamic conceptions of nature. C is not an Islamic Theologian but has a solid grasp on Islamic theology, and here provides the academic founding to this chapter. F and C met in 2009 when C was part of a small team of scholars researching diaspora communities' connections to grown food

C. Welch (✉)
University of Winchester, Winchester, UK
e-mail: Christina.Welch@winchester.ac.uk

F. B. Rahman
Independent Scholar, London, UK

© The Author(s) 2024
N. Finneran et al. (eds.), *Managing Protected Areas*,
https://doi.org/10.1007/978-3-031-40783-3_12

(Light & Welch, 2018). During the food and diaspora project, C learnt about F's personal connections with the natural world and saw how it tangibly connected her with her faith. However, with a focus on home-grown natural food, what fell outside the remit of the diasporic food project was Fahima's enthusiasm for fell walking and mountain climbing in the countryside that surround her home in the north of England, and in natural areas beyond.

In this chapter, it is F's experiences with the countryside that come to the fore and centre F's engagement with the natural world. They are however, set into wider Islamic eco-engagement, both globally and locally in England, to ensure a fuller picture of people and place in Islam. The chapter deliberately includes the recent rise in publicity about Muslim's consciously getting out into the British countryside in order to, quoting the tagline of MuslimHikers.com, 'champion diversity outdoors'; with spaces and places in White majority countries dominated by the majority population, engaging with them as a person of colour, especially a person of colour of the non-dominant religion (in rural Britain this would be Anglican Christianity) such engagement is a soft political as well as a potent spiritual act. And to ground the spiritual side, this chapter also places F's fell walking and other countryside pursuits within Islamic theology, and in particular Islamic environmentalism, arguably a tradition dating back to the reception of the Qur'an by Prophet Mohammed between 610 and 632 CE.

It is important to note that this chapter has an English focus in terms of personal Islamic engagements with the natural world, notably an adult female engagement. Explicitly stating this is not because Islamic engagements with the natural world are particularly country specific, but being a global religion, in some areas of the world, women's engagement with public spaces is restricted; this is far more cultural than religious per se, and not all Muslim-majority countries restrict female access to the public square. Such restriction is not the case in England, although in 2021 the countryside charity, Council for the Protection of Rural England (CPRE), published their report 'Access to Nature in the English Countryside' which stated that for Black, Asian and Minority Ethnic people (BAME), the countryside can be a space of exclusion for people of colour. An in-depth exploration of the reasons for this is beyond the scope of this chapter, but there are Muslims actively breaking down barriers, such as *Muslim Hikers* founded by Haroon Mota (2021). This community organisation includes Muslims from all over England; men, women and children from diverse backgrounds and includes women in *Hijabs* (head-coverings) and *Niqabs* (face-coverings).

Mota's idea for a Muslim walking group was sparked when in circa 2006, he was 'very surprised' to see women in *Hijabs* climbing Mount Snowdon, the highest peak in Wales located in Snowdonia National Park at 1085 metres (3560 ft) above sea level. It is of little surprise that he was taken aback, given that only one per cent of National Park visitors are from BAME backgrounds (Tailer, 2021). As the recent CPRE report states, in English countryside spaces, which are predominantly rural, close-knit, conservative and White (and we can add Anglican Christian in here), BAME people can feel 'as if they are objects of curiosity' (2021, 4), especially for Muslim women in *hijabs* or *niqabs* where National Parks and natural sites can be places of judgement rather than neutral spaces, or places for well-being, spiritual growth, and just 'being'. However, some Muslims in England, including women, have quietly been engaging with the English countryside for many years and this chapter centres Fahima's first-hand experiences as one of these pioneering British *Hijabi* hikers.

Before considering the experiences of F, and other female *Hijabi* hikers, exploring the nature-based ethos of Islam provides a religious and spiritual context for their stories because Islam is a way of life, with the Qur'an (the revealed word of Allah) at its core stating that the world, as Created (*Fitra*), should be cared for, protected and respected. It is important to note here though that the term nature is not theorised in this chapter. It is a contentious term (Ducarme & Couvet, 2020), but here nature is understood in the context of a non-urban space, typically a National Park or similar, not necessarily devoid of human habitation but slight in human population. It is, therefore, understood that a natural space is not an untouched wilderness, but a place cultivated by humankind (and indeed other-than-human kind), but largely full of natural stuff rather than cultural stuff. Further, the term Green in this chapter relates to environmentalism (Taylor, 2010) although it is often understood as the main colour associated with Islam.

Islam

Islam originated as a monotheistic faith in what is now northern Arabia as far back as the fifth century CE, and it is based upon the

Qur'an, its holy book and the teachings of the Prophet Mohammed (c 570–632 CE). Within a few hundred years Islamic armies had defeated the Byzantine Empire in the eastern Mediterranean and the Persian Sassanians and had established a large area of control (caliphate) over what is now north Africa, western Asia, Arabia and eastwards into Persia and central Asia. Over time Islamic communities became established as far east as China, and along the east African coast. Islamic communities grew in southern Europe too (especially in Spain). Over the last four or five hundred years Islamic communities have grown up across the globe and are found on every continent. In sub-Saharan African Islam is the fastest growing religion. In the UK and Europe, many Islamic communities are a relatively recent historical phenomenon. In France, for example, many Muslims are diasporic descendants of immigrants from France's north and west African colonies; in Germany, the mainly Turkish Islamic community has its origins in the Gastarbeiter (guest workers) who came as economic refugees from the 1950s. In the UK the majority of Muslims are descended from immigrants from the former West and East Pakistan (Pakistan and Bangladesh) from the 1960s. Muslims broadly belong to two main traditions, Sunni and Shi'a Islam but all Muslims express their faith through daily prayer, regular almsgiving and charitable works, cycles of fasting during holy festivals, and once in a lifetime go on pilgrimage (*Hajj*) to Mecca. It is estimated that globally there are some 1.9 billion Muslims, or approximately a quarter of the world's population.

12.2   ISLAM AND THE EARTH: FROM THEOLOGY TO PRACTICES

There are over 750 Suras (verses) in the Qur'an that are related to the natural world. Some Suras are named after animals (The Ant; The Cow), others after natural events (The Thunder; The Night) and in some, natural phenomena start a Sura and work as an oath; for example, the first words of Sura 89:1 are 'by the Daybreak', and Sura 95:1 starts 'by the fig and the olive'. Some Suras speak of the awareness that natural phenomena have of Allah, who created them; for example, Sura 17:44 states 'and there is not a thing but that it glorifies Him with His praise but you do not understand their glorification', and Suras 21:79 and 38:18 state that the mountains and the birds celebrate and praise Allah. For Muslims, the concept of *Tahweed*, the Oneness of God, is also the unity of all things and thereby ensuring the 'right relationship of harmony, peace and justice' with all things is an Islamic duty (Watling, 2009, 163); within Islam the whole of creation is Created with the ability to praise the Creator.

Praising God is central to the Islamic concept of the right relationship with Allah, and for humankind, to pray is a central Islam duty, with the Five Pillars of Islam specifying *Salat* (ritual prayer) five times each day. Before prayers, Muslims must do *Wudu*, a ritual washing with water. In Islam, water is understood as both 'pure and purifying' (Sura 25:48) so *Wudu* purifies the body enabling it to be ready for prayerful thoughts and deeds. But water is also understood to connect (unify) humanity with much of creation, and this *Tahweed* (unity) is specified in Sura 24:45 which states 'God created from water every animal that goes on its belly, on two legs and on four legs'; humankind then shares its method of creation with non-humankind. Further, some scholars of Islam understand the references to water in the Qur'an as metaphors for the knowledge and wisdom that Allah has gifted humanity. This particular exegesis interprets Suras 67: 30 and 72:16, which speak of flowing water and abundant rain respectively, as metaphors for the gifts of knowledge and wisdom that Allah gave humanity who uniquely amongst Creation have free-will and therefore the choice to abide by or stray from Allah's guidance (Shomali, 2008). As such the very act of *Wudu* can be seen to connect Muslim symbolically with other living beasts, and esoterically with the importance of learning (which increases human knowledge and wisdom), as well as it being an obligation before *Salat*.

Salat, the ritual prayer involves physical prostration, and thus a holistic submission, to Allah. Before mosques were created, and if mosques are not available, *salat* takes place outside on the earth/ground, and if water is not available for the purification of *Wudu*, then earth or earthy materials, such as sand, can be used in its stead. Earth as a substance then is more than just a place for prayer. Indeed, it is, along with water, the very stuff of creation. Sura 20:55 informs 'from the earth We created you and into it We shall send you back and from it We will raise you a second time'. As such the substance of earth connects humanity directly with their birth, death and resurrection into Paradise. But the earth also is a place for humanity to cultivate (cultivate food, and from observing Creation to cultivate the mind), and from which to create places to live (Sura 11:61). But Muslims are required to cultivate and create responsibly, and are to take responsibility for the land, and the animals that use it (Shomali, 2008, 5). Indeed, there is a hadith, a saying of the Prophet Mohammed who received the Qur'an and is deemed to be the embodiment of a perfect human, that asserts that even if a Muslim thinks the Day of Resurrection is about to come, if they are holding a seed in their hand, they should still plant it. This hadith speaks to the importance of the natural world, but also relates to another hadith that states that if a Muslim plants a seed and a bird, animal, or person, eats from it, that is seen as a *Sadaqah*, a charitable gift (Iqbal, 2020). There is in both hadiths an understanding that planting is social good.

In theology then Islam can be seen as an inherently 'Green' religion. Indeed, green is the colour most associated with Islam and for Muslims the colour typically signifies prosperity and a good life. The earth was made green by Allah (Sura 22:63) and Paradise is conceived of as a green harmonious space full of plants and water. Further, lush gardens are a well-known feature of Islamic countries (Watling, 2009, 166) with their traditional geometry (which is also found in Islamic calligraphy and architecture) understood as an expression of God-consciousness (*Ihan*), symbolising beauty and virtue, and the glory of Allah. Many hadiths relate to greening the earth; planting a fruit tree, watering a date tree are encouraged, and planting and farming in general are seen as being the most pleasant human occupations (Shomali, 2008, 6). The earth and all that is in it, the Qur'an states, was created for humanity (Suras 2:29, 14:32–34, 16:10–14, 31:20, 45:12, 55:10) to ensure the ongoing everyday survival of humankind, but also for its beauty which reflects the glory of Allah (16:5–6, 22:5, 27:60, 50:7); it is important to state here that in Islam, as noted, although the earth is created for humans, they are to treat it well, and treating the earth appropriately is considered an act of faith (Bagnied, 2016). This in essence means that for humankind, along with the great benefits of being gifted a Created Earth, comes great responsibilities (23:71, 33:72), and one of these responsibilities is not to squander what was given (7:31, 17:26–27), not to destroy (2:205), nor to do 'mischief on the earth' (7:56). Muslims are explicitly called to be *khalifah*, responsible stewards of Allah's Created world, and to live in balance

(*Mizan*) with the Created world (Khalid, 2002: 338), and as such there is a deep and strong eco-theology embedded in Islam.

Islam's explicit eco-theology is, as noted, based on the canonical sources of the Qur'an (Allah's directions for humanity given to the Prophet Mohammed) and the Hadiths (the saying and doings of the Prophet that act as tools for understanding the Qur'an), and form the basis of any Islamic environmental action. Such action has been termed an eco-Jihad; Jihad simply means a struggle, or striving to do something, and Islamic-centric ecological living is not a simple pursuit (Amri, 2014, 75; 79; Denny, 2010; Foltz et al., 2003). However, Fazlun Khalid prefers the term Green Jihad (2016), with his organisation, *The Islamic Foundation for Ecology and Environmental Sciences*, seeking to motivate and educate the Muslim world about their 'responsibilities as guardians' of Allah's Creation (IFEES, n.d). Khalid stresses that Islam is an inherently environmental spiritual lifeway and Muslims have a duty to live a green *Deen* (*Deen* is the Arabic term for a way of life) and extends Islamic eco-theology to fully embrace *Tawheed* as incorporating all life and all life-supporting systems on earth. He states that as all life is composed of water, all life is intimately connected, and further as water operates in a closed-system, water used today may have been water used 'by a Chinese farmer ten years ago, or...was the urine of an elephant in Africa 5 years ago'. Not only then do 'we all share the same water' (Khalid, 2016, 0.49–1.25 min) but that shared water is the very essence of all of us.

This shared relationship with the natural world has recently played out in the construction of an English Mosque. Opened in 2019 in Cambridge, the non-denominational Cambridge Central Mosque (CCM) had Islamic conceptions of nature, *Khalifa* and *Mizan* at its core with 'rainwater harvesting, air-source heat pumps and photovoltaic [solar panels] to minimize its carbon footprint' and speak to the 'deep green (faith)' that is Islam (en.shafaqna, 2019). The construction of such a place builds on decades of grass-roots English Islamic engagement with nature, conservation and sustainability, including the development of environmental groups locally, nationally (Gilliat-Ray & Bryant, 2010) and internationally (IFEES nd). But in England green mosques and green gardening have long been part of the Muslim response to the climate crisis (Gilliat-Ray & Bryant, 2011) and in 2011 construction started on the Hampshire Shi'a community's eco-friendly, Wessex Jamaat Al Mahdi Centre (Wessex Jamaat Al Mahdi Centre, n.d). This building consciously incorporated both green gardening with eco-architecture ensuring the building was warmed with a deep piped natural heat source and lit through solar panels. As Abdul-Matin has argued, green mosques are central to opening Muslim hearts up to *Tawheed* (unity) *Khalifah* (stewardship) and *Mizan* (balance) (Abdul-Matin, 2014, 289).

However, the concept of a Mosque as purely a building for regular community prayer is a narrow one. The earth itself is understood as a place of prayer; with the Hadith by Abu Sa'id al-Khudri (1057:4) confirming this; al-Khudri 'records the Prophet Mohammed saying: "Wherever you may be at the time of prayer, you may pray, for it [the earth] is all a masjid"… "the earth is a mosque"… "the mosque is sacred; therefore, the Earth Is sacred"' (in Apotsos, 2021, 137). For Muslims then, engaging with nature is deeply significant and a way for Muslims to connect with, and even live out, their spiritual path (*Deen*), and so, to F and her Islamic engagement with the earth.

12.3   Islamic Engagement with the Earth: Personal Practices

F is a female Muslim hiker. Her first big hike took place in 1997 when she was working with a youth group; the bug for hiking bit. Since then, F has climbed many British mountains as part of walking, hiking and/or climbing groups that comprise groups of Muslim-only women as well as groups made up of male and female Muslims and non-Muslims, and in every one of the events, in both types of groups, she has found mutual emotional, mental and physical support. Of course, whilst outdoor activity is good for keeping fit and aiding general well-being, it is the spiritual side that touches F most. As she walks, F recites '*SubhanAllahi wa bihamdihi subhanAllahil azeem*' (Glory is to Allah and all praise is to Him, glory is to Allah the great). This recitation is believed to have many spiritual benefits including bring the reciter closer to Allah and helping them clear sins and avoid evil, but recitation of this phrase is also considered 'one of the greatest acts of worship' (Qarabic, 2022). Walking in nature, reciting this verse, has helped Fahima deal with the grief of fractured relationships, of deaths in her family, and given her strength when life felt too tricky.

On one memorable occasion, on a winter family trek climbing up Mount Snowdon, she decided that, because it is a *Sunna* (an Islamic precedent) to walk back from the mosque a different way to the walk there, that it would be Islamically correct to walk back down Snowdon taking a different route. Having prayed in the snow at the summit already, some of the family decided, as the weather was poor, to walk back on a familiar path. But F and several others took up her challenge, and walked what turned out to be, a very difficult route back down. Their faith though gave them the courage to continue, and they eventually made it safely back. Faith too has played its part in her hikes during Ramadan where she has trekked in National Parks, and climbed a number of hills during the fast, not eating or drinking at all during the daylight hours. Although breaking the fast in wild places with her walking companions is a particular joy, she has found a very close spiritual connection to the nature world and Allah during Ramadan walks. It should be noted here that F would not suggest climbing Snowdon or engaging in similar strenuous activity during Ramadan without an appropriate level of fitness, and prior experience of arduous pursuits.

As noted previously, F has hiked and climbed as part of all-female Muslim groups, and she has found that these events connect her deeply to Islamic spirituality. One notable walk included her female companions praying on a mountain having first performing *Wudu* in a stream that ran down from its summit. On another hike, the experience was so moving for them all that one young member of the group shouted out '*Amu tumake bhalo bashi*' (I love you, in Bangali) from the mountaintop. The group then drank chai tea from tartan thermos flasks and ate biscuits and Kendall Mint Cake, in a wonderful mash-up of traditional Asian and British culture.

With countryside walking and hiking often perceived of as a very British activity, sadly going by the abuse some BAME people experience in the countryside, it is also understood by a vocal minority as a thing that should only be undertaken by very British people (Harley, 2022). In an interview with Aysha Sharif, a British Pakistani who has hiked alone in the Lake District National Park, she notes that as brown women wearing her *Hijab*, she received comments from other hikers that the countryside and the towns surrounding it, that the space was not a place for people like her; something she has not found when hiking as part of a group (Davis, 2022). Amira Patel, who wears a *Niqab*, has experienced the same intimidation whilst hiking alone. In response, and in large part due to the COVID-19 pandemic lockdown, she set up *Wanderlust Women*. Patel had started posting on social media to record her lockdown engagements with nature, and other Muslim women became interested in her activities and inspired by her confidence to be outside alone as a clearly observant Muslim woman.

Patel had previously set up a casual women's walking group and had been taken out on hikes as a child by her mother (also Muslim women of colour), but *Wanderlust Women* goes beyond encouraging Muslim women to walk in wild spaces, to supporting Muslim women to engage in outdoor activities such as kayaking, cycling, gorge walking and rock climbing. These activities are untypical for Muslim women in England, but Patel argues that 'it's only after you've stepped outside your comfort zone that you begin to change, grow and transform' (Patel, 2021). Like Fahima, she has broken her Ramadan fast with an *Iftar* meal outdoors and hiked during Eid; indeed, she states that she likes 'to bring that faith part' to engagements with the natural world as it is 'a nice way to engage and interact' (Patel in Lindon, 2021; Patel, 2021). It is clear that for many Muslim women, and indeed Muslims in general, being outdoors in nature, in places away from the hustle and bustle of daily life, is an authentic *Deen*, and given the eco-ethos of Islam, the countryside in general, and National Parks in particular, would do well to encourage more of this style of engagement.

12.4   Global Muslim Engagements with the Earth

Sadly, England is not alone regarding issues of countryside access amongst BAME people and its Muslim populations; America too has similar problems (Outdoor Foundation, 2021). However, just as in the UK, some Muslims in the US are seeking to address this issue by setting out into the great outdoors at home and overseas (Shervin, 2018). But as walking and hiking in natural spaces is not an option for every Muslim, this section of the paper will largely explore Muslim engagement with wider environmental issues. Eco-Islam is truly a globalised phenomenon (Schwencke, 2012) with many examples of green *Jihad*; green action that speaks to *Khalifa*, and *Mizan*. In some countries, eco-action is needed because climate change has, and continues, to fundamentally affect daily life. The Maldives is a case in point, as being low-lying, like Bangladesh, sea level rises are drowning useable land in these Muslim-majority countries (Atlantic Council, 2022). In the Maldives, environmental awareness and eco-care are embedded in the school curriculum (NIA, 2014), and education is a key principle in Islam. Indeed, as Mamoun Abuarqub notes in

Islamic Perspectives on Education, 'seeking education [in Islam] is obligatory', and knowledge is considered to be a path towards greater closeness with Allah (2009, 7). Unsurprisingly then there are numerous educational programmes that engage Muslims of all ages with nature and environmental activism. 'Green Muslims' (Green Muslims, n.d) operates out of Washington DC and has the tagline 'Living the environmental spirit of Islam'. The organisation is predominantly youth education focussed, but they also arrange hikes and engage in various recreational projects with the aim of inspiring young Muslims to understand the concept of *khalifa* and live a green *Deen*. In other parts of the world, there is a thriving Muslim permaculture movement with, 'Greening the Desert' operating in Jordan through 'Muslim Aid International', an Australian Muslim humanitarian-aid oriented NGO (Greening the Desert, n.d).

Permaculture is a way of utilising land which is ecologically sound, that is practical, sustainable and adaptable land use for people living anywhere they can grow produce, from city balconies through suburban allotments and community or educational spaces, to waste grounds and commercial premises. One of the aims of the Jordan project is to teach about permaculture use for arid areas, and with desertification one of the inevitabilities of climate change, this and similar projects are vital. However, they also speak to the nature-centric ethos of Islam. In seeking to green Jordan's part of the Wadi Rum desert, this Muslim-majority country has a project by a Muslim charity that draws on Muslim environmental principles, and the Muslim concept of *Rizq* which, often translated as sustenance, can be understood as 'anything that brings benefits or goodness' (Ahmed, 2019). In years to come, what people learn from the 'Greening the Desert' project will enable more Muslims to live their *Deen* through permaculture and bring *Rizq* to those living in lands where drought effects food production.

This notion of an environmentally engaged *Deen* is evident if we return to the Maldives where contemporary reef management principles have been driven by the concept of *Rizq*. The Maldives are reliant on their reefs for fishing, and the protection from the ocean waves that the low-lying series of islands requires. The concept of *Rizq* (physical and emotional sustenance) includes that resources should be for all people rather than just the preserve of one person, and that any person seeking their own *Rizq* must do so in a lawful (*Halal*) respectful manner. *Rizq* in the Maldives extended then beyond the individual fishermen to ensure respect for the non-human reef dwellers, as fish and coral are as Created by Allah as is humanity. And whilst the world in Islam theology was/ is Created for humanity, the concepts of *Mizan* and *Khalifa* must be respected in all actions. Mohammed, Gombay and Pirker assert that by bringing to the fore Muslim ways of engaging with the natural world, reefs like those in the Maldives can be saved from destruction (2019, 19). This application of reef management and *Rizq* was one that bore fruit in Zanzibar. Between 1999 and 2001 a project ran to bring Islamic environmental stewardship principles into marine conservation through, 'an Islamic oriented education programme amongst …fishing communities, religious leaders and government officials'. This programme led to the enhanced management and protection of a marine conservation area, stopping the destruction of coral which was depleting fish stocks, and disturbing turtle nesting grounds (IFEES Zanzibar, 2021).

A similar concept has been used recently in Indonesia although with a focus on repurposing land known as *Wadq*. *Wadq* land is a charitably donated piece of land where there is no intention of reclaiming it. As Luik, Fatoni and Ahmad note it is land 'still managed by traditional methods and bound to limited purposes, such as a mosques, graveyards, and madrasah' (2021, 60). In Indonesia such land is often abandoned, yet could be used to help the earth, and communities, prosper through the implementation of permaculture. As noted previously the earth in all its richness is central to Islam, and within this understanding, human agriculture is an important, even blessed, occupation. In the Hadith of Bukhari 2320, Mohammed is cited as saying 'there is none amongst the Muslims who plants a tree or sows seeds, and then a bird, or a person or an animal eats from it, but is regarded as a charitable gift for him' (in Luik et al., 2021, 62). And with paradise (*Jannah*) a lush garden of intertwining shady trees, flowing water and abundant orchards, permaculture can be seen to resonate intrinsically with Islam and concepts of a green jihad/ eco-Jihad. With no monocultural crop techniques, no harmful chemical pesticides or fertilisers, and a community and social responsibility ethos, utilising Indonesia's *waqf* land through permaculture practices, could assist with countering food insecurity and helping to green the country in a solidly Muslim way (Luik et al., 2021, 68). It should be noted however, that there are prohibitions on the use of graveyards beyond a place to bury the dead and commemorating them through graveside visits (Masjid al-Muslimiin nd). Additionally, it is highly unlikely many Muslims would wish to eat produce grown in such a *waqf*. However, other *waqf* spaces would provide useful productive ground in most Muslim-majority countries.

Wadq is not the only type of land suitable for permaculture. Islamic law divides land into three main categories: developed (*Amir*) land, undeveloped (*Mawat*) land and 'protective zones' (*Harim*). Although a somewhat simplistic explanation, the basic rule is that undeveloped land belongs to whoever brings it to life by one of the following methods: putting a hedge or wall around the land, irrigating dry land or draining wetland, digging a well, clearing the area of trees, breaking up stones, and levelling the ground, ploughing the land, planting crops or trees on it and/or erecting a building on it. *Harim* land is land which forms a protective zone, of any size, around every development.

The use of *Harim* land is forbidden to anyone but the owner. And a *Hima* is an area of undeveloped (*Mawat*) land that has been set aside for some reason, and often functions as a conservation area or wildlife reserve (Khalid, 2002, 336). It may be that permaculture might be suitable for *Amir* land with community groups taking the lead, and even some *Harim* land could be utilised in this manner if the owner were willing. However, a green *Deen* can extend to small individual acts such as those in the Lifemakers/IFEES guide that note everyday ways Muslims can help tackle climate change (IFEES, n.d). From eating locally produced produce to reduce food miles, minimising food waste, having showers rather than baths to save water and turning off lights to save power are all practical ways that Muslims can authentically engage with and respect the earth. Living out a green *Deen* can be expressed in actions big and small; it is taking those actions that is important.

12.5 CONCLUSION

Islam is not only a theologically environmentally connected religion, but many Muslims find ways to live out their faith through environmental action, or an eco-or green *Jihad*. The concept of *Rizq* has helped ensure reefs are protected and as such not only are the livelihoods and habitats of humans protected, but the habitats of sea creatures are too. The concept of *Tawheed* is pertinent here, the unity between Created things through the substance of water. But humankind and non-humankind are also theologically connected through earth, and as such Islamic permaculture embodies *Rizq* through ensuring all can benefit from produce grown in the earth. Utilising *Waqf* spaces for growing crops assists here too. As such in using the Created earth responsibility, Muslims live out their role of *Khalifa* of the planet. And on an individual level, as Fahima and the other Muslim walkers, hikers and followers of outdoor pursuits have demonstrated, being authentically Muslim in the landscape embodies a green *Deen*, with not even the daylight fast of Ramadan stopping some from engaging in countryside activities, and indeed for many, this can deepen their spiritual engagement with nature.

This chapter has demonstrated that for Muslims, natural spaces are imbued with religious significance. Rather than exploring theory, theology can be a way into understanding how people and place connect, and this stands for most people of faith, or with a spiritual lifeway; there are many books and academic articles on this topic with a wealth of information on from everything the global doctrinal religions to local traditional spiritual lifeways. Islam is a global religion, doctrinally focussed but also a spiritual way of life, and as such taking care of, being *Khalifa* for, the earth plays out in different ways in different countries. But from large NGO projects to everyday domestic acts, Muslims can authentically live a green *Deen*.

REFERENCES

Abdul-Matin, I. (2014). Green Mosques. In R. Bohannon (Ed.), *Religions and environments: A reader in religions, nature and ecology* (pp. 289–296). Bloomsbury.

Abuarqub, M. (2009). *Islamic perspectives on education*. Islamic Relief Worldwide. https://www.muslimplatformsd.org/wp-content/uploads/2017/03/Islamic-Perspectives-on-Education.pdf. Accessed 15 January 2023.

Ahmed, I. (2019, August 12). The concept of Rizq (Sustenance) in Islam. *The Siasat Daily*. https://www.siasat.com/concept-rizq-sustenance-islam-1593246/. Accessed 15 January 2023.

Amri, U. (2014). From theology to a praxis of "Eco-Jihad": The role of religious civil society organizations in combating climate change in Indonesia. In R. Veldman, A. Szasz, & R. Haluza-Delay (Eds.), *How the World's religions are responding to climate change: Social scientific investigations* (pp. 75–93). Routledge.

Apotsos, M. (2021). *The Masjid in contemporary Islamic Africa*. Cambridge University Press.

Atlantic Council. (2022). Rising sea levels and the climate change crisis in Bangladesh, the Maldives, and Sri Lanka. https://www.atlanticcouncil.org/blogs/southasiasource/rising-sea-levels-and-the-climate-crisis-in-bangladesh-the-maldives-and-sri-lanka/. Accessed 15 January 2023.

Bagnied, O. (2016, October 1). Caretakers of the Earth: An Islamic Perspective. *Khaleafa.com*. http://www.khaleafa.com/khaleafacom/caretakers-of-the-earth-an-islamic-perspective. Accessed 15 January 2023.

CPRE (Council for the Protection of Rural England). (2021). Access to nature in the English Countryside. https://www.cpre.org.uk/wp-content/uploads/2021/08/August-2021_Access-to-nature-in-the-English-countryside_research-overview.pdf. Accessed 15 January 2023.

Davis, H. (2022, February 25). For Muslim hikers, an empowering community makes all the difference. *National Geographic*. https://www.nationalgeographic.co.uk/adventure/2022/02/for-muslim-hikers-an-empowering-community-makes-all-the-difference. Accessed 15 January 2023.

Denny, F. (2010). Islam. In W. Jenkins (Ed.), *The Berkshire Encyclopedia of sustainability: The spirit of sustainability* (pp. 243–247). Berkshire Press.

Ducarme, F., & Couvet, D. (2020). What does "nature" mean? *Palgrave Commun, 6*(14). https://doi.org/10.1057/s41599-020-0390-y

En.shafaqna. (2019, December 10). Europe's First "Eco-Mosque" opens in the UK. *En.shafaqna.com*. https://en.shafaqna.com/126010/europes-first-eco-mosque-opens-in-uk/. Accessed 15 January 2023.

Foltz, R., Denny, F., & Baharuddin, A. (Eds.). (2003). *Islam and ecology: A bestowed trust*. Harvard University Press.

Gilliat-Ray, S., & Bryant, M. (2010). *Islamic Gardens in the UK: Dynamics of conservation, culture and communities*. Botanic Gardens Conservation International. https://www.cardiff.ac.uk/__data/assets/pdf_file/0009/2399499/islamic-gardens-in-the-uk.pdf. Accessed 15 January 2023.

Gilliat-Ray, S., & Bryant, M. (2011). Are British Muslims "Green"? An overview of environmental activism in Britain. *Journal for the Study of Religion, Nature and Culture, 59*(3), 284–306. https://doi.org/10.1558/jsrnc.v5i3.284

Green Muslims. (n.d). https://www.greenmuslims.org. Accessed 15 January 2023.

Greening the Desert. (n.d). https://www.youtube.com/watch?v=ycLbO02lb7w. Accessed 15 January 2023.

Harley, N. (2022, March 18). Muslim Hikers shrug off racist abuse to enjoy challenges of the Yorkshire Dales. *The National News*. https://www.thenationalnews.com/weekend/2022/03/18/muslim-hikers-shrug-off-racist-abuse-to-enjoy-challenges-of-the-yorkshire-dales/. Accessed 15 January 2023.

IFEES (Islamic Foundation for Ecology and Environmental Sciences). (n.d). https://www.ifees.org.uk. Accessed 15 January 2023.

IFEES. (2021). The application of Islamic environmental ethics to promote marine conservation in Zanzibar. https://www.ifees.org.uk/projects/islam-biodiversity/zanzibar/. Accessed 15 January 2023.

Iqbal, Z. (2020, September 29). Green Ahadith—Ecological Advice from Prophet Mohammed. *EcoMENA*. https://www.ecomena.org/green-aha dith/. Accessed 15 January 2023.

Khalid, F. (2002). Islam and the environment. In P. Timmerman (Ed.), *Social and economic dimensions of global environmental change* (pp. 332–339). John Wiley and Sons.

Khalid, F. (2016). A Green Jihad: Pioneering Islamic Environmentalism. https://www.youtube.com/watch?v=Vb6N4aT7Wn4. Accessed 15 January 2023.

Lifemakers/IFEES. (2008). *Muslim Green guide to reducing climate change*. https://ifees.org.uk/wp-content/uploads/2019/12/muslim_green_guide.pdf. Accessed 15 January 2023.

Light, A., & Welch, C. (2018). The cultural and spiritual aspects of growing edible plants: Testing for meaningfulness in Leeds, UK. In J. Zeunert & T. Waterman (Eds.), *The Routledge handbook of landscape and food* (pp. 570–582). Routledge.

Lindon, H. (2021, May 13). Leading a Hiking group for Muslim women. *ToGoMazagine.co.uk*. https://www.tgomagazine.co.uk/news/leading-a-hik ing-group-for-muslim-women/. Accessed 15 January 2023.

Luik, M., Fatoni, A., & Ahmad, M. (2021). Utilising permaculture to develop abandoned Waqf land in resolving food insecurity in Indonesia. *Social Business Journal, 2*(2), 60–70. https://doi.org/10.55862/asbjV2I2a007

Masjid al-Muslimiin. (n.d). *The Sunnah and prohibitions concerning grave-yards*. https://www.almasjid.com/content/sunnah_and_prohibitions_concer ning_graveyards. Accessed 15 January 2023.

Mohammed, M., Gombay, N., & Pirker, J. (2019). Development and the sacred; An account of Reef Resource Management in the Maldives. *International Journal of Social Research and Innovation, 3*(1), 1–22. https://doi.org/10. 55712/ijsri.v3i1.18

Mota, H. (2021). Muslim Hikers Inaugural Trek: Mini documentary. https:// www.youtube.com/watch?v=4aDEF606AMk. Accessed 15 January 2023.

NIE (National Institute of Education, Maldives) (2014). *Science in the National Curriculum. Ministry of Education, Maldives Government*. https://www.moe. gov.mv/assets/upload/Science_Key_Stage_2_Syllabus.pdf

Outdoor Foundation. (2021). Outdoor Participation Trends Report. *Outdoorindustry.org*. https://outdoorindustry.org/wp-content/uploads/ 2015/03/2021-Outdoor-Participation-Trends-Report.pdf. Accessed 15 January 2023.

Patel, A. (2021, June 1). Amira Patel—Founder of The Wanderlust Women Group encourages Hiking and Adventure for Muslim Women. *Toughgirlchallenges.com*. https://www.toughgirlchallenges.com/single-post/patel. Accessed 15 January 2023.

Qarabic. (2022). Subhanallah wa bihamdihi. https://qarabic.com/subhan-allah-wa-bihamdihi-meaning/. Accessed 15 January 2023.

Schwencke, A. (2012). *Globalized Eco-Islam: A survey of Global Islamic Environmentalism*. http://ppi.unas.ac.id/wp-content/uploads/2015/01/report-globalized-eco-islam-a-survey-schwencke-vs-24-february-2012-pdf.pdf

Shervin. (2018, June 2). Backpacking for Muslims can be hard but not Impossible—This is what you should know. *Mvslim.com*. https://mvslim.com/backpacking-for-muslims-can-be-hard-but-not-impossible-this-is-what-you-should-know/. Accessed 15 January 2023.

Shomali, M. (2008). *Aspects of Environmental Ethics: An Islamic perspective*. Al-Islam. https://www.al-islam.org/articles/aspects-environmental-ethics-islamic-perspective-mohammad-ali-shomali-0. Accessed 15 January 2023.

Tailer, N. (2021, December 31). The Muslim Hiker inspiring his community to hit the Hills. *Guardian*. https://www.theguardian.com/travel/2021/dec/31/the-muslim-hiker-inspiring-his-community-to-hit-the-hills. Accessed 15 January 2023.

Taylor, B. (2010). *Deep Green Religion: Nature spirituality and the planetary future*. University of California Press.

Watling, T. (2009). *Ecological imaginations in the World religions: An ethnographic analysis*. Continuum.

Wessex Jamaat al Mahadi Centre. (n.d). https://www.almahdi.org.uk. Accessed 15 January 2023.

CHAPTER 13

What Have We Learned from the Impact of the Pandemic on Our Relationship with Nature? The Importance of Views from Home

Marco Garrido-Cumbrera and Olta Braçe

13.1 Introduction. The Importance of Nature Viewed from Home

We cannot imagine a house without windows as they provide us with light, allow the air to enter inside the house, enable us to judge the weather and connect us with the nature outside. For some people, windows are an escape route that can soothe feelings of claustrophobia, monotony or boredom (Collins, 1975). Inside buildings, windows are

M. Garrido-Cumbrera (✉)
Health & Territory Research (HTR) and Department of Physical Geography and Regional Geographic Analysis, Universidad de Sevilla, Seville, Spain
e-mail: mcumbrera@us.es

O. Braçe
Health & Territory Research (HTR) and Department of Human Geography, Universidad de Sevilla, Seville, Spain
e-mail: obrace@us.es

N. Finneran et al. (eds.), *Managing Protected Areas*,
https://doi.org/10.1007/978-3-031-40783-3_13

the primary means of connection to the outdoors, and nature views even through a window may have restorative effects on the occupants (Ko et al., 2020). Furthermore, windows with views allow contact with the outside world and, when well designed, provide 'visual rest centres' (Markus, 1967). Views from windows are the most important in the everyday life of a person living in an urban area (Honold et al., 2016). Therefore, when people choose their home, they are concerned about the views they are going to experience from the windows of their home. Given its importance, real estate companies have always used views surrounding a building to attract the interest of buyers, even using them as an attraction to increase the price of a property (Bond et al., 2002).

13.2   THE INFLUENCE OF VIEWS ON MENTAL HEALTH AND WELL-BEING

Previous studies have shown that indirect exposure to nature through window views can positively influence stress reduction, increased attention span, mental health and overall well-being (Ko et al., 2020; Li & Sullivan, 2016). Viewing the sky and other features of the natural world can support psychological restoration, thus preserving health and well-being (Kaplan, 2001). Restoration implies the renewal of physical, psychological and/or social resources that have been exhausted in efforts to meet the demands of daily life (Hartig, 2021). Likewise, views of natural elements—including trees/woodland and lawn/mown grass—from office windows have been shown to influence employee well-being more positively than time spent outdoors during break periods (Gilchrist et al., 2015). Moreover, Dzhambov et al. (2018) in a study assessing university students found that those exposed to green and blue spaces views reported higher levels of concentration and better mental health.

A study by Kaplan (1983) demonstrated greater satisfaction with a neighbourhood when residents of blocks of flats or appartments could contemplate more natural than built environments. A few years later, the same author showed that being able to see natural elements from a window contributes substantially to neighbourhoods satisfaction and overall well-being in Ann Arbor, Michigan (USA) (Kaplan, 2001). Another study highlighted the restorative value of window view of the sky and greenness in a context of high urban density in Tehran (Iran) (Masoudinejad & Hartig, 2020). Elsadek et al. (2020) found how, in high-rise buildings in Shanghai (China), windows with green space views contributed to psychological well-being. While a study by Vemuri et al. (2011) found that residents of the Baltimore metropolitan area (USA) who viewed a greater number of tress from the windows of their residence reported higher life satisfaction. Additionally, Honold et al. (2016) reported that residents with views of vegetation from their home had significantly lower hair cortisol levels as a biomarker of chronic stress reduction in Berlin (Germany).

13.3   The Value of Home Views During the COVID-19 Pandemic

Although previous studies have demonstrated the influence and impact of views from home on health and well-being, the COVID-19 pandemic provided us with an unprecedented situation in which our entire society was confined to their homes, either totally or partially. The conditions created during the pandemic resulted in many cases of a lack of direct exposure to nature. During the COVID-19 pandemic, many countries took extremely strict measures as part of their public health strategies, not allowing people to leave their homes or, in the best cases, limiting their movements to a small area near their homes. Owing to these restrictive measures, physical barriers were established that did not allow people to interact with the natural environment, with consequences on their lifestyles.

One of the great losses that people experienced during the periods of home confinement was the lack of direct contact with the nature, owing to the dwelling being effectively converted into an opaque obstacle that prevented interaction between the environment and the human being. In this home confinement situation, the only option to mitigate this feeling of loss and establish contact with nature was to have access to views from home windows (Ugolini et al., 2021). In the case of home confinement, the only form of contact with nature was through indirect exposure, contemplating views from a window and thus not requiring any physical presence in nature. Thus, in a period of home confinement during the COVID-19 outbreak, when many occupants were able to benefit greatly from the restorative effects of window views from their home (Batool et al., 2021), the importance of assessing visual contact with nature has emerged.

13.4   THE GREENCOVID STUDY

Living this experience of pandemic and lockdown measures prompted an interdisciplinary group of researchers from the University of Seville (Spain), University of Winchester (England) and Maynooth University (Ireland), to join forces and to conduct the GreenCOVID study. This study addressed the perception, the relationship with nature and the health of the population at a special time—during the first wave of the COVID-19 pandemic—when contact with nature was limited or non-existent. The main objective of GreenCOVID was to assess the impact of the COVID-19 lockdown on access and exposure to nature and its impact on well-being and mental health of the general population in three European countries (Spain, England and Ireland) (Garrido-Cumbrera et al., 2023). We focused on assessing the value of home views and contact with nature as a protective factor against anxiety and depression during the first wave of the pandemic.

This cross-sectional study analysed information collected through an online survey that was disseminated to unselected adults over 16 years residing in Spain, England and Ireland. The questionnaire was designed by researchers from the Health and Territory Research (HTR) group at the University of Seville to gather information on sociodemographic characteristics, household/housing, behaviours and routines, outdoor contact, physical health, mental health and well-being as well as information on pre- and post-lockdown -closure perceptions and use of green and blue spaces. The GreenCOVID survey was disseminated through social media channels of the University of Seville and the Spanish Association of Geography (AGE). The survey remained open in Spain from April 7 to 25, 2020, using the SurveyMonkey platform.

Researchers from England and Ireland joined the initiative at a second stage. Translated versions of the survey were disseminated during the period from May 28 to July 24 2020 (Guzman et al., 2020), by the PeopleScapes Research Group, at the University of Winchester and from June 3 to July 1, 2020, by Maynooth University. As a result of this survey, 2500 responses were gathered in Spain, 500 in England and 261 in Ireland. Survey participants who did not respond to at least 70% of the questionnaire were excluded from the analysed sample for the purpose of obtaining a more robust sample with a reduced percentage of missing values. After data cleaning, 3127 respondents (2464 from

Spain, 420 from England and 243 from Ireland) who fulfilled the condition of having fully responded to the sociodemographic variables and the Hospital Anxiety and Depression, (HAD) scale were included in the analyses.

13.5 EVALUATION OF VIEWS FROM HOME

In the total sample, including the three GreenCOVID countries, 60.2% lived in a flat or apartment. This is due to the large proportion of people living in apartments in Spain (72.1%)—the country with the higher sample size—while in England it is more common to live in a house (51.7%) and in Ireland 47.3% of people live, respectively, in single family houses and 25.7% in semi-detached houses. Related to how being in/seeing/hearing the outdoors helped people cope with the pandemic crisis, in England and Ireland participants gave a high rating while for Spain this rating was significantly lower (Table 13.1). This is due potentially to the large number of people in Spain who live in a flat or apartment with difficulties accessing the outdoors.

It is remarkable that during the COVID-19 lockdown, respondents reported that their appreciation and interest in views of blue spaces, such as those to the sea, rivers, lakes, springs, or reservoirs increased significantly compared to their appreciation and interest before the pandemic. This increased interest in blue space views during the pandemic in Spain -compared to England and Ireland- may have been due to greater restrictions, including home confinement, and the fact that the majority of the population lives in flats or apartments. These forms of urban housing often hinder their exposure to nature. Consequently, this may make such views from home more desirable and appreciable in this situation of total lockdown, as was experienced during the first wave of the COVID-19 pandemic (Fig. 13.1).

Table 13.1
Assessment of the extent to which being/seeing/hearing the outdoors helps to cope during COVID-19 pandemic by countries (N = 3,080)

Country	N	Mean*	SD
Spain	2,464	5.7	3.4
England	379	8.5	2.1
Ireland	237	8.5	1.8
Total	3,080	6.3	3.4

*From 0 to 10

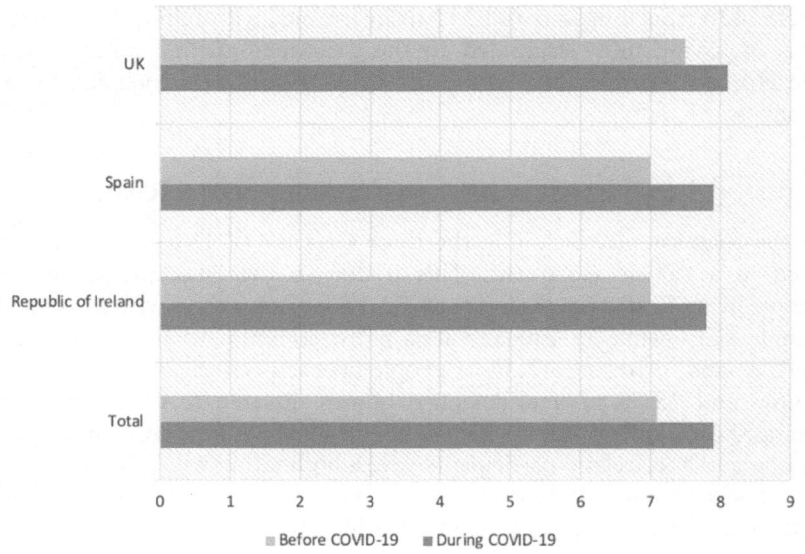

Fig. 13.1 Comparison of the value of views to the sea, rivers lakes, springs, or reservoirs before and during COVID-19 pandemic by countries (N = 3,083)

13.6   Contact with the Outdoors from Home

It should be noted that the situation of the Spanish population during the first wave of lockdown was quite different compared to the English and Irish population, since during the first wave of confinement they could not access the outdoors, a situation that was aggravated by the factor that most of the Spanish population lived in flats or apartments (Garrido-Cumbrera et al., 2021). Therefore, the only way of contact with nature during lockdown was through a window or balcony, and at the very best, having direct access to a yard or garden. Indeed, people who had been able to use outdoor spaces (courtyards or gardens) or had windows with outdoor views of natural environment reported that they have coped better with the lockdown than those who did not. In Spain, the most common element with outdoor views was the balcony and the communal roof terrace, whereas in England and Ireland, the element that most frequently allowed contact with the outdoors was the private garden or courtyard (Table 13.2).

Table 13.2 Type of elements in home that facilitate the contact with the outdoors during COVID-19 pandemic by countries (N = 2,953)

Country	N (%)			
	Spain	England	Ireland	Total
Balcony	1,340 (57.5)	21 (5.5)	17 (7.1)	1,378 (46.7)
Terrace[5]	–	20 (5.3)	2 (5.0)	32 (1.1)
Private rooftop	391 (16.8)	4 (1.1)	0 (0.0)	395 (13.4)
Communal rooftop terrace	531 (22.8)	0 (0.0)	5 (2.1)	536 (18.2)
Private garden/yard	338 (14.5)	332 (87.6)	204 (84.6)	874 (29.6)
Communal garden[1]	229 (9.8)	–	–	229 (7.8)
Private yard[2]	566 (24.3)	–	–	566 (19.2)
Communal yard[3]	348 (14.9)	–	–	348 (11.8)
Farmyard[4]	36 (1.5)	–	–	36 (1.2)
Other[6]	–	50 (13.2)	10 (4.1)	60 (2.0)
None of the above	312 (13.4)	25 (6.6)	18 (7.5)	355 (12.0)

[1,3,4]Communal garden, communal yard and farmyard were terms only used in Spain. [2]In Spain, the term private yard was defined separately from the private garden. [5,6]The categories 'terrace' and 'other' were not used in Spain

13.7 RATING OF VIEWS OF NATURE FROM THE HOME

Overall, according to the participants' opinions, the rating was 5.8 out of 10, which is due to the low score registered in Spain (5.6 out of 10). Whereas in England (6.5 out of 10) and Ireland (7.3 out of 10) the rating was higher. It should be noted that people tend to use outdoor spaces more frequently during the pandemic rather than indoor spaces with a view. However, the use of all spaces that can create a connection with nature was highly preferred by respondents in all countries (Table 13.3).

In Spain, the most frequent views from the living room or dining room were of streets with traffic (31.0%), followed by views of parks and gardens (26.7%). In contrast, in England and Ireland the most frequent views from the living room were of gardens (57.1% and 45.3%, respectively), followed by streets with little traffic (36.9% and 37.8%, respectively). This is because most people in England and Ireland live in houses that are equipped with courtyards or gardens (Table 13.4).

When the views contained greenery elements, it is worth noting that most respondents reported having views of plants or shrubs from their homes. In England and Ireland, views of more than ten trees were

Table 13.3 Use of any locations in home during COVID-19 pandemic (N = 3,067)

Country	n (%)			
	Spain	England	Ireland	Total
Outdoor spaces[1]	1,201 (48.7)	131 (36.0)	84 (35.3)	1,416 (46.2)
Spaces with views from windows	426 (17.3)	41 (11.3)	12 (5.0)	479 (15.6)
Both spaces	394 (16.0)	171 (47.0)	120 (50.4)	685 (22.3)
Not used them	146 (5.9)	5 (1.4)	5 (2.1)	156 (5.1)
Don't have any of them	212 (8.6)	18 (4.9)	15 (6.3)	245 (8.0)
Not allowed to use them	85 (3.4)	1 (0.3)	2 (0.8)	88 (2.9)

[1]Including rooftop, terrace, balcony, garden and yard

Table 13.4 Views from lounge or dining room by countries (N = 3,008)

Country	N (%)			
	Spain	England	Ireland	Total
No window	35 (1.9)	20 (5.5)	5 (2.2)	60 (2.0)
Inner courtyard	458 (24.4)	1 (0.3)	0 (0.0)	459 (15.3)
Street with traffic	749 (31.0)	4 (1.1)	0 (0.0)	753 (25.0)
Busy street with traffic[1]	–	34 (9.3)	17 (7.6)	51 (1.7)
Street with little traffic	518 (21.4)	135 (36.9)	85 (37.8)	738 (24.5)
Pedestrian area	282 (11.7)	48 (13.1)	32 (14.2)	362 (12.0)
Mostly houses[2]	–	124 (33.9)	75 (33.3)	199 (6.6)
Waste ground[3]	95 (3.9)	–	–	95 (3.2)
Industrial site/factories[4]	–	7 (1.9)	0 (0.0)	7 (0.2)
Offices[5]	–	2 (0.5)	0 (0.0)	2 (0.1)
Park[6]	–	22 (6.0)	15 (6.7)	37 (1.2)
Gardens[7]	–	209 (57.1)	102 (45.3)	311 (10.3)
Park or gardens[8]	645 (26.7)	–	–	645 (21.4)
Countryside[10]	–	92 (25.1)	43 (19.1)	230 (7.6)
Hills[9]	–	53 (14.5)	28 (12.4)	81 (2.7)
River or lake	25 (1.0)	10 (2.7)	0 (0.0)	35 (1.2)
Sea/beach/coastline	66 (2.7)	14 (3.8)	0 (0.0)	80 (2.7)

[1,2,4,5,9]These answers were not given in Spain. [3]This answer was only given for Spain. [6,7,8]In Spain, parks and gardens were not defined separately. [10]This term was translated as 'waste ground' in Spain

Table 13.5 View of nature elements from home by countries (N = 3,067)

Country	N (%)			
	Spain	England	Ireland	Total
Lawn	673 (27.3)	305 (84.3)	193 (80.1)	1,171 (38.2)
Plant or shrubs	1,340 (54.4)	330 (91.2)	211 (87.6)	1,881 (61.3)
1–3 trees	797 (32.3)	88 (24.3)	66 (27.4)	951 (31.0)
3–10 trees	654 (26.5)	108 (29.8)	59 (24.5)	821 (26.8)
More than 10 trees	686 (27.8)	171 (47.2)	113 (46.9)	970 (31.6)
None of these	1,498 (60.8)	2 (0.6)	9 (3.7)	1,509 (49.2)

common, while in Spain it was most common not to have any element of nature at all (Table 13.5).

13.8   Views as a Protective Factor Against Depression, Anxiety and for Well-Being

According to the GreenCOVID study, there was a higher prevalence of poor well-being in Spain (62.1%) than in England (47.2%) and Ireland (44.8%), as measured by the World Health Organisation-Five Well-Being Index (WHO-5) (Garrido et al., 2023). This may have been due to the greater impact of the pandemic and the more severe restrictions applied in Spain during the first wave of the pandemic (Garrido-Cumbrera et al., 2021), compared to the less restrictive environments of Ireland and England during the data collection period in those countries. In the overall sample including the three countries, poor well-being was specially associated with a younger age cohort, female respondents, students, individuals who were not physically active and those living in Spain. Poor well-being was also associated with poor quality of views from home and less perceived help from such views in coping with the lockdown.

Furthermore, in Spain a high proportion of people at risk of anxiety or depression was determined, measured using the Hospital Anxiety and Depression Scale (HADS). This result showed that the respondents did not have views of elements of nature from their home or the views they did have access to, were of poor quality. Because of this situation, they reported that they had not coped well with the effects of confinement compared with people who had views of nature. In addition, the risk of depression was higher in people who did not use outdoor spaces and window views. Thus, it can be affirmed that contact with nature and appreciation of nature contributed to reduce the risk of anxiety and depression among the general population in Spain during the COVID-19 pandemic.

These results are in line with the study conducted during the lockdown period by Fullana et al. (2020), in which performing activities of daily living such as staying outdoors or enjoying the view from the window at home were the best predictors of lower levels of depressive symptoms. The same findings were obtained in the study by Braçe et al. (2020), in which views of green spaces were associated with a lower risk of anxiety and depression, helping to cope with the impact caused by the COVID-19 pandemic, especially for those people who were under strict lockdown (Pouso et al., 2020). Another study showed that green views from windows were associated with higher levels of self-esteem, life satisfaction, perceived happiness and lower feelings of loneliness (Soga et al., 2021).

In Spain, those living mainly in flats or apartments, with no outdoor spaces other than windows and/or balconies, had no access to natural environments and a greater sense of poor well-being. Meanwhile, the situation in England and Ireland was different, as most of those living in houses had access to private gardens (Garrido-Cumbrera et al., 2022).

13.9   DISCUSSION

Nature is, therefore, important to the population and some natural elements such as trees, plants, shrubs, rivers, lakes or seas and wildlife, are particularly appreciated. This appreciation for nature has increased during the pandemic as people have more time to reflect and spend indoors. Apartment blocks are usually located in urban areas of higher population, building and traffic density, which makes it more important to enjoy visual access—through the windows of the home—to nature. In contrast, people who live in rural areas—surrounded by nature—or who have plants and gardens within their own homes are not as much in need of such a view of the outdoors.

In times of crisis such as the COVID-19 pandemic, when the population must stay at home for longer periods of time, it is necessary to assess the importance of views from home on their mental health and well-being. At the same time, the comparison of the situation in

Spain versus England and Ireland provides us with comparative evidence-based information. This confirms our initial hypothesis that the greater the restrictions and the less contact with nature, the worse the mental health and emotional well-being. In fact, during this unusual period of lockdown, nature surrounding the home may play a key role in mitigating adverse mental health outcomes (Soga et al., 2021). A similar study in Beijing (China) indicated that the population showed a greater appreciation for urban green spaces during the pandemic (Zhu & Xu, 2020). Likewise, an international survey conducted during the pandemic revealed that urban residents perceived an increased need for accessible urban green spaces, mainly for physical exercise, relaxation and nature observation (Ugolini et al., 2020).

GreenCOVID results have shown how during the first wave of the pandemic, participants in Spain—who mostly live in a flat or apartment, which greatly hindered their physical contact with nature—increased their appreciation for urban nature to a greater extent, compared to respondents in England and Ireland (Garrido-Cumbrera et al., 2022). In addition, GreenCOVID also revealed that the number of people living in flats or appartments in Spain was high (72.1%)—compared to only 12.7% in England and 12.9% in Ireland—reducing the possibilities for interaction with the outside world in this country. This explains the fact that in England/Ireland people had a greater access to private gardens inside the home (Garrido-Cumbrera et al., 2021). Home confinement meant that people living in flats, lacking outdoor spaces other than windows and balconies, missed access to natural environments more than those living in houses with access to private gardens (Garrido-Cumbrera, 2021). In fact, living in flats or apartments without views or with limited views of natural elements from their windows or balconies and poorer appreciation of views have had a significant impact on people's mental health and well-being during the pandemic. Furthermore, people who did not use outdoor spaces and window views and lacked contact with the outdoors were at far greater risk of anxiety or depression (Garrido-Cumbrera et al., 2022).

It should also be noted that the launch of the survey coincided with the most restrictive period of the first wave of the COVID-19 pandemic, when most people living in Spain were prevented from leaving home whereas there were far less restrictive measures in England and Ireland. In these cases, the population was allowed to move within certain zones of influence, close to their homes. In general, our survey also showed that the COVID-19 pandemic affected more women and young people (Garrido-Cumbrera et al., 2022), but did also impact those living alone and/or in nursing homes, such as the elderly and people with disabilities (Giri et al., 2021).

One of the strengths of the GreenCOVID study is the large sample size gathered in three European countries during the most severe period of the first wave of the COVID-19 pandemic. In addition to analysing factors associated with poor emotional well-being, sociodemographic, lifestyle and home environment characteristics were studied based on seven different factors. A limitation of the study was that the sample was mainly from Spain, compared with the UK, and Ireland, so the results cannot be extrapolated to the general population of these three countries. Another limitation had to do with the different survey periods in each country, as different timing and different lockdown policies adopted by the governments of each of the three countries may have affected the results. In short, the results of this study reaffirm the importance of visual contact with the natural environment from home in mitigating and coping with the negative effects on the mental health and well-being of the population. Visual contact with the natural environment is especially vital during periods of limited contact with the outside world, such as during the first wave of the COVID-19 pandemic.

Conclusions

House windows should not be seen as elements that only serve to allow light and air to enter, but as places that provide contact with the outside world, indirect exposure to nature and, ultimately, elements with capacity for recovery or restoration. Views from home have been fundamental in coping with the effects of the restrictive measures adopted to contain the spread of the SARS-CoV-2 virus. The value of views from home windows has been higher during times of home confinement and lower in places where there were no restrictions on walking and access to green or blue spaces.

Considering that window views are a fundamental aspect of the home, especially in crisis situations, improving people's views from their homes should become a priority as a key intervention to reduce mental health disorders and improve emotional well-being. This would involve increasing urban vegetation by adding plant species (trees, shrubs, herbs and flowering plants) on residential streets, as well as establishing urban planning and design guidelines that include the quality of views from residential dwellings. It is, therefore, incumbent upon urban planners and policy makers to design greener cities where more citizens can have access to views of nature and landscapes from their homes.

REFERENCES

Batool, A., Rutherford, P., McGraw, P., Ledgeway, T., & Altomonte, S. (2021). Window views: Difference of perception during the COVID-19 lockdown. *LEUKOS, 17*(4), 380–390. https://doi.org/10.1080/15502724.2020.185 4780

Bond, M., Seiler, V., & Seiler, M. (2002). Residential real estate prices: A room with a view. *Journal of Real Estate Research, 23*(1–2), 129–138. https://doi.org/10.1080/10835547.2002.12091077

Braçe, O., Garrido-Cumbrera, M., Foley, R., Correa-Fernández, J., Suárez-Cáceres, G., & Lafortezza, R. (2020). Is a view of green spaces from home associated with a lower risk of anxiety and depression? *International Journal for Environmental Research and Public Health, 17*(19), 1–9. https://doi.org/10.3390/ijerph17197014

Collins, B. (1975). *Windows and people: A literature survey. Psychological reaction to environments with and without windows.* Department of Commerce/National Bureau of Standards.

Dzhambov, A. (2018). Residential green and blue space associated with better mental health: A pilot follow-up study in university students. *Archives of Industrial Hygiene and Toxicology, 69*, 340–349. https://doi.org/10.2478/aiht-2018-69-3166

Elsadek, M., Liu, B., & Xie, J. (2020). Window view and relaxation: Viewing green space from a high-rise estate improves urban dwellers' wellbeing. *Urban Forestry and Urban Greening, 55*, 126846. https://doi.org/10.1016/j.ufug.2020.126846

Fullana, M., Hidalgo-Mazzei, D., Vieta, E., & Radua, J. (2020). Coping behaviors associated with decreased anxiety and depressive symptoms during the COVID-19 pandemic and lockdown. *Journal of Affective Disorders, 275*, 80–81. https://doi.org/10.1016/j.jad.2020.06.027

Garrido-Cumbrera, M., Foley, R., Braçe, O., Correa-Fernández, J., López-Lara E., Guzman V., Gonzalez-Marin, A., & Hewlett, D. (2021). Perceptions of change in the natural environment produced by the first wave of the COVID-19 pandemic across three European countries. Results from the GreenCOVID study. *Urban Forestry and Urban Greening, 64*, 127260. https://doi.org/10.1016/j.ufug.2021.127260

Garrido-Cumbrera, M., Foley, R., Correa-Fernández, J., González-Marín, A., Braçe, O., & Hewlett, D. (2022). The importance for wellbeing of having views of nature from and in the home during the COVID-19 pandemic. Results from the GreenCOVID study. *Journal of Environmental Psychology, 83*, 101864. https://doi.org/10.1016/j.jenvp.2022.101864

Garrido-Cumbrera, M., González-Marín, A., Correa-Fernández, J., Braçe, O., & Foley, R. (2023). Can views and contact with nature at home help combat anxiety and depression during the pandemic? Results of the GreenCOVID study. *Brain Behaviour, e2875*. https://doi.org/10.1002/brb3.2875

Gilchrist, K., Brown, C., & Montarzino, A. (2015). Workplace settings and wellbeing: Greenspace use, and views contribute to employee wellbeing at peri-urban business sites. *Landscape and Urban Planning, 138*, 32–40. https://doi.org/10.1016/j.landurbplan.2015.02.004

Giri, S., Chenn, L., & Romero-Ortuno, R. (2021). Nursing homes during the COVID-19 pandemic: A scoping review of challenges and responses. *European Geriatric Medicine, 12*(6), 1127–1136. https://doi.org/10.1007/s41 999-021-00531-2

Guzman, V., Garrido-Cumbrera, M., Braçe, O., Hewlett, D., & Foley, R. (2020). Health and Wellbeing under COVID-19: The GreenCOVID Survey. *Irish Geography, 53*(2), 157–162. https://doi.org/10.2014/igj.v53i2.1420

Hartig, T. (2021). Restoration in nature: Beyond the conventional narrative. In A. R. Schutte, J. C. Torquati & J. R. Stevens (Eds, *Nature and Psychology*. Nebraska Symposium on Motivation, vol. 67. Springer, Cham. https://doi.org/10.1007/978-3-030-69020-5_5

Honold, J., Lakes, T., Beyer, R., & Van der Meer, E. (2016). Restoration in urban spaces: Nature views from home, greenways, and public parks. *Environment and Behavior, 48*(6), 796–825. https://doi.org/10.1177/001391 6514568556

Kaplan, R. (1983). The role of nature in the urban context. In I. Altman & J. F. Wohlwill (Eds.), *Behavior and the natural environment* (pp. 127–162). Plenum.

Kaplan, R. (2001). The nature of the view from home: Psychological benefits. *Environment and Behavior, 33*(4), 507–542. https://doi.org/10.1177/001 3916012197311

Ko, W., Schiavon, S., Zhang, H., Graham, L., Brager, G., Mauss, I., & Lin, Y. (2020). The impact of a view from a window on thermal comfort, emotion, and cognitive performance. *Building and Environment, 175*, 106779. https://doi.org/10.1016/j.buildenv.2020.106779

Li, D., & Sullivan, W. (2016). Impact of views to school landscapes on recovery from stress and mental fatigue. *Landscape and Urban Planning, 148*, 149–158. https://doi.org/10.1016/j.landurbplan.2015.12.015

Markus, T. (1967). The function of windows—A reappraisal. *Building Science, 2*(2), 97–121. https://doi.org/10.1016/0007-3628(67)90012-6

Masoudinejad, S., & Hartig, T. (2020). Window view to the sky as a restorative resource for residents in densely populated cities. *Environment and Behavior,* 52(4), 401–436. https://doi.org/10.1177/00139165188072

Pouso, S., Borja, Á., Fleming, L., Gómez-Baggethun, E., White, M., & Uyarra, M. (2020). Contact with blue-green spaces during the COVID-19 pandemic lockdown beneficial for mental health. *Science of the Total Environment, 756,* 143984. https://doi.org/10.1016/j.scitotenv.2020.143984

Soga, M., Evans, M. J., Tsuchiya, K., & Fukano, Y. (2021). A room with a green view: The importance of nearby nature for mental health during the COVID-19 pandemic. *Ecological Applications, 31*(2), e2248. https://doi.org/10.1002/eap.2248

Ugolini, F., Massetti, L., Calaza-Martínez, P., Cariñanos, P., Dobbs, C., Ostoić, S. K., Marin, A. M., Pearlmutter, D., Saaroni, H., Šaulienė, I., & Simoneti, M. (2020). Effects of the COVID-19 pandemic on the use and perceptions of urban green space: An international exploratory study. *Urban Forestry and Urban Greening, 56,* 126888. https://doi.org/10.1016/j.ufug.2020.126888

Ugolini, F., Massetti, L., Pearlmutter, D., & Sanesi, G. (2021). Usage of urban green space and related feelings of deprivation during the COVID-19 lockdown: Lessons learned from an Italian case study. *Land Use Policy, 105,* 105437. https://doi.org/10.1016/J.LANDUSEPOL.2021.105437

Vemuri, A., Grove, J., Wilson, M., & Burch, W. (2011). A tale of two scales: Evaluating the relationship among life satisfaction, social capital, income, and the natural environment at individual and neighborhood levels in metropolitan Baltimore. *Environment and Behavior, 43,* 3–25. https://doi.org/10.1177/0013916509338

Zhu, J., & Xu, C. (2020). Sina microblog sentiment in Beijing city parks as measure of demand for urban green space during the COVID-19. *Urban Forestry and Urban Greening, 58,* 126913. https://doi.org/10.1016/j.ufug.2020.126913

Impacts and Lessons Learned from the COVID-19 Pandemic for Protected and Conserved Area Management

Mitali Sharma, Mariana Napolitano Ferreira, Rachel Golden Kroner, and Mohammad K. S. Pasha

M. Sharma (✉)
Independent Consultant for Environmental Organisations,
Singapore, Singapore
e-mail: mitalisharma.6776@gmail.com

M. N. Ferreira
WWF-Brasil, Brasília, Brazil
e-mail: marianaferreira@wwf.org.br

R. Golden Kroner
World Wildlife Fund US, Washington, D.C, USA
e-mail: rachel.goldenkroner@wwfus.org

M. K. S. Pasha
International Union for Conservation of Nature (IUCN), Asia Regional Office,
Bangkok, Thailand
e-mail: khalid.pasha@iucn.org

© The Author(s) 2024   243
N. Finneran et al. (eds.), *Managing Protected Areas*,
https://doi.org/10.1007/978-3-031-40783-3_14

14.1   INTRODUCTION

The Coronavirus Disease 2019 (COVID-19) pandemic, caused by the SARS-CoV-2 virus that emerged in December 2019 (Zhou et al., 2020), has had numerous devastating impacts worldwide, the most significant being the death of over 6.5 million people (Word Health Organization, 2022). One of the less discussed impacts is the effect that this pandemic has had on protected and conserved areas (PCAs) and how to implement any lessons learned for improved PCA management. Protected areas (PAs), which include national parks, nature reserves, and more, are defined as '...a clearly defined geographical space, recognized, dedicated and managed, through legal or other effective means, to achieve the long-term conservation of nature with associated ecosystem services and cultural values' (Dudley, 2008). These areas are highly important to consider because not only do they conserve nature and biodiversity, but they also provide economic value to many communities, and can improve physical and mental well-being (Hockings et al., 2020).

Additionally, PAs provide food, clean water and medicines, and can buffer the effects of climate change: it is estimated that PAs worldwide store at least 12% of carbon on land (IUCN WCPA, 2021). Aside from PAs, conserved areas, which may not have the same level of restrictions as PAs or a primary conservation objective, are also important for conservation and livelihoods and can provide similar benefits. The term 'conserved areas' could be used more generally or could include 'other effective area-based conservation measures' (OECMs), which were defined by the Convention of Biological Diversity (2018) as 'a geographically defined area other than a Protected Area, which is governed and managed in ways that achieve positive and sustained long-term outcomes for the in situ conservation of biodiversity, with associated ecosystem functions and services and, where applicable, cultural, spiritual, socioeconomic and other locally relevant values'. OECMs can cover a range of areas, such as Indigenous territories, fishing refuges and others, as long as they meet the IUCN World Commission on Protected Areas (WCPA) criteria for an OECM (IUCN WCPA, 2022) and are approved for addition in the UN Environment Programme World Conservation Monitoring Centre (UNEP-WCMC) Protected Planet database.

PCAs are also important for the 'One Health' approach that is growing in popularity, which is defined as 'a cross-sectoral and transdisciplinary approach that emphasizes the fundamental ways in which the health of humans, domestic and wild animals, fungi, plants, microbes, and natural and built ecosystems are interdependent' (Redford et al., 2022). This is particularly relevant in the context of COVID-19 as PCAs are an important form of protection against epidemics and pandemics as they maintain ecosystem integrity (Dobson et al., 2020) and can thereby suppress pathogen spillover—the process by which pathogens from animals 'jump' into humans (Reaser et al., 2021). It is estimated that 72% of zoonotic diseases have originated from wildlife (as compared to those from domestic animals) (Jones et al., 2008); PCAs act as a buffer and help reduce human exposure to emerging zoonotic infectious diseases (Ferreira et al., 2021) as they limit the contact between humans and other wild species that could transmit disease; conversely, land clearing leads to an increased risk of zoonotic disease transmission, especially in the tropics (Allen et al., 2017).

Given the clear value of PCAs, it is critical to monitor and assess the impacts of the COVID-19 pandemic on them. Insights from such an assessment can provide lessons to support strategic management decisions on PCA recovery and rebuilding, including improved management in the future. Assessments and recovery are needed across three categories: ecological, social and economic, each building on the one before. This approach will ensure that multiple key aspects of the issue are targeted for relief and recovery. Authorities, PCA managers, rangers, Indigenous peoples, local communities and other relevant stakeholders need to be involved at a site level in the rebuilding process to design adaptive and appropriate responses. Aside from the on-site support, knowledge generation on the lessons learned is also key to understanding how to prevent and address negative impacts in the future.

This chapter aims to piece together the expertise on this topic in the style of a review and further elucidate the effect of three major types of impacts from the COVID-19 pandemic—ecological, social and economic on PCAs (Fig. 14.1), including regional differences, where information was available. We draw insights from scientific articles and literature published by conservationists worldwide and from a dedicated IUCN WCPA 'COVID-19 and Protected Areas' Task Force. This chapter also discusses the lessons learned and their implications for public policies and improving PCA management, highlighting the IUCN Green List.

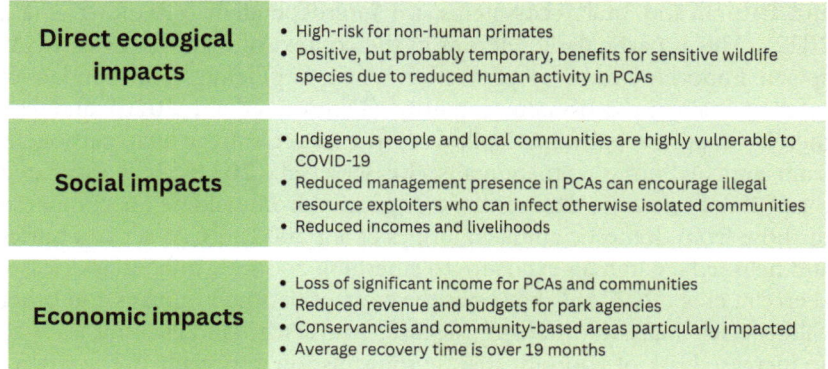

Fig. 14.1 Summary of the ecological, social and economic impacts of the COVID-19 pandemic on PCAs

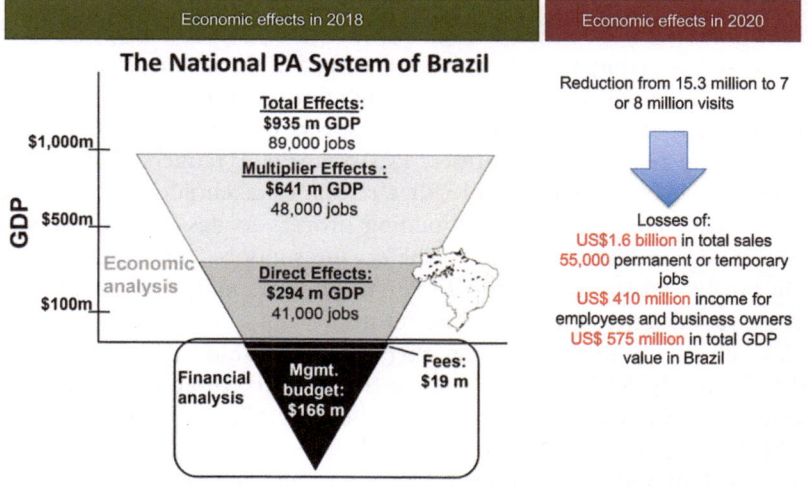

Analysis by Thiago Sousa, in Spenceley et al. 2021 in PARKS

Fig. 14.2 Economic impacts of the pandemic related to tourism in the Brazilian PA system

14.2 Ecological Impacts and Solutions

The COVID-19 pandemic has had several conflicting ecological impacts. Initially, it seemed as if nature could thrive again when lockdowns started occurring worldwide, as there had been many reports of wildlife re-occurring at sites and increases in species richness during these lockdowns (Manenti et al., 2020). Although there were positive effects including the increased breeding success of certain birds that are sensitive to human disturbance, and reduced road killings, there were also increases in invasive species and illegal hunting and fishing (Bennett et al., 2020; Manenti et al., 2020). This highlights the need to consider the various ecological implications of changes to human activities—especially for drastic changes such as during the lockdowns—on ecosystem functioning (Gilby et al., 2021). The negative ecological effects can be attributed to a reduction in patrolling, which greatly reduced the likelihood of detecting and responding to threats (Corlett et al., 2020). Researchers observed this globally, including with the illegal killing of birds on Italian islands (Manenti et al., 2020), and with wildlife poaching more than doubling in countries including Uganda and India during lockdowns (Athumani, 2020; Badola, 2020). In Bangladesh, poaching increased by 28 times during the 2020 lockdowns compared to 2019 (Rahman, 2021). However, there were reports that poaching decreased in other places, such as South Africa (Hockings et al., 2020).

Illegal land-use change was also a major ecological issue during the lockdowns. There were reports that illegal logging and natural resource extraction increased greatly in Nepal and Tunisia (Hockings et al., 2020). Countries with biodiversity-rich tropical forests experienced increased land clearing and mining as well (McNeely, 2021). Many parts of Asia, Africa and South America reported that deforestation increased during the pandemic (Fair, 2020), which has long-term implications for climate change; deforestation accounts for roughly ten per cent of anthropogenic greenhouse gas emissions (OECD, 2020) and even restoring these forests would not replace the carbon storage value of the older forests lost (Gibson et al., 2011). In the Brazilian Amazon, forest clearing increased by 28% in 2020 compared to the previous year (Escobar, 2020). Most of the land that had been cleared was changed into pastures for grazing cattle to support the beef industry in Brazil (McNeely, 2021).

In Bangladesh, these values were much higher—the number of forest loss alerts increased by 2,700% during the May 2020 lockdown period

compared to the same period in 2019 (Rahman, 2021); this under-
scores the importance of considering regional variations. These statistics
are concerning, as land-use change due to resource extraction or agriculture is the driving cause for zoonotic pathogen emergence (Ferreira
et al., 2021) and has caused more than 30% of new diseases reported
since 1960 (IPBES, 2020). Supporting PCAs and the achievement of
global targets such as 30 × 30 (to protect 30% of terrestrial and marine
spaces by 2030) would help limit land-use change if there are effective
and equitable measures in place to protect land.

New regulations on wildlife markets and the wildlife trade have resulted
in another ecological impact from the COVID-19 pandemic, as the virus
is thought to have originated in a market (Huanan Seafood Wholesale
Market) where live wild and farmed animals were traded (Worobey et al.,
2022). Markets with live animals, especially those under unregulated and
poor sanitary conditions, are an ideal location for pathogens to spread
because they contain stressed animals of different species from different
locations in stacked and overcrowded cages, all interacting with humans
(Aguirre et al., 2020). Even with the wildlife bans placed in China after
the pandemic started, people have found legal loopholes, as the medicinal
use of wildlife—which includes many species such as pangolins, bats and
tigers—is not covered by the ban (Wang et al., 2020). These species all
play unique roles in their ecosystems and impact the PCAs they live in.
For instance, bats act as biological and economical pesticides, and are
important pollinators and seed dispersers (Zhao, 2020). The suggestion
that the pandemic started from bats has hurt their reputation and placed
them at greater risk for actions such as mass slaughter and removal (Zhao,
2020), and it is unclear whether this will be a short or long-term impact.

That is why public education is important for both wildlife and
ecosystem conservation (Zhao, 2020); it is crucial for the media and other
platforms to improve their communication of the relationships between
nature, the pandemic and society—misleading narratives can place further
pressure on vulnerable ecosystems and species. Messages should be framed
with nature as the solution and not the problem, and the impacts of
human activities should be highlighted with clear calls to action (Gregg
et al., 2021). It is important to consider the wildlife trade when discussing
PCAs as the animals from this trade could be taken from PCAs, and if
another pandemic emerges due to the wildlife trade, PCAs will be at risk
again. PCA managers should work with authorities to establish strict legislation against illegally taking species from their premises. In addition, it is

important to have strong monitoring and enforcement systems in place, and consider means to avoid their disruption during a future pandemic.

Another group of animals that are at greater risk because of the pandemic are non-human primates such as apes, which are likely to be susceptible to many viruses that impact humans, such as Ebola and the SARS-CoV-2 virus (Melin et al., 2020). Primates play a key role in tropical biodiversity, forest regeneration and ecosystem health (Estrada et al., 2017). This is because many primates are frugivorous and can disperse seeds over long distances (Chapman et al., 2013). Apes that are habituated to humans, such as mountain gorillas, are at an even greater risk—an outbreak could devastate these gorillas and their ecosystem (Gillespie & Leendertz, 2020). Therefore, it is recommended that PCA managers add measures to limit or ban contact with great apes (Gillespie & Leendertz, 2020) with much greater caution and safety measures in place to protect them.

14.3   Social Impacts and Solutions

In terms of the social impacts, it is important to consider the people that are directly involved with PCAs, such as rangers, local communities and Indigenous peoples. Rangers play a critical role for PCAs as they are on the frontlines protecting these areas from threats including illegal logging and hunting (Singh et al., 2020). In some countries, the pandemic resulted in rangers getting fired due to budget reductions from tourism and other funding sources, which adversely affected their livelihoods and reduced protection for the areas they worked in (Hockings et al., 2020). These trends contributed to the stress of unemployment, increased anxiety from job insecurity and reduced ranger welfare (Singh et al., 2020; Smith et al., 2021). As for the rangers who have been working during the pandemic, a study on ranger welfare (Singh et al., 2020) revealed that a significant proportion of rangers believed that the pandemic increased threats to PCAs and negatively impacted their life and work—a job that already contained numerous challenges before the pandemic (Belecky et al., 2019; Singh et al., 2020). This is because rangers were required to work longer hours and spend less time with their families due to staff cuts and increased threats to PCAs (Singh et al., 2020). This could be a short-term impact if management measures and policies swiftly improve to support rangers. Therefore, it is recommended that greater emphasis should be placed on their well-being and funding.

Additionally, Appleton et al. (2022) recently found that personnel and ranger numbers are insufficient for global targets such as 30 × 30, which likely plays a major role in current management deficiencies; therefore, it is crucial to keep current staff on board where possible.

In the same survey on rangers (Singh et al., 2020), it should be noted that more than four out of five rangers in Asia, Africa and Latin America believe that their job success is dependent on the help of local communities, which were severely impacted by the pandemic. The communities that live near PCAs typically benefit from tourism in some way and rely on it for their livelihoods, such as by receiving a proportion of the PCA fees (Maekawa et al., 2015) or from their own businesses; as tourism was heavily impacted by the pandemic, local communities were negatively impacted due to a reduced source of income (Hockings et al., 2020). Local communities may also contribute to PCA management, thereby serving as critical stewards for conservation.

Before the pandemic, many local communities were already facing extreme poverty and other challenges including food security and human–wildlife conflict; the pandemic exacerbated these struggles (Bhammar et al., 2021). Food security and human–wildlife conflict are important to consider as local communities and Indigenous peoples could be driven to hunting and consuming wild animals, which would not only affect the conservation of certain species and, therefore, impact their surrounding habitats, but could also expose individuals to zoonotic diseases; one example is the Ebola virus outbreak from 2013 to 2016 that originated in Western Africa (Koh et al., 2021). These interactions could lead to further problems for health and livelihoods, even though wild meat is an important source of nutrition in rural areas (Friant et al., 2020). However, solutions to this should be considered carefully, as other foods could also contribute to threats to PCAs due to habitat loss from land-use change. For instance, communities in the African Congo consume around 5 million tons of wild meat annually—the same amount meat by cattle ranching would require converting up to 25 million hectares of forest into farmland, an area roughly the size of Great Britain (Cooney & Nasi, 2014). Since wild meat includes many types of animals, it could be worth improving education on which species are safer and more sustainable to consume.

Another way to support communities local to PCAs is by increasing benefit sharing. Not only can benefit-sharing arrangements increase success for local communities, but they can also help achieve conservation goals (Snyman & Bricker, 2019); benefit sharing increases the likelihood that communities will view PCAs positively and conserve their natural resources (Leung et al., 2018; Spenceley et al., 2017). Examples of benefit sharing include formalising revenue sharing, building capacity and skills, reducing human–wildlife conflict through mitigation or compensation, hiring local individuals for PCA management, increasing local sourcing for goods and offering grants to businesses (Bhammar et al., 2021). It is also important to include local communities when developing solutions for them so that they feel more empowered and motivated to protect these areas (Stolton et al., 2021).

Along with the local communities residing near PCAs, many PCAs overlap or share limits with Indigenous territories as well. Approximately 50% of Earth's lands are occupied by Indigenous peoples and local communities, and their lands have less deforestation and lower emissions than other spaces, with substantial biodiversity value (Garnett et al., 2018; Sze et al., 2022). Therefore, it is important to recognise Indigenous peoples' rights in these areas and their traditional knowledge systems. Indigenous peoples should be included in decision-making processes for their spaces as well, as they have been historically underrepresented and marginalised in conservation policy decisions (Forest Peoples Programme et al., 2020).

Aside from these aspects, the pandemic has also had a much broader social impact on mental health linked to access to natural spaces. Before the pandemic, people went to parks for recreation and education, but during the pandemic, people started visiting national parks and green spaces to maintain their mental and physical well-being (Kleinschroth & Kowarik, 2020; Miller-Rushing et al., 2021). For example, when restrictions were gradually lifted in European countries in the summer of 2020, visitor numbers increased significantly (McGinlay et al., 2020). This phenomenon demonstrates the importance of PCAs for people to manage stress and restore their mental and physical health during (and after) the COVID-19 pandemic (Mandić, 2021).

14.4 Economic Impacts and Solutions

The pandemic has had severe economic consequences for PCAs. Tourism is the most common use of PCAs and their largest financial contributor (Mandić, 2021; Spenceley et al., 2017); it contributes to gross domestic product (GDP), livelihoods, conservation funding (Snyman &

Bricker, 2019) and benefits local economies in numerous ways: money from tourists contributes to employment and businesses such as restaurants and tour services, which can enable individuals to learn new skills that can be applied to other industries as well (Leung et al., 2018). The World Travel & Tourism Council (WTTC) found that the COVID-19 pandemic caused a global loss of 62 million tourism jobs (not specific to PCAs) and $4.9 trillion USD from tourism's contribution to GDP in 2020, with some improvement in 2021 (WTTC, 2022). Historically, terrestrial PCAs received approximately 8 billion visits annually (Balmford et al., 2015). The reduction in tourists due to pandemic travel restrictions was particularly severe for places that depend on tourism in Africa and South America (Hockings et al., 2020; Spenceley et al., 2021; Fig. 14.2); monthly surveys of African safari tour operators revealed that over 90% of them had experienced more than 75% fewer bookings or had no bookings at all since April 2020, and the number of bookings still had not fully recovered in their last survey in May 2022, though it did improve (Beekwilder, 2022). Therefore, the impacts from this could be considered short to medium term, depending on when there is a full recovery in tourism there.

Funding is already a serious problem for PCAs, which results in poor management and issues with achieving conservation objectives (Bhammar et al., 2021). It is estimated that only 20% of PCAs are managed properly, despite their significant importance (Dasgupta, 2021). PCAs that are properly managed can advance social development in the form of fair employment, sustainable food production and safe drinking water access (Stolton et al., 2015). Therefore, increased funding for PCAs is required to reach societal goals; it is estimated that 140 billion USD annually could protect 30% of terrestrial and marine areas effectively by 2030, which is only 0.16% of the global GDP, and less than one-third of the subsidies provided to activities that harm nature (Waldron et al., 2020). Protecting 30% of these areas could generate up to 454 billion USD per year in revenue for four sectors (PAs/nature, agriculture, forestry and fisheries) by 2050, and the avoided-loss value of ecosystem services could be 170–534 billion USD per year by 2050 due to avoided flooding, climate change mitigation, soil loss prevention and storm protection (data based only on mangroves and forests; the value including the other biomes would be higher) (Waldron et al., 2020). Thus, the returns could be over seven times greater than the investment needed, including both the avoided loss and the revenue values.

In especially vulnerable regions such as Africa, nearly all PCAs lack proper funding; it is estimated that more than 1 billion USD is required annually to save the iconic species and habitats there (Lindsey et al., 2018). Increased funding can improve management effectiveness by hiring and training staff, increasing infrastructure investment and promoting outreach (Bhammar et al., 2021). Staff training is particularly important, not only in terms of the policies to effectively manage PCAs, but also in terms of their commercial expertise to ensure that business and financial requirements can be properly addressed (Bhammar et al., 2021; Stolton et al., 2021). In addition, increased funding would have much wider implications, including lowering the risk of future pandemics. It is estimated that the costs to monitor and prevent disease spillover across a ten-year period would be only two per cent of the estimated costs of the COVID-19 pandemic (Dobson et al., 2020). Thus, the benefits for society are estimated to significantly exceed the costs of increasing PCA funding (Waldron et al., 2020).

Despite the need for additional funding and regulatory support for PCAs, a recent analysis showed that 16 out of 20 major economies invested in activities that undermined environmental protection measures instead of supporting them, as part of their pandemic recovery efforts (Golden Kroner et al., 2021). Additionally, at least 22 countries rolled back or weakened their environmental protection for PCAs or reduced budgets. Rollbacks for PCAs are commonly due to new authorisations for activities such as new industrial plants or housing development, and they have been increasing over the past two decades, including protected area downgrading, downsizing, and degazettement (PADDD) (Golden Kroner et al., 2019). These rollbacks have occurred at times when the public cannot be consulted, including during the lockdowns, and the processes to justify PADDD lack rigour compared to those required to create PCAs (Pack et al., 2016; Golden Kroner et al., 2021). Nonetheless, there are still certain countries that are supporting PCAs during the pandemic; for example, Kenya pledged support for PCAs by promoting tourism, with the employment of 5,500 community scouts ($9.2 million USD) and 160 community conservancies ($9.2 million USD). Additionally, Pakistan created a Green Stimulus Initiative, which includes plans to expand PAs and add 15 national parks that cover 7,300 km^2 (supported with $24 million USD), create Pakistan's first National Parks Service, and around 5,000 new jobs (Golden Kroner et al., 2021).

Even though tourism will remain important for PCAs, it is important to reduce over-reliance on this sector and build resilience for PCAs. Therefore, more diverse and sustainable financing sources such as a combination of conservation trust funds, impact bonds and payments for ecosystem services are required in case one or more methods fail, especially in emergency situations such as the pandemic (Bhammar et al., 2021; Spenceley et al., 2021; Stolton et al., 2021). Another type of funding support is through official development assistance (ODA), defined as government aid that promotes and specifically targets the economic development and welfare of developing countries, which has proven to be a key resource in past emergencies and could be critical for protecting biodiversity-vulnerable nations (OECD, 2020). Additionally, subsidies that harm biodiversity and the environment, such as those for agriculture and fisheries, should be redirected to environmental conservation, including PCAs (Golden Kroner, 2021).

A promising approach to improve economic development linked to PCAs is through a collaborative relationship between communities, PCA managers and businesses (Stolton et al., 2021). Public–private partnerships (PPPs) and collaborative management partnerships (CMPs) between authoritative bodies, such as governments, and NGOs or other private bodies, have also shown to be effective tools for PCA management that have led to greater funding from increased donor confidence (Lindsey et al., 2021). It should be noted that each PCA exists in its own context and requires tailored approaches to increase economic development in its region (Stolton et al., 2021). In times of limited funding, resources should be maintained to continue to support staff to monitor and enforce protection and restoration, especially in places with high biodiversity and intact forests (Golden Kroner et al., 2021).

14.5   REGIONAL DIFFERENCES

It is clear that there were regional differences in the ecological, social and economic impacts of the pandemic on PCAs; Waithaka et al. (2021) found that, broadly, PCAs in wealthier nations have been able to manage the situation better than those in poorer nations. More specifically, the least affected regions were Europe, Oceania and North America, whereas the most severely affected PCAs were in Latin America and Africa. Eastern and Southern Africa were the most affected, and PCAs in Asia were moderately affected (Waithaka et al., 2021). However, this could also be due to underreporting from parts of Southeast and Central Asia. Based on the findings of Waithaka et al. (2021) and Hockings et al. (2020), the regions that appear to need the most immediate assistance are Latin America and Africa. Therefore, plans to help these regions should be prioritised.

These regions can be supported by increasing diverse sources of funding for resources and technology to support online platforms and remote work, as well as training and capacity building, as many countries lacked those; in particular, over 80% of the countries from Africa surveyed (Algeria, Benin, Cameroon, Chad, Ethiopia, Ghana, Guinea-Bissau, Madagascar, Malawi, Mozambique, Niger, Nigeria, Rwanda, Sao Tome and Principe, Seychelles, Somalia, Sudan, Swaziland and Uganda) indicated that their ability to cope with the pandemic was most affected by insufficient funding (Waithaka et al., 2021), which is why this aspect is especially important. In addition, emergency response guidelines and contingency plans need to be improved upon, which applies to all PCAs worldwide.

14.6   Green List Guidance in the Recovery of PCAs

One way that PCAs can begin to recover effectively from the pandemic is by using the Green List (IUCN & WCPA, 2017). The Green Listing mechanism enables PCAs to evaluate their challenges and bottlenecks and take corrective measures to remove impediments by fulfilling the criteria and components of the Green List framework. Combatting the impacts of COVID-19 on PCAs involves not only the elimination of high-risk factors, but also the adoption of ecological, social, economic safeguards. The Green List process validates and allows a monitoring mechanism for conservation efforts undertaken in areas and sites afflicted with high ecological, social and economic risks (Wells et al., 2016). There are four components in the Green List: Good Governance, Sound Design and Planning, Effective Management and Successful Conservation Outcomes, all of which can help improve the way PCAs are managed and enhance their contributions to conservation.

The Good Governance component within the Green List framework is the foundation towards building ecological, social and economic recovery pathways for PCAs. This component guarantees legitimacy and voice (criterion 1.1), achieving transparency and accountability (1.2), and enabling governance vitality and capacity to respond adaptively (1.3). The Effective Management component is also crucial when discussing PCA management; this component has seven criteria that include: developing and implementing a long-term management strategy (criterion 3.1), managing ecological condition (3.2), managing within the social and economic context of the site (3.3), managing threats (3.4), effectively and fairly enforcing laws and regulations (3.5), managing access, resource use and visitation (3.6), and measuring success (3.7). These, followed by the other two components, can enable successful conservation in PCAs and provide an international benchmark for quality that motivates improved performance and helps catalyse ecological, social and economic recovery in a global network of PCAs, as well as their revitalisation and expansion. Ensuring better safeguards in the future for PCAs through the robust Green List mechanisms and rebuilding sites could benefit both people and nature (Hockings et al., 2019; Wells et al., 2016).

14.7 Lessons Learned and Recommendations

It is critical that action is taken to safeguard PCAs, given their ecological, social and economic importance. The COVID-19 pandemic affected all three of these aspects of PCAs negatively in many parts of the world. Therefore, policies and management decisions to support PCAs and their rebuilding from COVID-19 are crucial and will have far greater implications beyond specific sites, such as supporting biodiversity, mitigating climate change, enhancing human health and supporting the economy (Kumar, 2010). It is important to consider a holistic 'One Health' approach (Osofsky et al., 2005) to PCA governance and management, which would require collaboration among experts from different fields, including human, animal and environmental health, to design and implement actions, policies and legislation that reflect research in this field (McNeely, 2021). It would also be beneficial for PCAs to consider using the Green List as a tool to help recover effectively (Hockings et al., 2019). Therefore, to summarise the key recommendations that were mentioned throughout this chapter, categorised as responses to the ecological, social and economic impacts of COVID-19 on PCAs:

Responses to Address the Ecological Impacts:

1. PCAs should be governed and managed effectively, supported by adequate regulations, institutions and funding, using a One Health approach. This approach could involve partnerships across sectors to monitor zoonotic diseases around PCAs and bans of certain visitor interactions with animals around PCAs, especially vulnerable species like great apes or those considered high risk, to reduce the chances of spillover. Tools including the IUCN Green List can support monitoring and encourage improvements to governance and management.

2. Legal protection measures, regulations and management to prevent illegal logging, poaching and wildlife being taken from PCAs (e.g. for live animal markets) must be strengthened; this will require the development of effective legislation and investment into management personnel and surveillance equipment. Community-led monitoring may also support this effort.

3. Habitat restoration and connectivity efforts should be intensified due to the increased deforestation during the pandemic, as well as the land-use change from intensive agriculture. Regions with the most land-use change should be prioritised. This will also help build ecological resilience and help prevent future pandemics.

Responses to Address the Social Impacts:

1. Working conditions for rangers and PCA staff should be improved, including humane policies (e.g. fair working hours). PCAs should try to retain as many staff members as possible, even during emergency situations like pandemics to prevent illegal activities in PCAs from increasing.

2. Benefit sharing should be increased for local communities by formalising revenue sharing in the law and creating more requirements to hire local staff in PCAs.

3. Indigenous peoples and local communities should be included in decision-making processes for PCAs that overlap with their territories for more inclusive governance. Their rights and Indigenous knowledge systems should also be respected more, with measures in place to uphold these.

Responses to Address the Economic Impacts:

1. Sustainable and diverse financing mechanisms should be in place, and emergency funds and ODA mechanisms should be created for PCAs that are heavily dependent on tourism, including those in Latin America, Africa and parts of Asia where tourism plays an integral role. Budget rollbacks that affect PCAs should also be avoided where possible.
2. A significant amount of funding needs to be provided to PCAs in Africa and Latin America to protect their unique and especially vulnerable wildlife and habitats, which should be prioritised based on the severe impacts the pandemic has had there.
3. There should be an annual investment from various sources of at least 140 billion USD into PCAs worldwide, as this could protect 30% of terrestrial and marine areas effectively by 2030, and the returns could be over seven times higher (including the projected revenues and avoided-loss values of $454 billion and $534 billion USD, respectively; Waldron et al., 2020).

Based on the ecological, social and economic impacts highlighted in this chapter, it is clear that each aspect plays a prominent role in the future of PCAs in different ways, and different management approaches should be taken to address each. Global policies, funding mechanisms and management plans should prioritise the more vulnerable regions first, namely, Latin America and Africa, to ensure that they receive adequate support to help their wildlife, communities and economies recover from the effects of the COVID-19 pandemic.

REFERENCES

Aguirre, A., Catherina, R., Frye, H., & Shelley, L. (2020). Illicit wildlife trade, wet markets, and COVID-19: Preventing future pandemics. *World Medical & Health Policy, 12*(3), 256–265. https://doi.org/10.1002/wmh3.348

Allen, T., Murray, K., Zambrana-Torrelio, C., Morse, S., Rondinini, C., Di Marco, M., et al. (2017). Global hotspots and correlates of emerging zoonotic diseases. *Nature Communications, 8*(1), 1124. https://doi.org/10.1038/s41467-017-00923-8

Appleton, M., Courtiol, A., Emerton, L., Slade, J., Tilker, A., Warr, L., et al. (2022). Protected area personnel and ranger numbers are insufficient to deliver global expectations. *Nature Sustainability, 5*, 1100–1110. https://doi.org/10.1038/s41893-022-00970-0

Athumani, H. (2020). *Wildlife poaching doubles in Uganda during COVID-19 Lockdown.* https://www.voanews.com/covid-19-pandemic/wildlife-poaching-doubles-uganda-during-covid-19-lockdown. Accessed 22 August 2021.

Badola, S. (2020). *Indian wildlife amidst the COVID-19 Crisis: An analysis of status of poaching and illegal wildlife trade.* https://wwfin.awsassets.panda.org/downloads/indian_wildlife_amidst_the_covid_19_crisis.pdf. Accessed 14 February 2023.

Balmford, A., Green, J., Anderson, M., Beresford, J., Huang, C., Naidoo, R., et al. (2015). Walk on the wild side: Estimating the global magnitude of visits to protected areas. *PLOS Biology, 13*, e1002074. https://doi.org/10.1371/journal.pbio.1002074

Beekwilder, J. (2022). *The impact of the Coronavirus pandemic on the safari industry.* https://www.safaribookings.com/blog/coronavirus-outbreak. Accessed 14 November 2022.

Belecky, M., Singh, R., & Moreto, W. (2019). *Life on the frontline 2019: A global survey of the working conditions of rangers.* https://www.worldwildlife.org/publications/life-on-the-frontline-2019-a-global-survey-of-the-working-conditions-of-rangers. Accessed 14 February 2023.

Bennett, N., Finkbeiner, E., Ban, N., Belhabib, D., Jupiter, S., Kittinger, J., et al. (2020). The COVID-19 Pandemic, small-scale fisheries and coastal fishing communities. *Coastal Management, 48*, 336–347. https://doi.org/10.1080/08920753.2020.1766937

Bhammar, H., Li, W., Molina, C., Hickey, V., Pendry, J., & Narain, U. (2021). Framework for sustainable recovery of tourism in protected areas. *Sustainability, 13*, 2798. https://doi.org/10.3390/su13052798

CBD (Convention on Biological Diversity). (2018, July 2–7). Protected areas and other effective area-based conservation measures. Twenty-second meeting Montreal. https://www.cbd.int/doc/c/9b1f/759a/dfcee171bd46b06cc91f6a0d/sbstta-22-1-02-en.pdf. Accessed 4 November 2022.

Chapman, C., Bonnell, T., Gogarten, J., Lambert, J., Omeja, P., Twinomugisha, D., et al. (2013). Are primates ecosystem engineers? *International Journal of Primatology, 34*, 1–14. https://doi.org/10.1007/s10764-012-9645-9

Cooney, D. & Nasi, R. (2014). Ebola and bushmeat in Africa: Q&A with leading researcher. *CIFOR Forests News.* https://forestsnews.cifor.org/23924/ebola-and-bushmeat-in-africa-qa-with-leading-researcher?fnl=. Accessed 22 August 2021.

Corlett, R., Primack, R., Devictor, V., Maas, B., Goswami, V., Bates, A., et al. (2020). Impacts of the coronavirus pandemic on biodiversity conservation. *Biological Conservation, 246*, 108571. https://doi.org/10.1016/j.biocon.2020.108571

Dasgupta, P. (2021). *The economics of biodiversity: The Dasgupta review. Abridged Version.* HM Treasury.

Dobson, A., Pimm, S., Hannah, L., Kaufman, L., Ahumada, J., Ando, A., et al. (2020). Ecology and economics for pandemic prevention. *Science, 369,* 379–381. https://doi.org/10.1126/science.abc3189

Dudley, N. (2008). *Guidelines for applying protected area management categories.* IUCN.

Escobar, H. (2020). Deforestation in the Brazilian Amazon is still rising sharply. *Science, 369,* 613–613. https://doi.org/10.1126/science.369.6504.613

Estrada, A., Garber, P. A., Rylands, A. B., Roos, C., Fernandez-Duque, E., Di Fiore, A., Nekaris, K. A. I., Nijman, V., Heymann, E. W., Lambert, J. E. and Rovero, F. (2017). Impending extinction crisis of the world's primates: Why primates matter. *Science Advances, 3*(1), p.e1600946. https://doi.org/10.1126/sciadv.1600946

Fair, J. (2020). COVID-19 lockdown precipitates deforestation across Asia and South America. *Mongabay Environmental News.* https://news.mongabay.com/2020/07/covid-19-lockdown-precipitates-deforestation-across-asia-and-south-america/. Accessed 22 August 2021.

Ferreira, M., Ellio, W., Golden Kroner, R., Kinnaird, M., Prist, P., Valdujo, P., & Vale, M. (2021). Drivers and causes of zoonotic diseases: An overview. *PARKS, 27,* 15–24. https://doi.org/10.2305/iucn.ch.2021.parks-27-simnf.en

Forest Peoples Programme, International Indigenous Forum on Biodiversity, Indigenous Women's Biodiversity Network, Centres of Distinction on Indigenous and Local Knowledge and Secretariat of the Convention on Biological Diversity (2020). *Local biodiversity outlooks 2: The contributions of Indigenous peoples and local communities to the implementation of the Strategic Plan for Biodiversity 2011– 2020 and to renewing nature and cultures. A complement to the fifth edition of Global Biodiversity Outlook.* Forest Peoples Programme. https://www.cbd.int/gbo/gbo5/publication/lbo-2-en.pdf. Accessed 16 April 2023.

Friant, S., Ayambem, W., Alobi, A., Ifebueme, N., Otukpa, O., Ogar, D., et al. (2020). Eating bushmeat improves food security in a biodiversity and infectious disease 'Hotspot.' *EcoHealth, 17,* 125–138. https://doi.org/10.1007/s10393-020-01473-0

Garnett, S., Burgess, N., Fa, J., Fernández-Llamazares, Á., Molnár, Z., Robinson, C., et al. (2018). A spatial overview of the global importance of Indigenous lands for conservation. *Nature Sustainability, 1*(7), 369–374. https://doi.org/10.1038/s41893-018-0100-6

Gibson, L., Lee, T., Koh, L., Brook, B., Gardner, T., Barlow, J., et al. (2011). Primary forests are irreplaceable for sustaining tropical biodiversity. *Nature, 478*(7369), 378–381. https://doi.org/10.1038/nature10425

Gilby, B., Henderson, C., Olds, A., Ballantyne, J., Bingham, E., Elliott, B., et al. (2021). Potentially negative ecological consequences of animal redistribution on beaches during COVID-19 lockdown. *Biological Conservation, 253*, 108926. https://doi.org/10.1016/j.biocon.2020.108926

Gillespie, T., & Leendertz, F. (2020). COVID-19: Protect great apes during human pandemics. *Nature, 579*, 497–497. https://doi.org/10.1038/d41586-020-00859-y

Golden Kroner, R. (2021). *Nature-based COVID-19 recovery: Investing in Protected and Conserved Areas (PCAs) for planetary hesalth. IUCN WCPA Technical Note Series No.4.* IUCN WCPA.

Golden Kroner, R., Qin, S., Cook, C., Krithivasan, R., Pack, S., Bonilla, O., et al. (2019). The uncertain future of protected lands and waters. *Science, 364*(6443), 881–886. https://doi.org/10.1126/science.aau5525

Gregg, E., Kusmanoff, A., Garrard, G., Kidd, L., & Bekessy, S. (2021). Biodiversity conservation cannot afford COVID-19 communication bungles. *Trends in Ecology and Evolution*, S0169534721001919. https://doi.org/10.1016/j.tree.2021.07.003

Hockings, M., Dudley, N., & Elliott, W. (2020). Editorial Essay: COVID-19 and protected and conserved areas. *PARKS, 26*(1), 7–24. https://doi.org/10.2305/IUCN.CH.2020.PARKS-26-1MH.en

Hockings, M., Hardcastle, J., Woodley, S., Sandwith, T., Wilson, J., Bammert, M., & Miranda Lodoño, J. (2019). The IUCN green list of protected and conserved areas: Setting the standard for effective conservation. *PARKS, 25*(2), 57–66. https://doi.org/10.2305/IUCN.CH.2019.PARKS-25-2MH.en

IUCN and WCPA. (2017). *IUCN Green List of protected and conserved areas: Standard, Version 1.1.* IUCN.

IUCN WCPA. (2021). *World Commission on Protected Areas (WCPA) information note for UNFCCC COP26: Role of Protected and Conserved Areas (PCAs) in attaining the Paris Agreement Target.* IUCN. https://www.iucn.org/sites/default/files/2022-07/pcasclimatechangecop26_1.pdf

IUCN WCPA. (2022). *Site-level tool for identifying other effective area-based conservation measures (OECMs). Version 2.0.* IUCN World Commission on Protected Areas.

IPBES (Intergovernmental Platform on Biodiversity and Ecosystem Services). (2020). *Workshop report on biodiversity and pandemics of the intergovernmental platform on biodiversity and ecosystem services.* IPBES Secretariat. https://doi.org/10.5281/zenodo.414317.

Jones, K., Patel, N., Levy, M., Storeygard, A., Balk, D., Gittleman, J., & Daszak, P. (2008). Global trends in emerging infectious diseases. *Nature, 451*(7181), 990–993. https://doi.org/10.1038/nature06536

Kleinschroth, F., & Kowarik, I. (2020). COVID -19 crisis demonstrates the urgent need for urban greenspaces. *Frontiers in Ecology and the Environment*, *18*, 318–319. https://doi.org/10.1002/fee.2230

Koh, L., Li, Y., & Lee, J. (2021). The value of China's ban on wildlife trade and consumption. *Nature Sustainability*, *4*, 2–4. https://doi.org/10.1038/s41893-020-00677-0

Kumar, P. (Ed.). (2010). *The economics of ecosystems and biodiversity: Ecological and economic foundations*. Earthscan.

Leung, Y.-F., Spenceley, A., Hvenegaard, G., & Buckley, R. (2018). *Tourism and visitor management in protected areas: Guidelines for sustainability*. IUCN.

Lindsey, P., Baghai, M., Bigurube, G., Cunliffe, S., Dickman, A., Fitzgerald, K., et al. (2021). Attracting investment for Africa's protected areas by creating enabling environments for collaborative management partnerships. *Biological Conservation*, *255*, 108979. https://doi.org/10.1016/j.biocon.2021.108979

Lindsey, P., Miller, J., Petracca, L., Coad, L., Dickman, A., Fitzgerald, K., et al. (2018). More than $1 billion needed annually to secure Africa's protected areas with lions. *Proceedings of the National Academy of Sciences*, *115*, E10788–E10796. https://doi.org/10.1073/pnas.1805048115

Maekawa, M., Lanjouw, A., Rutagarama, E., & Sharp, D. (2015). Mountain gorilla ecotourism: Supporting macroeconomic growth and providing local livelihoods. In H. Young, L. Goldman, & J. Egeland (Eds.), *Livelihoods, natural resources, and post-conflict peacebuilding* (pp. 191–210). Taylor and Francis.

Mandić, A. (2021). Protected area management effectiveness and COVID-19: The case of Plitvice Lakes National Park, Croatia. *Journal of Outdoor Recreation and Tourism*, *100397*,. https://doi.org/10.1016/j.jort.2021.100397

Manenti, R., Mori, E., Di Canio, V., Mercurio, S., Picone, M., Caffi, M., et al. (2020). The good, the bad and the ugly of COVID-19 lockdown effects on wildlife conservation: Insights from the first European locked down country. *Biological Conservation*, *249*, 108728. https://doi.org/10.1016/j.biocon.2020.108728

McGinlay, J., Gkoumas, V., Holtvoeth, J., Fuertes, R., Bazhenova, E., Benzoni, A., et al. (2020). The impact of COVID-19 on the management of European protected areas and policy implications. *Forests*, *11*, 1214. https://doi.org/10.3390/f11111214

McNeely, J. (2021). Nature and COVID-19: The pandemic, the environment, and the way ahead. *Ambio*, *50*, 767–781. https://doi.org/10.1007/s13280-020-01447-0

Melin, A., Janiak, M., Marrone, F., Arora, P., & Higham, J. (2020). Comparative ACE2 variation and primate COVID-19 risk. *Communications Biology*, *3*, 641. https://doi.org/10.1038/s42003-020-01370-w

Miller-Rushing, A., Athearn, N., Blackford, T., Brigham, C., Cohen, L., Cole-Will, R., et al. (2021). COVID-19 pandemic impacts on conservation research, management, and public engagement in US national parks. *Biological Conservation*, *257*, 109038. https://doi.org/10.1016/j.biocon.2021.109038

OECD (Organisation for Economic Co-operation and Development). (2020). Biodiversity and the economic response to COVID-19: Ensuring a green and resilient recovery. https://www.oecd.org/coronavirus/policy-responses/biodiversity-and-the-economic-response-to-covid-19-ensuring-a-green-and-resilient-recovery-d98b5a09/. Accessed 22 August 2021.

Osofsky, S. A., Koch, R. A., Koch, M. D., Kalema-Zikusoka, G., Grahn, R., Leyland, T., & Karesh, W. (2005). *Building support for protected areas using a 'one health' perspective*. IUCN.

Pack, S., Ferreira, M., Krithivasan, R., Murrow, J., Bernard, E., & Mascia, M. (2016). Protected area downgrading, downsizing, and degazettement (PADDD) in the Amazon. *Biological Conservation*, *197*, 32–39. https://doi.org/10.1016/j.biocon.2016.02.004

Rahman, M., Saidur, A., Salekin, S., Belal, M., & Rahman, M. (2021). The COVID-19 pandemic: A threat to forest and wildlife conservation in Bangladesh? *Trees*. *Forests and People*, *5*, 100119. https://doi.org/10.1016/j.tfp.2021.100119

Reaser, J., Tabor, G., Becker, D., Muruthi, P., Witt, A., Woodley, S., et al. (2021). Land use-induced spillover: Priority actions for protected and conserved area managers. *PARKS*, *27*, 161–178. https://doi.org/10.2305/IUCN.CH.2021.PARKS-27-SIJKR.en

Redford, K., da Fonseca, G., Gascon, C., Rodriguez, C., Adams, J., Andelman, S., et al. (2022). Healthy planet healthy people. *Conservation Letters*, *15*(3), e12864. https://doi.org/10.1111/conl.12864

Singh, R., Gan, M., & Barlow, C. (2020). What do rangers feel? Perceptions from Asia. Africa and Latin America. *PARKS*, *26*(1), 63–76. https://doi.org/10.2305/IUCN.CH.2020.PARKS-26-1RS.en

Smith, M., Smit, I., Swemmer, L., Mokhatla, M., Freitag, S., Roux, D., & Dziba, L. (2021). Sustainability of protected areas: Vulnerabilities and opportunities as revealed by COVID-19 in a national park management agency. *Biological Conservation*, *255*, 108985. https://doi.org/10.1016/j.biocon.2021.108985

Snyman, S., & Bricker, K. S. (2019). Living on the edge: Benefit-sharing from protected area tourism. *Journal of Sustainable Tourism*, *27*, 705–719. https://doi.org/10.1080/09669582.2019.1615496

Spenceley, A., McCool, S., Newsome, D., Báez, A., Barborak, J., Blye, C.-J., et al. (2021). Tourism in protected and conserved areas amid the COVID-19 pandemic. *PARKS, 27*, 103–118. https://doi.org/10.2305/IUCN.CH. 2021.PARKS-27-SIAS

Spenceley, A., Snyman, S. & Eagles, P. (2017). *Guidelines for tourism partnerships and concessions for protected areas: Generating sustainable revenues for conservation and development.* IUCN.

Spenceley, A., Snyman, S., & Rylance, A. (2019). Revenue sharing from tourism in terrestrial African protected areas. *Journal of Sustainable Tourism, 27*, 720–734. https://doi.org/10.1080/09669582.2017.1401632

Stolton, S., Dudley, N., Avcıoğlu Çokçalışkan, B., Hunter, D., Ivanić, K.-Z., Kanga, E., et al. (2015). Values and benefits of protected areas. In G. Worboys, M. Lockwood, A. Kothari, S. Feary, & I. Pulsford (Eds.), *Protected area governance and management* (pp. 145–168). Australian National University Press.

Stolton, S., Timmins, H. & Dudley, N. (2021). *Making money local: Can protected areas deliver both economic benefits and conservation objectives? Technical series 27.* Montreal, Secretariat of the Convention on Biological Diversity, Montreal.

Sze, J., Carrasco, L., Childs, D., & Edwards, D. (2022). Reduced deforestation and degradation in Indigenous Lands pan-tropically. *Nature Sustainability, 5*(2), 123–130. https://doi.org/10.1038/s41893-021-00815-2

Waldron, A., Adams, V., Allan, J., Arnell, A., Asner, G., Atkinson, S., et al. (2020). *Protecting 30% of the planet for nature: Costs, benefits and economic implications.* Campaign for nature. https://helda.helsinki.fi/handle/10138/326470. Accessed 14 February 2023.

Walker, W., Gorelik, S., Baccini, A., Aragon-Osejo, J., Josse, C., Meyer, C., et al. (2020). The role of forest conversion, degradation, and disturbance in the carbon dynamics of Amazon indigenous territories and protected areas. *Proceedings of the National Academy of Sciences, 117*(6), 3015–3025. https://doi.org/10.1073/pnas.1913321117

Wang, H., Shao, J., Luo, X., Chuai, Z., Xu, S., Geng, M., & Gao, Z. (2020). Wildlife consumption ban is insufficient. *Science, 367*, 1435–1435. https://doi.org/10.1126/science.abb6463

Waithaka, J., Dudley, N., Álvarez, M., Arguedas Mora, S., Chapman, S., Figgis, P., Fitzsimons, J., Gallon, S., Gray, T.N., Kim, M. and Pasha, M.K.S. (2021). Impacts of COVID-19 on protected and conserved areas: A global overview and regional perspectives. https://doi.org/10.2305/IUCN.CH. 2021.PARKS-27-SIJW.en

Wells, S., Addison, P., Bueno, P., Costantini, M., Fontaine, A., Germain, L., et al. (2016). Using the IUCN green list of protected and conserved areas to promote conservation impact through marine protected areas. *Aquatic*

Conservation: Marine and Freshwater Ecosystems, 26, 24–44. https://doi.org/10.1002/aqc.2679

Winter, S. (2020) Waldverlust in Zeiten der Corona-Pandemie. Holzeinschlag in den Tropen. Berlin, WWF. https://www.wwf.de/fileadmin/fm-wwf/Publikationen-PDF/WWF-Analyse-Waldverlust-in-Zeiten-der-Corona-Pandemie.pdf. Accessed 14 February 2023.

World Health Organization. (2022). WHO Coronavirus (COVID-19) Dashboard. https://covid19.who.int. Accessed 28 October 2022.

World Travel & Tourism Council (WTTC). (2022). Economic impact 2022: Global trends. https://wttc.org/research/economic-impact. Accessed 14 November 2022.

Worobey, M., Levy, J., Serrano, L., Crits-Christoph, A., Pekar, J., Goldstein, S. et al. (2022). The huanan seafood wholesale market in Wuhan was the early epicenter of the COVID-19 pandemic. *Science*, 951–959. https://doi.org/10.1126/science.abp8715.

Zhao, H. (2020). COVID-19 drives new threat to bats in China. *Science, 367*, 1436. https://doi.org/10.1126/science.abb3088.

Zhou, P., Yang, X.-L., Wang, X.-G., Hu, B., Zhang, L., Zhang, W., et al. (2020). Addendum: A pneumonia outbreak associated with a new coronavirus of probable bat origin. *Nature, 588*, E6–E6. https://doi.org/10.1038/s41586-020-2012-7

Tourism and Visitor Management in Protected Areas Post-pandemic: The English Context

Denise Hewlett, Richard Gunton, Debra Gray, Ainara Terradillos, Sheela Agarwal, Natalia Lavrushkina, and Danny Byrne

15.1 INTRODUCTION

Across the planet, the network of protected areas provides the backbone of biodiversity conservation. These geographically defined areas, protected by legal or other means, enable biodiversity conservation by maintaining key habitats and allow if not facilitate species migration

D. Hewlett (✉)
PeopleScapes Research & Knowledge Exchange Centre, Department of
Responsible Management, University of Winchester, Winchester, UK
e-mail: denise.hewlett@winchester.ac.uk

Bournemouth University, Dorset, UK

R. Gunton
Winchester Business School, University of Winchester, Winchester, UK
e-mail: Richard.Gunton@winchester.ac.uk

N. Finneran et al. (eds.), *Managing Protected Areas*,
https://doi.org/10.1007/978-3-031-40783-3_15

267

and movement (Dudley, 2008). Where protected areas are sustainably managed, as promoted within the UN Sustainable Development Goals and by the International Union for the Conservation of Nature (IUCN), such areas can also act as buffers against the transfer of zoonotic diseases, such as COVID-19. Moreover, they provide for multiple additional benefits including balancing natural landscape processes and enabling ecosystem services, both of which are essential for human survival. They additionally provide opportunities for engaging with nature-based solutions to some of the most pressing twenty-first-century challenges we are experiencing, for example, ameliorating impacts of climate change, supporting water and food security, providing for carbon sequestration, and contributing to improvements in our air quality.

To support the purposes of protected and conserved areas, government and political will is essential. In many cases, this is demonstrated, albeit often prescriptively, in documents generated by global institutions and national governments, in their legislative and policy frameworks. Yet, the 'devil is in the detail' (Graham et al., 2003, (ii). Political frameworks are key for providing funding mechanisms and essential conservation programmes. Often underpinning these activities is the promotion of sustainable tourism products such as nature or heritage tourism, which ultimately has been shown to safeguard, not only the protected and conserved area network and its range of biodiversity and ecosystem

---

D. Gray
PeopleScapes Research & Knowledge Exchange, University of Winchester, Winchester, UK
e-mail: Debra.Gray@winchester.ac.uk

A. Terradillos
Universidad de Sevilla, Andalusia, Spain

S. Agarwal
Department of Tourism Management, University of Plymouth Business School, Plymouth, UK
e-mail: s.agarwal@plymouth.ac.uk

N. Lavrushkina
Faculty of Management, Bournemouth University, Poole, UK
e-mail: nlavrushkina@bournemouth.ac.uk

D. Byrne
Independent Consultant, Hampshire, UK
e-mail: byrne729@btinternet.com

services but also provide support to local economies and rural communities. A substantial global study on the economic value of protected area tourism, published in 2015, estimated that protected area tourism could be exceeding US$600 billion in direct in country expenditure annually: a notable amount that way exceeds what was also estimated at that time as < US$10billion required to safeguard these areas (Balmford et al., 2015). Mindful also of increasing understanding for the health and wellbeing benefits that public access to protected and conserved areas and other forms of green spaces can provide, increasing reliance is being placed on these areas for safeguarding not only environmental and ecosystem wellbeing, but undoubtedly enhancing, if not additionally safeguarding human health benefits.

Such is the potential for positive interactions between humans, protected area settings, and other urban forms of green and blue spaces, that the *One Health* initiative was driven by the IUCN in 2021 (IUCN, 2021). *One Health* recognises how interlinked people and place actually are, and that 70% of zoonotic diseases transfers to humans from wild animals and livestock (Allen et al., 2017; Jones et al., 2008). The initiative emphasises the prioritisation of solutions at national levels that are needed in order to prevent and mitigate impacts from the potential for pandemics in the future, and advocates the application of 'a coordinated, collaborative, multidisciplinary, transboundary and cross-sectoral approach to address risks that originate at the animal–human–ecosystem interface' (Hockings et al., 2020, 8).

15.2 IMPACTS OF COVID

The importance for recognising the interconnectivity of people and place was emphasised during the pandemic. Worldwide, our respective experiences of the pandemic have demonstrated to academics and practitioners of protected area management alike, the importance for coordination, collaboration, and communication across sectors. These experiences additionally demonstrated requirements for multidisciplinary approaches to research and management approaches required to address socioeconomic-environmental issues in and around protected areas worldwide. COVID-19 rocked global systems, institutions, national economies, and social confidence, not least in terms of tourism and leisure activities contributing to the quality of life that many across the world had previously enjoyed pre-pandemic through their access to protected areas, including nature reserves and coastlines, and to rural and urban green and blue spaces. COVID-19, and how the pandemic was managed, has impacted, and is continuing to impact many of these protected areas greatly.

Various strategies to prevent the spread of COVID-19 were undertaken by governments worldwide. Some of these were more restrictive than others. These strategies have included lockdowns and restricting public movement and interactions and were fundamentally inclusive of preventing/restricting overseas travel. Between January and May 2020, every global destination-imposed travel restrictions, and 45 per cent either totally or partially closed their borders to tourists (UNWTO, 2020). This strategy had impacts on local economies, and on communities dependent on tourism expenditure (Hockings et al., 2020). It resulted in the long-term closure of some protected areas, which in itself meant job losses for staff, and therefore with what fundamentally became unmanaged areas, had the potential to promote the risk of transferring additional zoonotic diseases. Combined, these factors alone have disrupted 'decades of conservation effort' (Hockings et al., 2020, 8).

Government budgets are increasingly being stretched, directed towards supporting public health strategies and concurrently dealing with the global socio-economic crises including cost-of-living increases, effects of war in Ukraine, and fuel price hikes for example. As a result, any funding originally directed to support the planning and management of natural environments is continuing to be questioned if not at risk of being discontinued, with huge consequences for progressing protected area management activities (Spenceley et al., 2021). Also, the likelihood of 'rollbacks' is increasingly overt, whereby governments' previously planned commitments for environmental protection, conservation, and direction towards sustainable economic growth are at risk (Kroner, 2020).

15.3   THE CASE OF THE UK

Restrictions imposed across the UK due to COVID-19 have left a lasting legacy in terms of 'a decline in mental health' (ONS, 2021). The situation drove people to engage with their personal and public spaces in new, if not alternative ways. Remote working from home has become commonplace, around one in six (17% of) businesses intended to move to homeworking permanently, with 61% of these pointing to improved staff wellbeing as a reason for the change (ONS, 2021). Bedrooms, kitchens have been converted into office spaces and naturalistic spaces including public and pocket parks in urban areas and protected areas in rural regions, have become of increased focus and attraction for many people, including for those whose interest in such spaces, prior to the pandemic, was negligible at best. Emerging research is demonstrating that for many—especially in urban areas and cities, a sense of freedom was sought, and nature became of increasing value in terms of enhancing health and wellbeing (c.f. Garrido et al., 2021, 2022). Indeed, nature experiences have proved to be considered as a 'source of solace for many' (ONS, 2021). The result of this situation for the UK's protected areas has been a surge in visitors: a phenomenon equally reported across other protected areas across the EU and North America for example (McClanahan, 2020; Rose, 2021). Whilst a totally comprehensive reporting of the situation in the UK remains outstanding, examples of overwhelming public popularity for these areas includes visits to parks in Cornwall rising by 280% during the summer of 2020, with similar situations experienced in Devon and Norfolk, South England (ONS, 2021).

This mass exodus of people escaping from cities to experience the freedom of the open countryside has created additional challenges for protected area agencies and government bodies. Issues include reports of extensive overcrowding, claims of new visitor profiles with implications for managing new demands from visitors, extensive experiences of antisocial behaviour, rising not least to conflicts amongst different user groups (McGinlay et al., 2020). The situation took its toll on capacities to cope with increased and extensive footfall, and multiple incidences of environmental and social detrimental impacts in areas that are designated for their cultural and environmental values. For example, the Peak District National Park Authority is reported as spending on average £38,000 per year collecting litter: this expenditure was expected to double post-COVID (Pidd, 2021). Equally in Snowdonia, the second most visited National Park in Europe (Statista, 2021), it was reported that the numbers of visitors tackling the highest mountain in Wales, had risen by 40% compared to those in 2018. Unprecedented scenes of hundreds of people walking up Snowdon were considered to result in 'the busiest visitor day in living memory' according to National Park Authority representatives (BBC, 2021a). Such challenges have required effective, proactive, adaptive, and reactive management both to deal with issues as they surface, but also to work towards minimising risks of further excessive and unplanned use of our protected areas (Snow, 2021).

15.4   MANAGEMENT FRAMEWORKS, GUIDELINES, AND TOOLKITS: PRO-ACTION AND ADAPTATION

All forms of tourism will have some impact on natural environments: this has long been recognised and reported upon (c.f. Mathieson & Wall, 1982). Equally, a number of planning approaches, management frameworks, and best practice guidelines (See example in Box Case Study) have been promoted in academic and grey literature to manage tourism and visitor impacts. These work to avoid exceeding environmental and social carrying capacities of a given area, and to direct effective decisions on the management, monitoring, and development activities in protected areas guided by sustainability principles (Europarc, 2021).

Europarc Federation Sustainable Tourism in Protected Areas: Technical Guidelines (Europarc 2021)
- '**Giving priority to protection:** A fundamental priority for the development and management of sustainable tourism should be to protect the area's natural and cultural heritage and to enhance awareness, understanding, and appreciation of it.
- **Contributing to sustainable development**: Sustainable Tourism should follow the principles of sustainable development which means addressing all aspects of its environmental, social and economic impact in the short and long term.
- **Engaging all stakeholders**: All those affected by sustainable tourism should be able to participate in decisions about its development and management, and Partnership working should be encouraged.
- **Planning sustainable tourism effectively**: Sustainable Tourism development and management should be guided by a well-researched plan that sets out agreed objectives and actions.
- **Pursuing continuous improvement**: Tourism should be managed in such a way as to continuously reduce the negative impacts while improving visitors' satisfaction, economic performance, local prosperity, and quality of life. Regular monitoring and reporting of progress and results should be part of the process'.

Guided by the European Landscape Convention (ELC), at the heart of protected area management is sustainable development and landscape planning (Dejeant-Pons, 2007). This takes a forward-looking stance on how a landscape is managed, it considers the integration of distinct land uses, be they designated for development and/or for conservation, and it emphasises approaches that involve multiple stakeholders and public views in decisions taken: demonstrating appropriate governance and government of an area (Hewlett, 2015; Hewlett & Brown, 2018; Hewlett & Edwards, 2013). The emphasis is to be proactive, which calls for visionary planning and exploratory planning scenarios: the former 'as a bottom-up approach for managing complex social–ecological systems in response to multiple system stresses, the climate emergency and competing policy priorities'; and the latter, exploratory scenarios, that provide for alternative considerations for what might happen in the future, for example in relation to coastal erosion, climate change, or other potential disasters, be they of natural or of human origin—as experienced during and since the COVID-19 pandemic (Lo et al., 2021, 446). Through visioning activities, actions on site can be guided, and enable policymakers to identify opportunities for facilitating any changes that might be needed.

To facilitate resource protection, tourism management and accommodate visitors, management frameworks such as limits of acceptable change (LAC), recreation opportunity spectrum (ROS), and carrying capacities, whether focused on social or environmental capacity, can provide structures to deploy and at least in theory, provide for opportunities to prevent/minimise environmental impacts. A commonly deployed planning tool is zonation. The demarcation of zones within a protected area allows for multiple and seemingly incompatible uses within it, by restricting or promoting visitor access according to the sensitivity of those zones. Thus, an area, at least in principle, can be configured to have the potential to accommodate multiple purposes including the development of tourism and leisure activities, tourism infrastructure, i.e. transportation hubs, car parking facilities, and/or creating zones for conservation and research purposes.

Yet as promising as these frameworks can appear to be, they can be highly challenging to implement in practice. This is especially due to the number of stakeholders involved in decision-making (Borrini-Feyerabend et al., 2013; Hewlett & Edwards, 2013) and the fact that a considerable investment in time, money and in diversity of skills on the part of rangers and other staff, is needed to work with these frameworks. For example, environmental carrying capacity, which at its most basic, means determining just how many visitors can be accommodated by any one area without destroying or degrading the environment, can present particular challenges. Carrying capacity relies upon a quantifiable measurement, a threshold of usage by which an environment is understood to be pejoratively impacted. This threshold is notoriously difficult to measure. Its construction fundamentally draws upon a range of stakeholders' views, (as discussed further) and in terms of the environment itself, carrying capacity should be also drawing upon our knowledge of ecosystems, which by themselves will be a dynamic factor to consider. Consequently, the construction of carrying capacity, in terms of staff knowledge and skills, requires expertise in human behaviour **and** ecology, warranting that a multidisciplinary and interdisciplinary team should be enabled to guide the process. Ultimately, in the case of the EU, in line with principles of the ELC, constructing carrying capacities, additionally and fundamentally, requires the involvement of **all** stakeholders, including protected area staff, government representatives, residents, and the wider public, all of whom will present different values of an area, ideas, preferences, and invariably agendas on what they do or do not consider is too much or not enough tourism being attracted to a given area.

Alternative, and 'softer', approaches to visitor management (Ling Kuo, 2002; Mason, 2005) include the use of human guides, interpretation-information opportunities, and visitor communication strategies, all of which can have the potential to help mitigate and minimise negative impacts of visitors in environmentally sensitive areas (c.f. Mason, 2005 for advantages and challenges of approaches). Likewise, visitor communication strategies (e.g. media campaigns, leaflets, posters, digitally disseminated information on site) can be designed to convey information on the environmental and cultural values of a protected area and purposed to encourage pro-environmental and prosocial visitor behaviours. Such communications strategies have the potential of providing information to a far wider audience of visitors, as well as to residents.

What each of these tools has in common is the need for preparatory time for development and implementation, and the need to take into consideration visits forecast, to prepare for what impacts might be expected on site, and what tourists needs might be. They are often also influenced in practice, by far wider national policy agendas. Data is essential, and information will need to be sourced from a number of key stakeholders, landowners, environmental agencies, and tourism/leisure providers. If coastal areas are included, this will necessitate an additional array of marine-specific expertise, stakeholders, policies, and legislation to navigate. For historical context, benchmarks of environmental data, footfall, and visitor usage for example, are important to evaluate the longitudinal context of usage and impacts. To advise planning and management processes, information is also required, for example: on visitor numbers, visitors' views on their experiences, visitors' behaviours on site, where the most and least popular routes for walking and cycling might be, visitor dispersion behaviours from key tourism transportation hubs/hotspots, and monitoring for any changes are fundamental amongst many other factors considered in evaluations to inform planning and management processes (Eagles et al., 2002).

The importance of these aspects is widely known, yet such tools can be hugely challenging to implement. For example, fundamental to all management strategies is funding, for both staff time on projects and potentially for the appropriation of existing or installation of new infrastructure. A diverse range of skills will be required and these together with the political will to provide support through funding mechanisms, legislative and policy frameworks, are fundamental to ensure that these vast spaces of highly regarded environmental and often cultural value are supported. Moreover, although such adaptive management strategies are unquestionably warranted as a key principle for both planning and implementing activities ahead of potential challenges (Eagles et al., 2002; Hewlett & Edwards, 2013; Spenceley et al., 2021) not all factors can necessarily be forecast: the pandemic and the surge in visitors experienced in protected areas is a prime example. Consequently, an ability to adapt if not react, informedly, to address 'unexpected disruptions' (Lo et al., 2021, 446) is also essential.

15.5 The Pandemic, Protected Areas, Management Strategies, and Outcomes: Case of Dorset, UK

One area that was overwhelmed with visitors during the post-lockdown period of the pandemic, was Dorset in Southern England. The county encompasses the Dorset Area of Outstanding Natural Beauty, classified of international importance as a Category V IUCN protected landscape/seascape. Covering over 1,129 square kilometres, the Dorset AONB covers approximately 42% of the county of Dorset and stretches from Lyme Regis in the west, along the coast to Poole Harbour in the east, and north to Hambledon Hill near Blandford Forum (Fig. 14.1). It includes three Ramsar Sites; wetlands of global importance; nine Special Areas of Conservation (SACs) of international importance for habitats and species, and three Special Protection Areas (SPAs) of international importance for birds. Together, SACs and SPAs form a network of 'Natura 2000' sites—European sites of the highest value for rare, endangered, or vulnerable habitats and species. National designations include nine National Nature Reserves (NNRs) which lie wholly within the AONB: 67 Sites of Special Scientific Interest (SSSIs) of national importance for their wildlife and/or geological interest; 646 Sites of Nature Conservation Interest (SNCIs); 1,581 hectares of Ancient Semi-Natural Woodland; of the eastern two-thirds of the Dorset and East Devon Coast World Heritage Site ('Jurassic Coast') and Marine Protected Areas in the sea adjacent. Access to the county is served by an international airport, a number of motorways linking major cities including London to Dorset, and relatively good rail links. It is one of the major tourism destinations in the south, and one of the most popular destinations. Such is its popularity and ease of access that in the surge of visitors experienced in 2020, Google mobility data recorded a 500% increase in visitations during this time (Google, 2021).

With visitors came multiple forms of impacts, some traditional and some new. These were especially evident in key tourism hotspots, where reports of exceeding capacity of tourism infrastructure was evident, even in the neighbouring town of Bournemouth where authorities announced a national emergency in an attempt to manage the overwhelming numbers of visitors to the south coast (BBC, 2020a). The social impact on communities included increases in illegal parking, the indiscriminate disposal of litter, human, and dog waste near residences and other forms of antisocial behaviour, including graffiti spray painted along parts of the coastal cliff face (BBC, 2020b; BBC, 2021b; BBC, 2021c; Fitch, 2020, 2021). Reports of fires and arson events were experienced, requiring Fire and Rescue Services intervention. Additionally, tourist activity was concentrated in well-known places, often promoted through Instagram and other online promotions that seemingly had replaced the use of literature provided at visitor centres. Yet although visitors focused on key known hotspots, their apparent quest for freedom and exploration led to their dispersal across vast areas of open space, resulting in an identifiable pattern of footpath erosion and concerns for new pathways being cut into the landscape by the increased footfall (Fig. 15.1).

Fig. 15.1 Location of COVIM (*Sources* © Crown copyright and database rights 2022 Ordnance Survey [100025252] and Dorset Council)

15.6 MANAGEMENT RESPONSES: THE CASE OF COVID VISITOR AND IMPACT MANAGEMENT STUDY (COVIM)

Such events called for urgent responses. At the international level, launched by the IUCN, the Global Taskforce on COVID-19 and Protected areas, led discussions amongst members representative of governing bodies, academic institutions, and protected area managing agencies worldwide, on impacts of the pandemic, potential for strategic responses and sharing of best practice. This resulted in members taking recommendations to protected area agencies and other stakeholders in their respective countries.

In the UK case, and working at the regional level of Dorset, governing bodies, area authorities, and other stakeholders including emergency services were engaged in the continuation of their management practices (c.f. Fitch, 2021), namely scenario scoping and visioning practices, that resulted in adaptive management plans, that are guided by landscape planning and regional land-use approaches. To supplement the strategic working amongst a broad range of key stakeholders including custodian public sector officers of the Jurassic Coast and Dorset AONB, landowners, NGOs, and through governance consultative mechanisms with the wider public, a visitor communication campaign, 'Promise to Love Dorset' was developed by Dorset Council using Visit Dorset promotional platforms. The Council procured an external agency (Alive) to design, deliver the campaign, and supply all visuals and content. It was funded by a central government emergency fund created as part of the national strategy designed to support economic recovery post-COVID. This ran from April through to September 2021 and was primarily disseminated via social media and website platforms. Additional tools used included e-newsletter campaigns, on-site signage/banners at key car parks, radio and podcast advertisements, and roadside electronic billboards. Visit Dorset, the tourism marketing arm of Dorset Council, management organisations, the Lulworth Estate, the National Trust, Natural England, and NGOs, as well as town and parish councils, shared the campaign through their own channels to increase the messages to their audiences. The NGO Litter Free Dorset ran a focused multimedia campaign alerting people to the risks of lighting campfires or having barbecues (Snow, 2021), the campaign's key objective was to influence visitor behaviour in terms of reducing what was considered as four major impacts: (a) the production of litter, (b) the number of fires caused by Barbecues or campfires, (c) illegal parking and camping, and (d) antisocial behaviour including vandalism, graffiti, and disposal of human and dog waste. A public safety message was emphasised to inform visitors to take care of eroding cliffs, a volatile coastline, to be aware of strong water currents and take caution with open water swimming, kayaking, and paddle-boarding, increasingly popular activities, to reduce the number of people getting into difficulty and requiring emergency services.

The visitor communication campaign was successful in terms of its reach on social media. An evaluation conducted by CAN Digital showed that between July and September 2021, 18,815,682 impressions (number of digital views or engagements) were recorded and 281,629 people engaged with the campaign (reported as interactions recorded as click throughs, swipe ups, likes, shares, video views over a ten-second period). Additional data gathered from advertising platforms showed that the campaign appeared on the devices of 3.7 million people and that various elements of the campaign were interacted with (clicked, liked, shared) over 550,000 times (Snow, 2021).

Alongside this evaluation of the campaign on social media, the Covid Visitor and Impact Management study (COVIM) , was designed to: (1) evidence the suspected environmental impacts of increased visitors' numbers; and (2) evaluate the Promise to Love Dorset campaign's effectiveness and persuasiveness in influencing visitors' attitudes and behaviours on the ground. To date there has been little research that has looked at how effective or persuasive visitor communication strategies are in practice, and this study was intended to address this key gap. An overall aim of the project was to provide recommendations to inform existing management strategies and inform the design of an impact and visitor management framework. COVIM was led by Nature Based Collaborative Solutions, a Community Interest Company comprised social and environmental scientists, in partnership with the teams of Dorset AONB, Visit Dorset, additional landowners and managing agencies, and NGOs. COVIM was funded in part by a grant from the National Geographic Society, with supplementary support in terms of information and secondary data provided from governing bodies, NGOs, landowners and tourist organisations, and fire and rescue services in the area. COVIM commenced in May 2021 and was completed in July 2022.

15.7 Case Study Areas

Four tourism hotspots on the Dorset coast and along the Jurassic World Heritage coastline, were selected as case studies areas: (1) **Studland** a headland at the southern side of the entrance to Poole Harbour; it is characterised by beach, heathland, woodland, and protected sand dunes. (2) **Old Harry Rocks** an east-facing promontory around 2 miles to the south of Studland, and popular viewpoint. Around 15 miles to the west, (3) **Durdle Door** is centred around a natural sea arch of limestone. Adjoining it to the east, (4) **Lulworth Cove,** a natural south-facing cove featuring a popular beach and village (Fig. 15.1).

15.8   Environmental and Visitor Studies

COVIM was based on two stages that ran concurrently. Firstly, an environmental study, the aim of which was to evaluate environmental impacts of visitors in each of the four case study areas. Four key indicators pertaining to visitor impacts on the natural environment were determined in consultation with our partners: fires, off-road parking, litter production, and footpath dynamics. Fires and off-road parking are primarily drivers of environmental degradation, whereas litter production and path erosion are impacts of visitors that may also serve as proxies for harder-to-measure ecological impacts. Data were collected from various sources to cover the period 2014 to 2021. The focal period was the years 2020 and 2021, whilst the earlier years enabled us to place the most recent observations in a temporal context, revealing the extent to which recent levels of impacts are atypical, and what a return to longer-run normality would entail.

Concurrently, we conducted a series of visitor's studies, looking at visitors' behaviour, attitudes, and experiences in some of the case study sites, with specific focus on the effectiveness of the Promise to Love Dorset campaign on encouraging pro-environmental behaviour and attitudes of visitors. It was intended that this study would contribute to an enhanced understanding of the factors influencing visitors' responses to messages that the campaign sought to convey, and to identify a set of 'practices' amongst tourists who share similar habits, thereby shedding light on how the design and distribution of these messages may be more persuasive. To meet these aims, three different visitor studies were conducted: (1) a visitor survey, (2) an on-site visitor observation study, and (3) on-site mini-interviews with visitors. Across all studies, data were collected in two periods: Summer (June–August) 2021 and Easter (April) 2022 to capture peak holiday periods over an 11-month period.

15.9   Key Findings

The environmental study sought to establish what impacts during 2020 could be discerned in comparison with previous years. As such a benchmark of such impacts, were provided to our partners, that could inform management strategies, especially for use in times where an informed reactive management approach to existing adaptive strategies may need to be deployed in the future. The key overall finding was that the year of the pandemic, 2020, was a year of unusually intense visitor impacts on the natural environment. Although the temporal resolution is only annual for most of our environment data, the existence of a national lockdown throughout the spring of 2020 implies that the heavy impacts of this year were additionally concentrated into a shorter window of recreational time. In addition, environmental damage appears to have been increasing gradually but exponentially across the study area over the period 2014–2021. This is evidenced by average trends of 16% more fires per year, 27% more parking fines per year, and around 5% more footpath area per year, along with increasing annual fluctuations. The season of 2020 (despite being shortened by the pandemic), saw elevated impacts in all the features studied, whilst footpath erosion appears to have risen dramatically since the summer of 2020.

There were some marked disparities between the sites, however. In the Studland area (including Old Harry Rocks), fires were dramatically more common (an 85% increase demonstrated through number of calls received identifying incidences: Fig. 14.2) in 2020 than expected based on the seven-year trend, whereas the Lulworth area showed no such increase. Litter collected also increased strongly in 2020 (by around 100% above the seven-year rising trend: Fig. 14.3) at Studland, but not (as far as collection records show) at Lulworth. In the Lulworth area (including Durdle Door: Fig. 14.4), on the other hand, footpath erosion was stronger, apparently reaching 40% per year in the 2021–2022 period (Fig. 15.2).

In 2021, most environmental impacts appear to have decreased compared to 2020. Fire incidence at Studland dropped by 46% from the 2020 level (Fig. 15.2) to return to a value that was not significantly different from its seven-year trend ($t = 1.3$, $P = 0.1$: 2-tailed test). Parking fines at Studland dropped by 70% from the 2020 level (Fig. 9.2) to return to a value that was not significantly different from their seven-year trend ($t = -1.8$, $P = 0.07$: 2-tailed test). Our data were insufficient to detect changes in rates of litter production or footpath widening in 2021 (Fig. 15.3).

With regard to the *visitor studies*, Google mobility data recorded a 500% increase in 2020 on the previous year's visitations. In terms of who was on site, our data analyses showed that most visitors were day and short-break holiday-takers, two-thirds had visited the sites before, and most reported their intentions to return to international travel when restrictions were lifted. Perhaps unsurprisingly, given that our data were

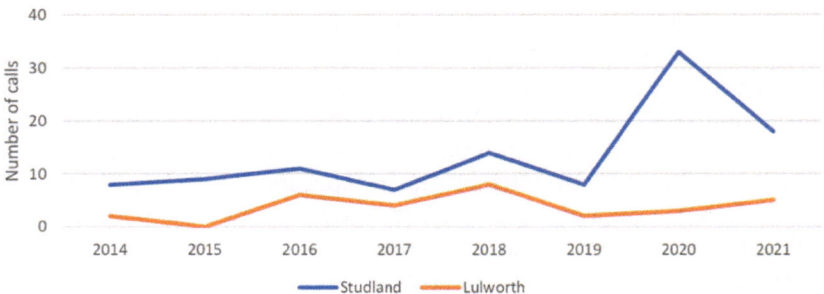

Fig. 15.2 Numbers of emergency calls likely related to fires for the two main areas studied, from 2014 to 2021. The Studland area experienced a significant anomaly (nearly twice the number of calls expected from an exponential regression line) in the lockdown year 2020

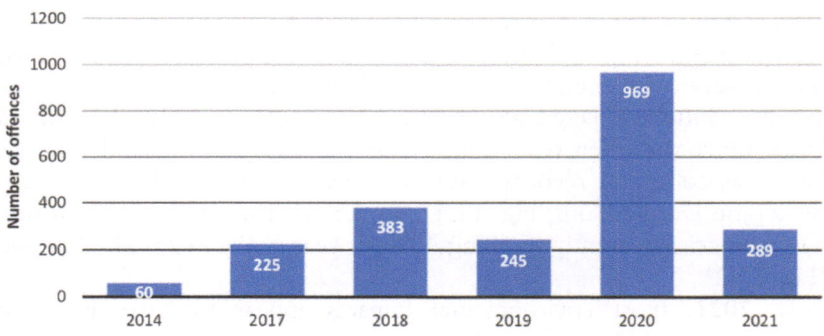

Fig. 15.3 Fly parking offences recorded in 2014 and 2017 onwards around the Studland site. There is a significant anomaly (about 40% above the number expected from an exponential regression line) in 2020

collected in peak school holiday periods, most visitor groups comprised of families with children. There were relatively small numbers of older adult visitors across all sites. This was particularly so in the summer season of 2021, where only seven per cent of visitors fell in this category, but even in Easter 2022 they made up just 10% of visitors observed. This could be COVID related, particularly in Summer 2021 when COVID restrictions were just being relaxed. Yet, this could also be due to the terrain that would need to be navigated by visitors. This situation was particularly noted at Old Harry Rocks, a footpath that led along rough terrain, was open to nature and had few sitting points: the second example that involved navigating a < 600 m footpath which leads to 145 steps down to the beach. In terms of steepness, this varies on different sections of the footpath, however, the steepest section has been approximated of ranging between 10 and 15 degrees.

There is little infrastructure in place to aid visitors. Here despite the footpaths being in keeping with the landscape, they were frequently criticised by older and/or less mobile adult visitors as being inaccessible. In terms of visitor attitudes and behaviours, most visitors evinced a holiday/ tourist motivation for visiting these locations, where nature was seen as an important background to (rather than foreground to) their reasons for being there. The key motivation for most visitors was simply to get away from it all, to relax and to spend time away with family and friends, particularly post the peak of COVID as most had been unable to go away on holidays for some time. Most visitors showed a strong environmental orientation, though this did not necessarily translate into varying levels of environmental behaviour being observed. For example, it seems clear that where behaviours required a high degree of 'effort', they were less likely to happen, e.g. taking litter home rather than putting it into a recycling bin on site.

Relatedly, most visitors seemed happy to obey the rules, but they did not always know what the rules were. Instead, they had their own ideas about what was appropriate on site that was related to what they had done in the past when on holiday, or what they saw as important to their own experience on site. Barbecues are a good example of this, as some visitors did not see these as being problematic, but instead thought they were an important part of what they did on the beach. Through the interviews it was clear, a tension did exist between the different 'groups' of tourists who were at a site in terms of their expectations and experiences. In particular, this was noted between those who want more touristic infrastructure such as bars and eateries and those who wanted less and who reported experiencing an over-commercialism of areas steeped in nature and heritage. This situation presents a significant challenge to those responsible for tourism and visitor management in terms of how to reconcile these competing demands.

In relation to the visitor communications campaign, of the 336 participants that were surveyed or interviewed on site, less than ten per cent, reported seeing the campaign. Thus, whilst the reach of the visitor campaign was extensive in terms of social media presence, and its design was based on conveying positive messages to visitors, it would seem that its impact in influencing visitors' attitudes and behaviours on the ground was limited, and the campaign did not appear to change their understanding of the unique and diverse habitat, rules, and dangers presented by these locations. Fly parking was still evident, albeit 60% less in 2021 than would be expected from the seven-year trend. This might have been attributable to the campaign, although fire incidence was 40% higher than expected from the seven-year trend. It must be said that this is not entirely unusual in terms of social media campaigns, as there is considerable evidence that informational visitor campaigns are insufficient to achieve pro-environmental outcomes (e.g. see Abrahamsen et al., 2005). Research has indicated that to truly influence behaviour, it is important to fully incorporate behavioural change theory, and importantly to design campaigns that incorporate the views of those whose behaviours you are trying to change (c.f. Green et al., 2019). Our data on site provides some evidence for the real value of such an approach, with many participants expressing the environmental value of the area and their interest in doing more as individuals and groups. Overall, this does indicate that there is huge potential, as also identified by Visit Dorset (Snow, 2021), to change visitors' behaviours, where campaigns are designed with this in mind.

15.10   Discussion

Adaptive and proactive management, scenario scoping and landscape planning practices are at the core of best practice in protected area management across the EU, and in the UK. Based on sustainable principles, in keeping with balancing use and conservation in these areas, their vision for longevity, and the legacy they provide for current and future generations is emphasised. Ongoing management assessment processes using indicators that help authorities know if they are or are not achieving their goals should be evident and improve protected area management, either directly through on-the-ground proactive and adaptive management or indirectly through improvement of national or international conservation approaches: all of which require funding. Yet, this process is all well and good when the status quo of an area is relatively well maintained.

During the pandemic, management processes including the availability of staff could not however, have perhaps forecast the speed with which COVID-19, was able to consume governments time, attention, funding mechanisms, etc., worldwide, and the activities warranted to make urgent decisions for public safety have had a knock-on pejorative effect on protected areas. The unfolding situation benefitted natural areas indirectly by the absence of public use, yet it also challenged their management through the sheer and sudden surge in visitor numbers. The situation was unprecedented. As evidenced by the case of Dorset, best practice management principles are demonstrated across the area, including in terms of increasing organisations' insights into the tourism attracted to the area, through engaging with the COVIM research team, which additionally enabled an evaluation of the authorities' own visitor campaign. Yet multiple impacts continued to be demonstrated and visitations could feasibly be considered to have exceeded social and environmental capacities.

Key indicative impacts recorded in our study from 2020 related to footfall, in that main footpaths widened by an average of up to 40% per year following the release from lockdown, whilst fires increased by up to 85%, litter production by 100%, and illegal parking by up to 150%. Thus, visitor numbers impacted extensive tracts of managed footpaths, and cut new footpaths into the landscape, whilst inappropriate use of barbecues resulted in a number of fires, one of which on the outskirts of our case study areas destroyed two acres of forest: replicating a fire in the same forest the previous year that damaged 220 hectares (BBC, 2021d). The amount of litter grew in 2020, (Pidd, 2021) and clear examples of antisocial behaviour were regularly reported by the media.

The visitor communications campaign was designed to ameliorate these negative impacts. Importantly, in line with best practice and the management of protected areas and green spaces (Leung et al., 2018; EUROPARC, 2021), although it is recognised that evaluations of techniques used can be difficult (Slaymaker, 2016), an evaluation of the campaign on the ground was welcomed. As discussed above, the COVIM evaluation showed that despite the reach of the campaign, fewer than ten per cent of those surveyed reported that they had seen the campaign and of those that could, none were able to recall its messages. This meant that COVIM was unable to make any quantifiable determination of how the campaign influenced visitors when on site. The researchers observed and reported on clear abuses of safety precautions in Durdle Door for example, in relation to just how close the public were to crumbling cliff edges. The media reported on a number of incidences along the cliffs, where visitors ignored messages as to their vulnerability along cliff edges and incidences of visitors in distress in the sea resulted in their needing help from rescue services (examples Klein, 2021; Ping, 2022) (Fig. 15.4).

The visitor studies provided insights into their experiences and how tourism and visitor management was progressed in the case study areas. As discussed above, in all areas, many tourists complained about the lack of bins made available, and some expressed concerns during the heat of the summer, for taking their litter home, including dog waste, with them when they left. Additional complaints were conveyed as to how public transportation to the sites could be improved in terms of frequency and location to case study areas, resulting in long walks often over make-shift walkways. Related to inaccessibility reported by visitors, and specifically at Durdle Door, concerns were additionally conveyed for what was considered to be only exclusive access being provided for those able, relatively young and physically fit. At each of the case study areas, there was also a clear distinction, if not conflicting views amongst visitors as to those who wanted more tourism infrastructure, i.e. shops, cafes, increased car parking facilities and those who wanted less.

Despite years of underfunding, political involvement in the management of these areas, and a multitude of stakeholders involved in the

Fig. 15.4   Visitors at cliff edge above Man O' War Beach, Durdle Door (Denise Hewlett)

governance and management of our protected areas, best practices in the management of such protected areas are being demonstrated. Yet as with so many other protected areas in Europe (McClanahan, 2020), the unprecedented nature of the pandemic and surge in visitors to our green and blue spaces, meant that management in situ was weakened. Additionally, this situation is exacerbated as in the UK's case, there is a striking inability to constrain visitor numbers, many of our spaces are not boundaried, and there are no central hubs of transportation with which to count numbers entering/leaving an area—this is fundamentally important to be able to inform carrying capacity evaluations and monitoring visitor numbers. Additionally, our managing agencies in Dorset have for years received limited government funding, impacting staffing in terms of numbers and skills, and impacting capital asset budgets required to instal infrastructure that can enhance zonation strategies for example, and direct visitors to areas identified for usage, and away from areas designated for conservation and/or research purposes. The situation is still, at this time of writing, very questionable. The UK, having left the European Union, is increasingly facing political turmoil with huge and impactful deficits in our national economy, in part created by a context derived from the post-pandemic outfall, political and economic instability, and war in Ukraine. Moreover, with major impacts on our economy, recession, fuel hikes, and inflation, the UK national debt, is currently forecast to tip £2.45 trillion (Ukpublicspending, 2023): an eye watering figure which will challenge public expenditure budgets for decades to come. It is suspected that as with other protected areas worldwide (Kroner, 2020), a 'rolling back' of funding and of political support, may well be coming towards the UK protected area network.

Acknowledgements Sincerest thanks to the National Geographic Society who awarded the COVIM team with funding that made this research possible. Our thanks especially go to Julia Luthringer, our funding officer at the National Geographic who provided support and guidance throughout our collaborations with the organisation. We are also hugely grateful to the teams of the Dorset Area of Outstanding Natural Beauty, of Visit Dorset, of the National Trust Studland Team, of the GIS officer of the Dorset Fire and Rescue Service, and thanks to the Lulworth Estate and the Lulworth Rangers: all of whom provided the research team with access permissions, and who provided secondary data sets that enabled comparative research to be completed.

REFERENCES

Abrahamsen, W., Steg, L., Vlek, C., & Rothengatter, T. (2005). A review of intervention studies aimed at household energy conservation. *Journal of Environmental Psychology, 25,* 273–291. https://doi.org/10.1016/j.jenvp.2005.08.002

Andreasen, A. (2006). *Social marketing in the 21st century.* Sage.

Allen, T., Murray, K., Zambrana-Torrelio, C., Morse, S., Rondinini, C., Di Marco, M., et al., (2017). Global hotspots and correlates of emerging zoonotic diseases. *Nature Communications, 8* Article 1124. https://doi.org/10.1038/s41467-017-00923-8.

Balmford A., Green J., Anderson M., Beresford J., Huang C., Naidoo R., et al., (2015). Walk on the wild side: Estimating the global magnitude of visits to protected areas. *PLoS Biol, 13*(2). https://doi.org/10.1371/journal.pbio.1002074.

BBC News. (2020a, June 25).Coronavirus: It's a national emergency not a national holiday. *BBC News.* https://www.bbc.co.uk/news/av/uk-england-dorset-53179224. Accessed 10 May 2021.

BBC News (2020b, June 2). Jurassic Coast beach crowds 'showed shocking disregard for area'. *BBC News.* https://www.bbc.co.uk/news/uk-england-dorset-52890608. Accessed 16 February 2023.

BBC News. (2021a, March 22). Cornoavirus: 'unprecedented' crowds in Wales despite warnings. *BBC News.* https://www.bbc.co.uk/news/uk-wales-51994504. Accessed 15 February 2023.

BBC News. (2021b, February 24). Durdle Door beach owners call for 'basic manners' code. *BBC News.* https://www.bbc.co.uk/news/uk-england-dorset. Accessed 15 February 2023

BBC News. (2021c, March 16). March 16. Covid: Bournemouth prepares for post lockdown visitor surge. *BBC News.* https://www.bbc.co.uk/news/uk-england-dorset-56412367. Accessed 15 February 2023.

BBC News. (2021d, September 7). Wareham Forest: More heathland destroyed as fire breaks out. *BBC News.* https://www.bbc.co.uk/news/uk-england-dorset-58479215. Accessed 1 October 2021d.

Blake, J. (1999). Overcoming the 'value-action gap' in environmental policy: Tensions between national policy and local experience. *Local Environment, 4,* 257–278. https://doi.org/10.1080/13549839908725599

Borrini-Feyerabend, G., Dudley, N., Jaeger, T., Lassen, B., Pathak Broome, N., Phillips, A., & Sandwith, T. (2013). *Governance of protected areas: From understanding to action. Best practice protected area guidelines series No. 20.* IUCN.

Bouman, T., & Steg, L. (2019). Motivating society-wide pro-environmental change. *One Earth, 1*(1), 27–30. https://doi.org/10.1016/j.oneear.2019.08.002

Corral-Verduga, V., & Armendáriz, L. (2010). The 'new environmental paradigm' in a Mexican community. *The Journal of Environmental Education, 3*, 25–31. https://doi.org/10.1080/00958960009598642

Dejeant-Pons, M. (2007). The European landscape convention. *Landscape Research, 31*(4), 363–384. https://doi.org/10.1080/01426390601004343

Dorset Council (2021). *Evaluation report—Promise to Love Dorset behaviour change campaign.* https://moderngov.dorsetcouncil.gov.uk/documents/s27 819/Appendix%20C%20-%20Promise%20to%20Love%20Dorset%20Evaluat ion.pdf. Accessed 2 August 2022.

Dorset Echo. (2021, February 24). The shocking impact of 'hit-and-run' visitors on Durdle Door. *Dorset Echo.* https://www.dorsetecho.co.uk/news/ 19112639.shocking-impact-hit-and-run-visitors-durdle-door/. Accessed 16 February 2023.

Dorset Tourism Association. (2019). Market Characteristics of Dorset Holiday Makers. https://www.dorsetlep.co.uk/userfiles/files/Detailed%20a nalysis%20-%20Experian%20V9%20-%20FINAL.pdf. Accessed 9 August 2022.

Dudley, N. (Ed.). (2008). *Guidelines for applying protected area management categories.* IUCN.

Eagles, P., McCool, F. & Haynes, C. (2002). *Sustainable tourism in protected areas: Guidelines for planning and management.* IUCN.

Europarc. (2021). *Sustainable tourism guidelines.* https://www.europarc.org/sus tainable-tourism/. Accessed 1 August 2022.

Fitch, H. (2020, August 28). Increase in anti-social behaviour like wild camping as Dorset reaches "full capacity". *Swanage News.* https://www.swanage. news/increase-in-anti-social-behaviour-like-wild-camping-as-dorset-reaches-full-capacity/. Accessed 6 August 2022.

Fitch, H. (2021, March 20). Plans to prevent staycationer hell, rolled out by Dorset Council. *Swanage News.* https://www.swanage.news/plans-to-prevent-staycationer-hell-rolled-out-by-dorset-council/. Accessed 13 October 2022.

Flint, R. (2021, July 25). Wales' tourist image 'could be harmed' by over-loaded infrastructure. *BBC News.* https://www.bbc.co.uk/news/uk-wales-57898837. Accessed 6 August 2022.

Getmapping. (2020). *High resolution (25cm) vertical aerial imagery.* https:// www.getmapping.com/. Accessed 16 February 2023.

Garrido-Cumbrera, M., Braçe, O., Hewlett, D., & Foley, R. (2022). Health and wellbeing under COVID-19: The green COVID Survey. *Irish Geography, 53*(2), 157–162. https://doi.org/10.2014/igj.v53i2.1420

Garrido-Cumbrera, M., Foley, R., Brace, O., Correa-Fernandez, J., Lopez-Lara, E., Guzman, V., & Hewlett, D. (2021). Perceptions of change in the natural environment by the first wave of the COVID-19 pandemic across three European Countries. Results from the GreenCOVID survey. *Urban Forestry and Urban Greening*. https://doi.org/10.1016/j.ufug.2021.127260.

Google. (2021). *Covid-19 community mobility*. COVID-19 Community Mobility Reports. https://www.google.com/covid19/mobility/. Accessed 25 March 2021.

Graham, J., Amos, B., & Plumptre, T. (2003). *Governance principles for protected areas in the 21st Century*. Prepared for the Fifth World Parks Congress Durban, South Africa-June 2003. IOG and Parks Canada. https://www.files.ethz.ch/isn/122197/pa_governance2.pdf. Accessed 16 April 2023.

Green, K., Crawford, B., Williamson, K., & DeWan, A. (2019). A meta-analysis of social marketing campaigns to improve global conservation outcomes. *Social Marketing Quarterly, 25*(1), 69–87.

Hewlett, D. (2015). Broadly engaging with tranquillity—more than a feeling. http://wherewelivenow.com/2015/12/03/making-sense-of-the-place-in-which-we-live-more-than-a-feeling/. Accessed 13 January 2021.

Hewlett, D., & Brown, L. (2018). Planning for tranquil spaces in rural destinations through mixed methods research. *Tourism Management, 67*, 237–247. https://doi.org/10.1016/j.tourman.2018.01.011

Hewlett, D., & Edwards, J. (2013). Beyond prescription: Community engagement in the planning and management of national parks as tourist destinations. *Tourism Planning and Development, 10*(1), 45–63. https://doi.org/10.1080/21568316.2012.723041

Hockings, M., Dudley, N., Elliott, W., Napolitano-Ferreira, M., MacKinnon, K., Pasha, M., et al. (2020). Covid-19 and protected and conserved areas. *Parks, 26* (1). https://parksjournal.com/wp-content/uploads/2020/06/Hockings-et-al-10.2305-IUCN.CH_.2020.PARKS-26-1MH.en_-1.pdf. Accessed 16 April 2023.

Imran, S., Alam, K., & Beaumont, N. (2014). Environmental orientations and environmental behaviour: Perceptions of protected area tourism stakeholders. *Tourism Management, 40*, 290–299. https://doi.org/10.1016/j.tourman.2013.07.003

Jackson, N. (2013). *Promoting and marketing events: Theory and practice*. Routledge.

Jackson, T. (2004). *Motivating sustainable consumption. A review of evidence on consumer behaviour and behavioural change. A report to the sustainable development research network, as part of the ESRC sustainable technologies programme, Centre for environmental strategy, University of Surrey, Guildford*. https://timjackson.org.uk/wp-content/uploads/2018/04/Jackson.-2005.-Motivating-Sustainable-Consumption.pdf. Accessed 16 April 2023.

Jones, K., Patel, N., Levy, M., Storeygard, A., Balk, D., Gittleman, J., & Daszak, P. (2008). Global trends in emerging infectious diseases. *Nature, 451,* 990–9993. https://doi.org/10.1038/nature06536

Klein, J. (2021, September 4). Man injured in landslide at Kimmeridge Bay. *Bournemouth Echo.* https://www.bournemouthecho.co.uk/news/19559524. man-injured-landslide-kimmeridge-bay/. Accessed 4 June 2022.

Kollmuss, A., & Agyeman, J. (2002). Mind the Gap: Why do people act environmentally and what are the barriers to pro-environmental behaviour? *Environmental Education Research, 8,* 239–260. https://doi.org/10.1080/13504620220145401

Kotler, P., Roberto, N., & Lee, N. (2002). *Social marketing: Improving the quality of life.* Sage.

Kroner, R. (2020). Rolling back environmental protections under cover of the Pandemic. *Scientific American Newsletter.* http://scientificamerican.com/art icle/rolling-back-environmental-protections-under-cover-of-the-pandemic/. Accessed 6 August 2022.

Leung, Y.-F., Spenceley, A., Hvenegaard, G., & Buckley, R. (Eds.). (2018). *Tourism and visitor management in protected areas: Guidelines for sustainability. Best practice protected area guidelines series No. 27.* IUCN.

Ling Kuo, I. (2002). The effectiveness of environmental interpretation at resource-sensitive tourism destinations. *International Journal of Tourism Research, 4*(2), 87–101. https://doi.org/10.1002/jtr.362

Lo, V., Lopez Rodriguez, M., Metzger, M., Osteros Rozas, E., Cebrián-Piqueras, M., Ruiz-Mallén, I., et al. (2021). How stable are visions for protected area management? Stakeholder perspectives before and during a pandemic. *People and Nature, 4,* 445–461. https://doi.org/10.1002/pan3.10292

Manfredo, M., Teel, T., Berl, R., Bruskotter, J., & Kitayama, S. (2020). Social value shift in favour of biodiversity conservation in the United States. *Nature Sustainability, 4,* 323–330. https://doi.org/10.1038/s41893-020-00655-6

Mason, G. (2005). Visitor management in protected areas: From 'hard' to 'soft' approaches? *Current Issues in Tourism, 8*(2–3), 181–194. https://doi.org/10.1080/13683500508668213

Mathieson, A., & Wall, G. (1982). *Tourism: Economic, physical and social impacts.* Longman.

McClanahan, P. (2020, December 10). The newest challenges for Europe's parks: a surge of new nature lovers. *New York Times.* https://www.nyt imes.com/2020/12/10/travel/european-parks-pandemic.html. Accessed 9 August 2022.

McCool, S. (2006). Managing for visitor experiences in protected areas: Promising opportunities and fundamental challenges. *Parks, 16*(2), 3–9.

McGinlay, J., Gkoumas, V., Holtvoeth, J., Armas Fuertes, R., Bazhenova, E., Benzoni, A., et al. (2020). The impact of COVID-19 on the management of

European protected areas and policy implications. *Forests, 11*, 1214. https://doi.org/10.3390/f11111214

McEachan, R., Taylor, N., Harrison, R., Lawton, R., Gardner, P., & Conner, M. (2016). Meta-analysis of the reasoned action approach (RAA) to understanding health behaviours. *Annals of Behavioural Medicine, 50*, 592–612. https://doi.org/10.1007/s12160-016-9798-4

Moscardo, G. (1996). Mindful visitors. *Annals of Tourism Research, 23* (2), 376–397.https://doi.org/10.1016/0160-7383(95)00068-2.

Moscardo, G. (1999). *Making visitors mindful: Principles for creating quality sustainable visitor experiences through effective communication*. Sagamore Publishing.

Moscardo, G. (2008). Understanding tourist experience through mindfulness theory. In M. Kozak & A. Decrop (Eds.), *Handbook of tourist behaviour* (Vol. 16, pp. 99–115). Routledge.

ONS (Office of National Statistics). (2021). How has lockdown changed our relationship with nature? https://www.ons.gov.uk/economy/environmentalaccounts/articles/howhaslockdownchangedourrelationshipwithnature/2021-04-26. Accessed 6 August 2022.

Pidd, H. (2021, January 1). The litter was a shock. *Guardian*. https://www.theguardian.com/environment/2021/jan/01/the-litter-was-a-shock-2020-covid-rush-on-uk-national-parks. Accessed 12 July 2022.

Ping, S. (2022, May 19). *Anti-social behaviour orders' slapped on Dorset beaches and forests for this summer*. Dorset Live. https://www.dorset.live/news/dorset-news/8-anti-social-behaviour-orders-7101469. Accessed 6 August 2022.

Rose, A. (2021, June 13). National park visitors surge as Covid-19 pandemic restrictions wane. CNN. https://edition.cnn.com/travel/article/national-park-visitors-surge/index.html. Accessed 3 July 2022.

Slaymaker, B. (2016). *Visitor behaviour and best practice visitor services in European protected areas*. Europarc federation & Alfred Toepfer natural heritage scholarship https://www.europarc.org/wp-content/uploads/2015/02/ATS-2016_Visitors-in-European-protected-areas_BThomson.pdf. Accessed 16 April 2023.

Steg, L., & Vlek, C. (2009). Encouraging pro-environmental behaviour: An integrative review and research agenda. *Journal of Environmental Psychology, 29*, 309–317. https://doi.org/10.1016/j.jenvp.2008.10.004

Snow, K. (2021). *Evaluation report—Promise to Love Dorset behaviour change campaign*.https://moderngov.dorsetcouncil.gov.uk/documents/s27819/Appendix%20C%20-%20Promise%20to%20Love%20Dorset%20Evaluation.pdf. Accessed 6 August 2022.

Spenceley, A., McCool, S., Newsome, D., Baez, A., Barborak, J., Blye, C., et al. (2021). Tourism in protected and conserved areas amid the Covid-19 pandemic. *Parks*, *27* (Special Issue March 2021). https://parksjournal.com/wp-content/uploads/2021/03/Spenceley_et_al10.2305-IUCN.CH_.2021. PARKS-27-SIAS.en_.pdf. Accessed 22 April 2023.

Statista. (2021). Leading national parks in Europe. https://www.statista.com/statistics/1058601/leading-national-parks-europe/. Accessed 16 February 2023.

Tung, V., & Ritchie, J. (2011). Exploring the essence of memorable tourism experiences. *Annals of Tourism Research*, *38*(4), 1367–1386. https://doi.org/10.1016/j.annals.2011.03.009

UK Government. (nd.). *Fire statistics definitions*. https://assets.publishing.service.gov.uk/government/uploads/system/uploads/attachment_data/file/610453/fire-statistics-definitions.pdf. Accessed 11 August 2022.

Ukpublicspending. (2023). *UK National Debt*. https://www.ukpublicspending.co.uk/. Accessed 2 January 2023.

UNWTO (United Nations World Tourism Organisation). (2020). *Covid-19 travel restrictions*. https://www.unwto.org/covid-19-travel-restrictions. Accessed 2 February 2022.

CHAPTER 16

Climate Change—Protected Areas as a Tool to Address a Global Crisis

Zachary J. Cannizzo, Elise M. S. Belle, Risa B. Smith, and Tom P. Mommsen

16.1 INTRODUCTION

The interconnected crises of biodiversity loss and climate change are having profound effects on human well-being and the natural world on which society depends (Newbold, 2018). Addressing these crises

Z. J. Cannizzo (✉)
National Marine Protected Areas Center, National Oceanic and Atmospheric Administration Office of National Marine Sanctuaries, Washington, DC, USA
e-mail: zac.cannizzo@noaa.gov

E. M. S. Belle
WCMC Europe, Bruxelles, Belgium
e-mail: Elise.Belle@wcmc-europe.eu

R. B. Smith
IUCN/World Commission On Protected Areas, Gland, Switzerland
e-mail: risa.smith.wcpa.iucn@gmail.com

T. P. Mommsen
Canada University of Victoria, University of Victoria, Victoria, BC, Canada
e-mail: tpmom@uvic.ca

N. Finneran et al. (eds.), *Managing Protected Areas*,
https://doi.org/10.1007/978-3-031-40783-3_16

will require leveraging and integrating natural processes more fully into management and policy actions. Protected areas (PAs) are increasingly recognised as a tool to protect, restore, and enhance critical natural processes. Traditionally established for biodiversity conservation, PAs also serve as tools for climate change adaptation and mitigation by protecting and maintaining carbon sinks and reservoirs, natural infrastructure that protects human communities, species that are threatened by climate change, and the ecological processes and services that humans depend on in an increasingly variable world (Dinerstein et al., 2019; IPCC, 2019; Rockström et al., 2017).

While this 'traditional' role addresses the biodiversity crisis, the ability of PAs to be designed for and act to address climate change is a recently recognised and increasingly implemented management model. In fact, the recent sixth assessment report (AR6) of the Intergovernmental Panel on Climate Change (IPCC) recognises ecosystem-based adaptation through area-based conservation, protection, and restoration of ecosystems as an effective strategy to reduce the vulnerability of biodiversity and humans to climate change (Pörtner et al., 2022). This interplay of humans and nature is integral to successfully addressing the climate and biodiversity crises. Rather than requiring the exclusion of people, the role of PAs in climate adaptation, and the need for greater protection of biodiversity and ecosystems services require both intentional, thoughtful governance and the integration of human and natural systems. Globally, 37% of remaining natural lands are traditionally owned, managed, or occupied by indigenous peoples (Garnett et al., 2018). The traditional conservation and management practices implemented in these areas are integral to the cultural practices of these communities and are increasingly recognised for the important benefits they bring to biodiversity conservation, as well as climate change adaptation and mitigation (Dinerstein et al., 2019; Schuster et al., 2019). These and other co-benefits make PAs a powerful tool to enhance climate mitigation and the adaptation of ecological and human communities.

16.2   Impacts of Climate Change on Protected Areas

From the most visited national park to the most remote marine reserve, there is no corner of the globe that is untouched by climate change. The release of greenhouse gases (GHGs) from burning fossil fuels and other human activities has led to increasing global air and sea temperatures, ocean waters that are more acidic and lower in oxygen, alterations to atmospheric and oceanic circulation, and countless changes to ecosystems and species. Although some areas have enjoyed more climate stability than others, resulting in refugia, these areas are now also threatened (Brown et al., 2020; Kocsis et al., 2021). Climate change is now one of the top five drivers of biodiversity loss globally (Diaz et al., 2019) and thus a prevalent threat to PAs, including World Heritage sites (Osipova et al., 2020). In addition, non-climate stressors such as habitat fragmentation, pollution, direct damage to ecosystems, and unsustainable management further increase the vulnerability of ecosystems to climate change (Pörtner et al., 2022).

Most climate change impacts on PAs are a direct or indirect result of an increase in temperature. Perhaps most noticeably, gradual warming encourages species to move in geographic space, generally poleward, upslope, or to deeper waters to track preferred conditions (Pinsky et al., 2020; Poloczanska et al., 2013). Such range shifts have already been observed in plant and animal species found in terrestrial, freshwater, and marine habitats (Feeley et al., 2020; Pinsky et al., 2020; Poloczanska et al., 2013) and can have profound ecological impacts regardless of biome. One widespread example is the over-grazing of temperate macroalgae and seagrass habitats by poleward expanding tropical herbivorous fishes in marine PAs (MPAs), from Japan (Nagai et al., 2011) and Australia (Vergés et al., 2016) to the Mediterranean (Vergés et al., 2014b) and the United States (Fodrie et al., 2010; Vergés et al., 2014a). Further, range shifts are predicted to result in declines in terrestrial species abundance within the European PA network (Araújo et al., 2011). Such climate-driven changes in ecosystem interactions and species composition and abundance can prevent PAs from meeting their conservation objectives (Araújo et al., 2011; Johnston et al., 2013).

Species movements and changes to ecosystem structure are also influenced by other factors. In the ocean, warming-driven changes to ocean currents (Peng et al., 2022) and oxygen levels (Laffoley & Baxter, 2019) affect species movement and ecosystems. Further, increasing temperatures are reducing the thickness and extent of Arctic sea ice, threatening ice-dependent species and the people who depend on these unique ecosystems. On land, warmer and drier climates are leading to a loss of alpine communities, even in the large PA complexes of North America (Holsinger et al., 2019), and are causing tundra ecosystems to be overtaken by trees and shrubs (Myers-Smith et al., 2015). Changes in precipitation and atmospheric circulation can have profound impacts on terrestrial and freshwater species and ecosystems. Decreases in precipitation can lead to local species extinctions, especially for species dependent on freshwater. In particular, the interaction of water loss and temperature increases makes amphibian species extremely vulnerable to climate change (Lertzman-Lepofsky et al., 2020). The drying of wetlands has already caused the decline of several amphibian species within Yellowstone National Park (McMenamin et al., 2008).

In addition to gradual changes, extreme events are increasing in frequency and intensity. Flooding and droughts are already altering freshwater wetlands, including those in PAs. In the Pripyat River in Eastern Europe, large areas of flood plains are drying out, which could result in reduced plant recruitment and an ecological shift (Moomaw et al., 2018). Further, a mass die-off of mangrove forests in Australia's Gulf of Carpentaria Marine Park has been attributed to El Niño-driven drought and sea level anomalies (Abhik et al., 2021; Duke et al., 2017), while an unprecedented localised mortality event in the U.S.'s Flower Garden Banks National Marine Sanctuary may have been triggered by a flood-induced pulse of river water (Johnston et al. 2018). Droughts have also been associated with an increase in tree mortality globally, as well as a loss of threatened bird species and outbreaks of woodboring insects (Ruthrof et al., 2018). Compounding this effect, warmer, drier air can cause landscape-scale fires (Liu et al., 2010) that devastate landscapes and release carbon into the atmosphere. Because of their slow generation time, trees are highly vulnerable to rapid changes in climate and there are concerns that existing forests might not persist in the future (Brodribb et al., 2020). Increasingly frequent and severe heatwaves (Frölicher et al., 2018; Perkins-Kirkpatrick & Lewis, 2020) also exacerbate the impacts of gradual warming such as species range shifts (Lonhart et al., 2019), drought (Ruthrof et al., 2018), fire (Liu et al., 2010), ecosystem transformation (Beas-Luna et al., 2020), and coral bleaching (Day et al., 2020). A 2016 marine heatwave compressed upwelling in California's Monterey Bay National Marine Sanctuary closer to shore. Humpback whales followed the associated prey shoreward, encountering fishing gear from a Dungeness crab fishery delayed due to a harmful algal bloom (Santora et al., 2020). This confluence of events driven by the heatwave led to record levels of whale entanglement (Santora et al., 2020).

MPAs are also experiencing additional climate impacts. Warmer waters have a lower capacity to hold oxygen and warming may have already led to a two per cent decrease in global ocean oxygen since 1960, with the potential loss of an additional three to four per cent by 2100 (Laffoley & Baxter, 2019; Stramma & Schmidtko, 2019), increasing rates of hypoxia (Altieri & Gedan, 2015; Chan et al., 2019) and threatening benthic ecosystems and species (Diaz & Rosenberg, 1995). Further, rising sea levels driven by warming-induced thermal expansion of sea water and melting ice caps threaten to drown coastal ecosystems like mangroves and salt marshes that provide critical nursery habitat and coastal protection. Ecosystems can also be damaged by tropical cyclones and other extreme storms, which are becoming stronger and wetter as the ocean warms (Knutson et al., 2019). In 2018, Hurricane Walaka passed through Papahānaumokuākea Marine National Monument, erasing an islet important to endangered Hawaiian monk seals and devastating a nearly pristine coral reef (Pascoe et al., 2021). Independent of warming, the absorption of increasing atmospheric CO_2 by ocean waters has raised the acidity of the ocean by at least 30% since the 1800s (Doney et al., 2009). Ocean acidification reduces the ability of organisms such as shellfish and corals to make and maintain their stony shells and skeletons (Hofmann et al., 2010) and can threaten the survival and growth of larvae and plankton (Bednarsek et al., 2020; Bednaršek et al., 2017). Reductions in these populations can create cascades through the food chain that threaten higher trophic levels (Hodgson et al., 2018; Piatt et al., 2020).

16.3 Protected Areas as a Tool for Climate Adaptation

Buoyed by strong policy and management, PAs represent one of our most effective tools to adapt to a changing climate. Furthermore, they have been shown to provide a thermal buffer against climate change (Xu et al., 2022). The role of PAs in aiding natural resources, and the human communities that depend on them, adapt to climate change is increasingly being recognised. The IPCC recently highlighted that 'maintaining the resilience of biodiversity and ecosystem services at a global scale depends on effective and equitable conservation of approximately 30 to 50% of Earth's land, freshwater and ocean areas' (Pörtner et al., 2022). By enhancing the resilience of biodiversity and ecosystems, PAs contribute to the adaptation of natural and human communities.

Ecological Adaptation

Well-managed and well-designed PAs minimize non-climate stressors such as habitat destruction, pollution, and invasive species, providing an environment that gives organisms their best chance to adapt to climate changes. This adaptation benefit is demonstrated by the slower rate of decline of some species, especially birds (Virkkala et al., 2019), within PAs. In many instances, PAs are expected to retain better climatic suitability compared to unprotected areas (Araújo et al., 2011; Johnston et al., 2013). Targeting areas for protection where changes are occurring more slowly, known as 'spatial climate refugia', can give organisms the time they need to adapt to changing conditions. For example, British birds and butterflies that have shifted their distributions towards the poles survive better within PAs (Gillingham et al., 2015). Similarly, protecting areas where organisms are already adapted to conditions similar to expected future conditions, known as 'adaptive climate refugia', can ensure species have the genetic diversity to adapt to environmental change (Boyd et al., 2016; Dawson et al., 2011).

The size and geographical characteristics of a PA can also play a critical role in its ability to provide climate adaptation benefits. Species often have a restricted distribution and uneven population densities within PAs and, in the absence of barriers, can often shift their distributions and abundance patterns within a PA of sufficient size (Thomas & Gillingham, 2015). For instance, when temperatures increase, species can adapt by moving uphill or to deeper waters, shifting to slopes facing towards the poles, or moving into denser vegetation or other microhabitats (Scridel et al., 2018; Suggitt et al., 2011). In mountain areas, if PAs are large enough, species can adjust to warmer temperatures by reaching higher elevations, as observed with small mammals in Yosemite National Park (de la Fuente et al., 2022; Moritz et al., 2008). Temperature gradients related to the aspect of hill slopes (Thomas & Gillingham, 2015), vegetation cover (Lenoir et al., 2017; Suggitt et al., 2018), localised upwelling, and other features can also provide a variety of microclimates for species to adapt locally, without having to move out of a PA. However, small and isolated PAs are likely to be more vulnerable to changing climatic conditions (Loarie et al., 2009) and, in some cases, less likely to allow species to adapt locally. In addition, species that already occupy the coolest microclimates within a PA will not be able to shift their distribution locally into more suitable conditions. When species do undergo large geographic shifts, it is important to consider the whole PA network rather than individual PAs and seek to enhance its suitability for species to adapt to climate change.

When integrated into well-designed networks, PAs that are connected to each other and integrated with other land use plans can facilitate species' adaptation to climate change and act as stepping stones for range shifts (Littlefield et al., 2019; Lopoukhine et al., 2012; MacKinnon et al., 2020). PA networks that incorporate considerations of both connectivity and climate change in their design and management provide protected routes and safe 'landing places' for shifting species, ensuring they remain protected even as they move in geographic space (Roberts et al., 2017). Empirical evidence of PAs facilitating species dispersion have been found in Great Britain for butterfly and odonate (dragonfly) species (Gillingham et al., 2015), as well as other invertebrate species and birds (Thomas et al., 2012). Furthermore, ecological networks and corridors between PAs allow species to adapt by shifting their ranges to more suitable habitats and climates (Hilty et al., 2020). Such connectivity considerations are particularly important to the conservation of highly migratory species, which may move across jurisdictional boundaries, as well as species with limited dispersal ability, which may require targeted protection.

Effective protection, facilitated by adaptive and intentional PA design, is a powerful tool to increase the adaptation of ecological communities to climate change. However, sometimes PAs may also need more direct management actions to facilitate the adaptation of ecosystems most at risk from climate change. For example, Florida Keys National Marine Sanctuary is working with partners to identify and grow genetic strains of coral resistant to warming and disease to be used during reef restoration. Further, managers in multiple PAs have been able to accelerate the adaptation of coastal wetlands to sea level rise by adding sediment (Berkowitz et al., 2017), restoring hydrological regimes (White & Kaplan, 2017), or providing space for these habitats to move inland (Wigand et al., 2017).

Human Adaptation

Effectively managed PAs provide essential ecosystem services such as food, clean water, recreation, economic opportunities, cultural practices, and disaster protection, that can help people adapt to the negative impacts of climate change (Dudley et al., 2010; Ivanić et al., 2017). Importantly, many terrestrial PAs contribute to the conservation and enhanced management of natural ecosystems that provide freshwater supplies (Harrison et al., 2016; MacKinnon et al., 2019) for agriculture, domestic use, and management of natural resources. For example, Cambodia's Tonle Sap lake, which has been designated as a PA, provides a critical supply of water and other ecosystem services to millions of people (Neugarten et al., 2020). PAs can also contribute to human health by protecting habitats that provide access to traditional medicine and protect the genetic material on which modern medicines depend (MacKinnon et al., 2019). Further, by protecting intact natural ecosystems, PAs play a key role in limiting the spread of zoonotic diseases (Terraube et al., 2017). Following the COVID-19 pandemic, the importance of PAs in buffering against novel disease outbreaks by maintaining ecosystem integrity is being increasingly recognised (McNeely, 2021; Terraube & Fernández-Llamazares, 2020). Finally, PAs can also protect biodiversity that is important for agriculture, including crop wild relatives that facilitate crop breeding and pollination services (Dudley et al., 2010), which are currently better represented in PAs than in seed banks (Wambugu & Henry, 2022).

In addition to the direct provision of services, PAs often protect habitats such as terrestrial and mangrove forests, tidal marshes, and coral reefs that provide protection to communities from climate impacts and natural disasters. This 'green infrastructure' stabilises slopes from earth and snow movement (Dudley et al., 2010) and reduces the effects of waves (Möller et al., 2014), flooding (Narayan et al., 2017), storm surges (Krauss et al., 2009; Zhang et al., 2012), and wind damage (Das & Crépin, 2013). These ecosystems provide benefits on the order of billions of dollars annually (Storlazzi et al., 2019) and save lives during extreme events like cyclones and tsunamis (Bayas et al., 2011). PAs also often preserve more pristine ecosystems, such as primary forests, that are less susceptible to wildfire and other disasters (Adrianto et al., 2019). Similarly, PAs that conserve grasslands, dry forests, and healthy desert vegetation can contribute to halting desertification (Dudley et al., 2015), protect watersheds, increase soil water retention, and reduce grazing pressure.

16.4   PROTECTED AREAS AS A TOOL
FOR CLIMATE MITIGATION

Attaining the Paris Climate Agreement will require rapid and significant reductions in GHG emissions supplemented by maintaining and enhancing the ability of natural ecosystems to draw down carbon, preventing the release of stored carbon, and large-scale permanent removals of carbon dioxide (CO_2) from the atmosphere. According to the IPCC, the removal of 200 to 800 Gt of atmospheric CO_2 will be required to meet the 2050 target of net zero (IPCC, 2022), yet none of the technological solutions for CO_2 removals are operational on the scale and time frame needed. However, nature can provide part of the solution and biological sinks already absorb up to 59% of anthropogenic GHG emissions (IPCC, 2021). Besides absorbing CO_2 from the atmosphere, ecosystems also store carbon as organic matter in the vegetation and soils of forests, peatlands, freshwater wetlands, grasslands, tidal marshes, mangroves, seagrass beds, and on the ocean floor (Fig. 16.1) (Smith et al., 2020). If released, much of the carbon stored within ecosystems is considered irrecoverable within a timeframe relevant for attaining the Paris Climate Agreement targets. About 23% of this irrecoverable carbon is found within PAs, with about half of it concentrated in just 3.3% of the planet's land (Noon et al., 2021).

International opinion is converging on the importance of ensuring that Natural Climate Solutions (NCS) have benefits for biodiversity, human well-being, and climate change, while avoiding perverse incentives and outcomes that have only modest climate change benefits (Bradfer-Lawrence et al., 2021; Donatti et al., 2022; Leclère et al., 2020; Pascual et al., 2022; Pörtner et al., 2021; Smith et al., 2019). PAs can play a key role towards achieving multiple positive outcomes and are increasingly recognised as one of the most effective solutions for addressing several societal issues, particularly the twin crises of climate change and biodiversity loss (Cook-Patton et al., 2021). In fact, the role of PAs in climate change mitigation has already been demonstrated. In Asia, carbon emissions in PAs are 61% lower than outside PAs, and biodiversity hotspots (measured as species richness) overlap with carbon-dense hotspots in 38% of mapped areas (Graham et al., 2021). Similarly, the 364 refuges in the U.S. National Wildlife Refuge Systems store 16.6 Gt of carbon, with the refuges created earlier storing more carbon per unit area than refuges

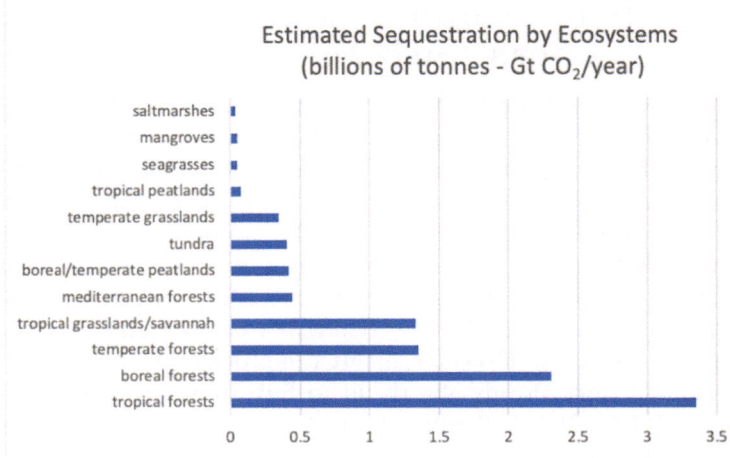

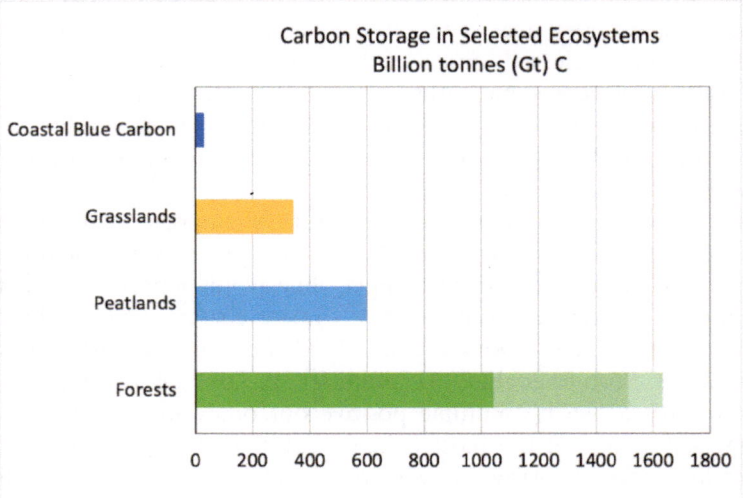

Fig. 16.1 The estimated values of carbon sequestration for different ecosystems are derived from Supplementary material in (Taillardat et al., 2018). Below, estimates of global carbon storage in selected ecosystems are from several sources: Macreadie et al. 2021 for coastal blue carbon; Lorenz and Lal [2018] for grasslands; Strack et al. [2022] for peatlands; Woods Hole Climate Research Center et al. [2020] for forests. For forests, the bar is split into the estimates for boreal forests (left, darkest green), tropical forests (centre, mid-shade green), and temperate forests (right, lightest green)

created more recently, and with more carbon stored within the wildlife refuge system than outside it (Zhu et al., 2022).

Forests

The vast extent of forests globally (~40 million km^2, 25% of the earth's land surface), make them essential for climate change mitigation. Since 1870, 26% of CO_2 emissions have come from deforestation and forest degradation (Watson et al., 2018) while, in contrast, PAs were responsible for a 29% reduction in greenhouse gas emissions from tropical deforestation between 2000 and 2012 (Bebber & Butt, 2017). Primary forests are particularly important for climate change mitigation and are estimated to have carbon stocks consisting of 49–53% of all tropical forest carbon, as well as CO_2 drawdown rates equivalent to 8–13% of annual global anthropogenic emissions (Mackey et al., 2020).

Grasslands

Grasslands currently cover 31–43% of the Earth's terrestrial area (excluding Greenland and Antarctica) and make key contributions to climate change mitigation (O'Mara, 2012). However, mainly due to land conversion, grasslands appear to have switched from being a GHG sink to a source of 1.8 ± 0.7 Gt CO_2 each year (Gomez-Casanovas et al., 2021). Remaining grassland soil stocks are estimated to be about 343 Gt C (Lorenz & Lal, 2018) and increasing the representation of grasslands in PAs can play an important role in preventing further loss (Griscom et al., 2017).

Freshwater Wetlands

At present, freshwater wetlands are considered net carbon sinks, drawing down about 6% of the current annual increase in CO_2 (Moomaw et al., 2018). Wetlands continue to accumulate carbon in their substrate over centuries to millennial time scales—making their carbon stores irrecoverable (Taillardat et al., 2020). When they are disturbed or warmed, however, wetlands can become carbon sources by releasing three major greenhouse gases—carbon dioxide, methane, and nitrous oxide. Due to this effect, different types of freshwater wetlands have different roles in climate change mitigation. Peatlands remove CO_2 from the atmosphere and store it in deep layers of organic soil, which build up over hundreds to thousands of years. They cover about 3% of the earth's land area (~4.23 million km^2), but are estimated to store up to 30% of all soil carbon (~600 Gt C) and act as a carbon sink with the potential to absorb 1.1 to 2.6 Gt CO_2/year globally by 2030. Again, negative feedbacks from warming temperatures could turn peatlands into sources of greenhouse gases. Indonesia's tropical peatlands cover 24,000 km^2 (Astuti, 2021) and store approximately 55 Gt of carbon—20 times more than non-peat containing tropical rainforest of the same size (Jaenicke et al., 2011). Conversion of Indonesian peatlands to agriculture and other human activities has resulted in the release of significant amounts of this stored carbon, but PAs have been shown to be effective in avoiding deforestation of these areas, while also reducing carbon emissions (Graham et al., 2021).

While at least 75% of the remaining peatlands are relatively undisturbed (Strack et al., 2022), they are poorly represented in PA networks. For example, the 350,000 km^2 Hudson Plains peatland complex in Northern Canada stores about 30 Gt C (Packalen et al., 2014) and captures 74.6 Mt of CO_2 per year (Bergeron & Fenton, 2012), but only 12% is protected (Abraham & McKinnon, 2011). Similarly the Cuvette Centrale, in the Congo Basin, possibly the most extensive tropical peatland complex in the world, covers 145,500 km^2 and stores ~ 30.6 Gt of carbon below ground (Dargie et al., 2017). Only 11% of it is protected (Avagyan et al., 2017).

The protection of freshwater mineral wetlands, wetlands that do not produce peat, has received less attention than other ecosystems as carbon sinks and stores. This is likely because the extent to which freshwater mineral wetlands capture and store, or produce and emit, GHGs is tied to microbial soil populations, which in turn respond to hydrological regimes. The balance between microorganisms that release the potent GHG methane (methanogens) and microorganisms that consume methane (methanotrophs) determines whether a mineral wetland is a net source or sink of greenhouse gases (Maietta et al., 2020).

Blue Carbon

Oceanic blue carbon refers to non-coastal marine habitats and processes that contribute to carbon drawdown and storage. The ocean covers 71% of the surface of the Earth, has absorbed about 90% of the heat generated by rising GHG emissions, and captured about 40% of all anthropogenic carbon dioxide released into the atmosphere in the last century (DeVries, 2022; UNFCCC, 2021). In addition, the ocean stores about 38,000 Gt C in its sediments, which has accumulated over millennia, and absorbs about 1.4 Gt C (=5.13 Gt CO_2) from the atmosphere each year (Bollmann et al., 2010). Because of complex relationships between marine biota, ocean chemistry, and climate change, about 1% of absorbed CO_2 ends up as stored carbon, largely in deep ocean sediments (DeVries, 2022; Henson et al., 2022). Organisms living in the upper ocean are key players in the global carbon cycle through absorption of atmospheric CO_2 and eventual carbon storage in sediment. However, the ocean has experienced devastating impacts (i.e., acidification, warming, oxygen depletion) due to its function in climate change mitigation, including increasing risks to marine life and coastal communities (UNFCCC, 2021). The role of marine systems in climate change mitigation differs between the vegetated coastal zone, or coastal blue carbon (mangroves, tidal marshes, seagrass meadows), and the open ocean.

Coastal blue carbon habitats, covering an estimated 410,000 km^2 (Taillardat et al., 2018) to 490,000 km^2 (Herr et al., 2017), are found on every continent except Antarctica, and have carbon drawdown rates more than 10 times greater (on a per-area basis) than those of terrestrial ecosystems (McLeod et al., 2011). Although coastal blue carbon ecosystems are the most efficient per-area natural carbon sinks, their limited global scale minimises their climate change mitigation potential. However, the most recent IPCC report recognises the importance of coastal blue carbon in mitigating climate change by 2050 (Nabuurs et al., 2022). Coastal blue carbon can contribute to GHG emission reduction targets for countries with large coastlines and small GHG emissions. For example, in 2014, mangroves mitigated more than 1% of national fossil fuel emissions in Bangladesh, Colombia, and Nigeria (Taillardat et al., 2018). Some permanent effects of increased CO_2 in the atmosphere have been blurred by the ability of the ocean to remove CO_2. It is predictable that this process is finite and reversible, especially as ocean acidification and increased surface temperatures are starting to counteract the ability of the ocean to remove atmospheric CO_2 and the role of biota is not yet well understood.

16.5   Policy and Action to Enhance the Role of Protected Areas as Tools to Address Climate Change

While PAs provide a wealth of benefits that enhance climate mitigation and the adaptation of both human and ecological communities, to successfully achieve these benefits, strong PA policy and management are necessary to reduce other threats and integrate climate change proactively into management (Gross et al., 2016).

Expanding and Enhancing the Global Protected Areas Network

Increasing the coverage of PAs can significantly enhance community resilience to climate change (Lehikoinen et al., 2021). In 2010, the UN Convention on Biological Diversity (CBD) set a target to protect 17% of the land and 10% of the ocean by 2020 and, as of August 2022, 15.8% of the land and 8.1% of the ocean were protected (Bingham et al., 2021). The Kunming-Montreal Global Biodiversity Framework, adopted in 2022, is even more ambitious. It set a target to protect at least 30% of terrestrial, inland water, and coastal and marine areas by 2030 (see Chapter 1). In addition to coverage, for PAs to efficiently address climate change, it will be crucial to improve (1) PA management effectiveness to combat other threats, such as habitat loss and degradation due to land conversion, and the overexploitation of natural resources; and (2) the connectivity and integration of PAs in the surrounding landscape and seascape (MacKinnon et al., 2020).

Incorporating Protected Areas into National and International Climate Strategies and Agreements

NCS are also increasingly recognised as crucial to attaining the Paris Climate Agreement goal to limit the mean global temperature to well below two degrees Celsius, and preferably one point five degrees Celsius above pre-industrial levels. More than a third of countries identified PAs in their nationally determined contributions towards the Paris Climate Agreement as a means of attaining their adaptation and mitigation goals, with half of these intending to expand their PA coverage (Hehmeyer et al., 2019). However, almost half of the countries did not mention PAs and almost a third did not consider NCS. Thus, while PAs are imperative to achieve climate change mitigation and adaptation targets, there is the potential for them to play an even greater role as countries increase their climate change ambitions.

Protecting areas that are rich in biodiversity and important carbon reservoirs provides multiple benefits (Roberts et al., 2020). While PAs are effective in maintaining carbon stores and enhancing carbon drawdown, they have rarely, if ever, been created for this specific purpose (Graham et al., 2021; Shi et al., 2020). Global mapping shows a 38% overlap between hotspots for carbon and areas with both high biodiversity and intactness. Yet only 12% of these areas of overlap are in PAs (Fig. 16.2) (Soto-Navarro et al., 2020). Although areas rich in biodiversity and those high in carbon do not always overlap, in cases where they do, vast potential exists for creation of new types of PAs, variously labelled as 'carbon stabilization areas' (Dinerstein et al., 2019), 'strategic carbon reserves' (Law et al., 2018), or 'carbon stewardship areas' (Wilson & Hebda, 2008).

Planning and Managing Protected Areas for Climate Change Mitigation and Adaptation

Reinforcing PA management and policies to better take climate change into account is key to enhancing the role PAs play in climate change adaptation and mitigation. PA managers will increasingly have to manage for change, rather than focus on maintaining existing systems. This will involve reviewing and revising management plans and taking climate change into account during PA designation, management, and expansion (Gross et al., 2016; Lawler et al., 2020). Furthermore, it is increasingly important that a diversity of climatic conditions are represented within PAs (Elsen et al., 2020).

Funding affects the quality of PA management, and hence their ability to adapt to climate change (Coad et al., 2019). By linking funding for climate change, biodiversity conservation, and human wellbeing, new funds can be leveraged to provide multiple benefits. Some of these funding mechanisms include Payments for Ecosystem Services (PES), Reducing Emissions from Deforestation and forest Degradation (REDD +) under the UN Framework Convention on Climate Change (UNFCCC), and voluntary carbon credits or regulatory offsets. The value of these programmes has increased in recent years (e.g., the monetary value of voluntary carbon markets has tripled since 2016) (Sreekar

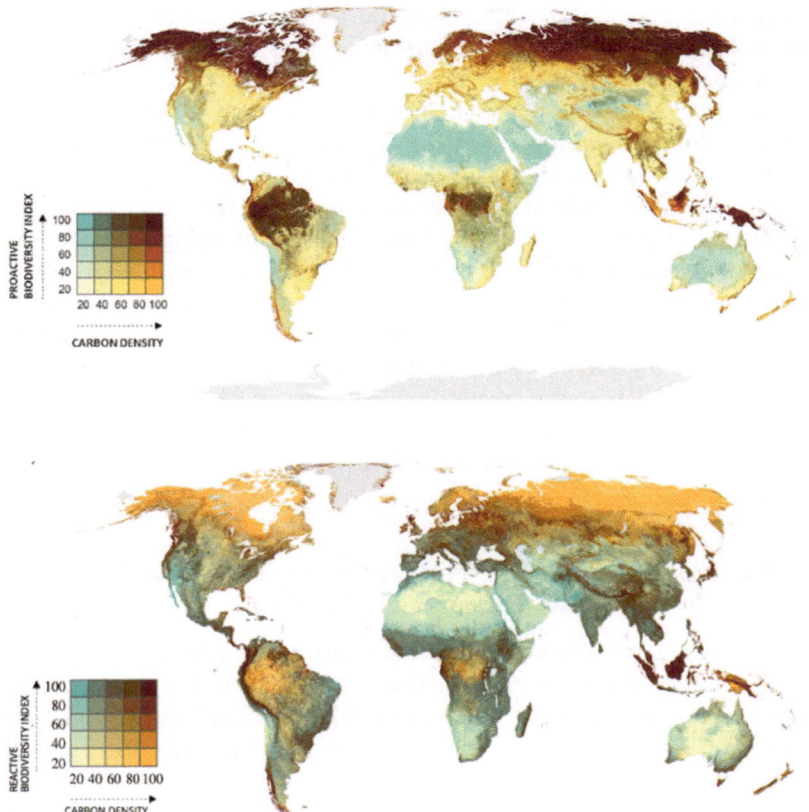

Fig. 16.2 (top) Area of overlap between a 'proactive biodiversity index' (areas of high local biodiversity [high species richness, range-size rarity], high local intactness, and high average habitat condition) with carbon density (carbon in biomass and soil organic carbon to one metre depth). The dark brown depicts areas of highest overlap. (below) Area of overlap between a 'reactive biodiversity index' (areas of high local biodiversity but low average habitat condition and high threats) with carbon density. The dark brown depicts areas of the highest overlap (Soto-Navarro et al., 2020) (*Source* Printed with permission)

et al., 2022). However, the monetisation of nature also has an increasing number of detractors, in part because evidence shows that many long-standing carbon offset programmes overestimate the carbon benefits (Badgley et al., 2022; Hook & Laing, 2022; West et al., 2020). Strong commitments are needed across national borders to not only better manage PAs in the face of climate change, but entire landscapes and seascapes. PAs provide a key instrument to address these related issues by offering an efficient way to integrate climate change mitigation, adaptation, and related development goals (Roberts et al., 2020). They are a powerful tool to address the twin crises of climate change and biodiversity loss, but strong national and international policy is necessary to ensure these benefits are maintained and leveraged to their fullest extent.

References

Abhik, S., Hope, P., Hendon, H., Hutley, L., Johnson, S., Drosdowsky, W., et al. (2021). Influence of the 2015–2016 El Niño on the record-breaking mangrove dieback along northern Australia coast. *Scientific Reports, 11*, 1–12. https://doi.org/10.1038/s41598-021-99313-w

Abraham, K. & McKinnon, I. (2011). Hudson Plains ecozone+ evidence for key findings summary. In Ministers, C. C. O. R. (Ed.), *Canadian Biodiversity: ecosystem status and trends 2010*. https://publications.gc.ca/site/eng/412444/publication.html. Accessed 16 April 2023.

Adrianto, H., Spracklen, D., Arnold, S., Sitanggang, I., & Syaufina, l. (2019). Forest and land fires are mainly associated with deforestation in Riau province. *Indonesia. Remote Sensing, 12*. https://doi.org/10.3390/rs12010003

Altieri, A., & Gedan, K. (2015). Climate change and dead zones. *Global Change Biology, 21*, 1395–1406. https://doi.org/10.1111/gcb.12754

Araújo, M., Alagador, D., Cabeza, M., Nogués-Bravo, D., & Thuiller, W. (2011). Climate change threatens European conservation areas. *Ecology Letters, 14*, 484–492. https://doi.org/10.1111/j.1461-0248.2011.01610.x

Astuti, R. (2021). Governing the ungovernable: The politics of disciplining pulpwood and palm oil plantations in Indonesia's tropical peatland. *Geoforum, 124*, 381–391. https://doi.org/10.1016/j.geoforum.2021.03.004

Avagyan, A., Baker, E., Barthelmes, A., Velarde, H., Dargie, G., Guth, M. et al. (2017). *Smoke on Water: Countering Global Threats from Peatland Loss and Degradation. a UNEP rapid response assessment* (edited by J. Crump). https://www.unep.org/resources/publication/smoke-water-countering-global-threats-peatland-loss-and-degradation-rapid. Accessed 16 April 2023.

Badgley, G., Freeman, J., Hamman, J., Haya, B., Trugman, A., Anderegg, W. & Cullenward, D. (2022). Systematic over-crediting in California's forest carbon offsets program. *Global Change Biology, 28,* 1433–1445. https://doi.org/10.1111/gcb.15943

Bayas, J., Marohn, C., Dercon, G., Dewi, S., Piepho, H., Joshi, L., et al. (2011). Influence of coastal vegetation on the 2004 Tsunami wave impact in West Aceh. *Proceedings of the National Academy of Sciences, 108,* 18612–18617. https://doi.org/10.1073/pnas.1013516108

Beas-Luna, R., Micheli, F., Woodson, C., Carr, M., Malone, D., Torre, J et al. (2020). Geographic variation in responses of kelp forest communities of the California current to recent climatic changes. *Global Change Biology, 26,* 6457–6473. https://doi.org/10.1111/gcb.15273

Bebber, D., & Butt, N. (2017). Tropical protected areas reduced deforestation carbon emissions by one third from 2000–2012. *Scientific Reports, 7,* 1–7. https://doi.org/10.1038/s41598-017-14467-w

Bednaršek, N., Feely, R., Beck, M., Alin, S., Siedlecki, S., Calosi, P., et al. (2020). Exoskeleton dissolution with mechanoreceptor damage in larval Dungeness crab related to severity of present-day ocean acidification vertical gradients. *Science of the Total Environments,* 136610. https://doi.org/10.1016/j.scitotenv.2020.136610

Bednaršek, N., Feely, R., Tolimieri, N., Hermann, A., Siedlecki, S., Waldbusser, G., et al. (2017). Exposure history determines Pteropod vulnerability to ocean acidification along the US west coast. *Scientific Reports, 7,* 4526. https://doi.org/10.1038/s41598-017-03934-z

Bergeron, Y., & Fenton, N. (2012). Boreal forests of eastern Canada revisited: Old growth, non-fire disturbances, forest succession, and biodiversity. *Botany, 90,* 509–523. https://doi.org/10.1139/b2012-034

Berkowitz, J., Vanzomeren, C., & Piercy, C. (2017). Marsh restoration using thin layer sediment addition: Initial soil evaluation. *Wetland Science and Practice, 34,* 13–17.

Bingham, H., Lewis, E., Belle, E., Stewart, J., Klimmek, H. et al. (2021). *Protected planet report 2020: Tracking progress towards global targets for protected and conserved areas.* https://livereport.protectedplanet.net. Accessed 16 April 2023.

Bollmann, M., Bosch, T., Colijn, F., Ebinghaus, R., Froese, R., Guessow, K. et al. (2010) *World ocean review 2010: Living with oceans.* https://agris.fao.org/agris-search/search.do?recordID=XF2015007946. Accessed 16 April 2023.

Boyd, P., Cornwall, C., Davison, A., Doney, S., Fourquez, M., Hurd, C. et al. (2016). Biological responses to environmental heterogeneity under future ocean conditions. *Global Change Biology, 22,* 2633–2650. https://doi.org/10.1111/gcb.13287

Bradfer-Lawrence, T., Finch, T., Bradbury, R., Buchanan, G., Midgley, A. & Field, R. (2021). The potential contribution of terrestrial nature-based solutions to a national 'net zero' climate target. *Journal of Applied Ecology, 58,* 2349–2360. https://doi.org/10.1111/1365-2664.14003

Brodrib, T., Powers, J., Cochard, H., & Choat, B. (2020). Hanging by a thread? forests and drought. *Science, 368,* 261–266. https://doi.org/10.1126/science.aat7631

Brown, S., Wigley, T., Otto-Bliesner, B., Rahbek, C., & Fordham, D. (2020). persistent quaternary climate refugia are hospices for biodiversity in the Anthropocene. *Nature Climate Change, 10,* 244–248. https://doi.org/10.1038/s41558-019-0682-7

Chan, F., Barth, J., Kroeker, K., Lubchenco, J., & Menge, B. (2019). The dynamics and impact of ocean acidification and hypoxia. *Oceanography, 32,* 62–71. https://doi.org/10.5670/oceanog.2019.312

Coad, l., Watson, J., Geldmann, J., Burgess, N., Leverington, F., Hockings, M., et al. (2019). Widespread shortfalls in protected area resourcing undermine efforts to conserve biodiversity. *Frontiers in Ecology and the Environment, 17,* 259–264. https://doi.org/10.1002/fee.2042

Cook-Patton, S., Drever, C., Griscom, B., Hamrick, K., Hardman, H., Kroeger, T., et al. (2021). Protect, manage and then restore lands for climate mitigation. *Nature Climate Change, 11,* 1027–1034. https://doi.org/10.1038/s41558-021-01198-0

Dargie, G., Lewis, S., Lawson, I., Mitchard, E., Page, S., Bocko, Y., et al. (2017). Age, extent and carbon storage of the central Congo basin peatland complex. *Nature, 542,* 86–90. https://doi.org/10.1038/nature21048

Das, S., & Crépin, A.-S. (2013). Mangroves can provide protection against wind damage during storms. *Estuarine, Coastal and Shelf Science, 134,* 98–107. https://doi.org/10.1016/j.ecss.2013.09.021

Dawson, T., Jackson, S., House, J., Prentice, I., & Mace, G. (2011). Beyond predictions: Biodiversity conservation in a changing climate. *Science, 332,* 53–58. https://doi.org/10.1126/science.1200303

Day, J., Heron Scott, F., & Markham, A. (2020). Assessing the climate vulnerability of the world's natural and cultural heritage. *Parks Stewardship Forum, 36.* https://doi.org/10.5070/P536146384

De la Fuente, A., Krockenberger, A., Hirsch, B., & Cernusak, l. & Williams, S. (2022). predicted alteration of vertebrate communities in response to climate-induced elevational shifts. *Diversity and Distributions, 28*(6), 1180–1190. https://doi.org/10.1111/ddi.13514

Devries, T. (2022). The ocean carbon cycle. *Annual Review of Environment and Resources, 47.* https://doi.org/10.1146/annurev-environ-120920-111307

Diaz, R., & Rosenberg, R. (1995). Marine benthic hypoxia: A review of its ecological effects and the behavioural responses of benthic macrofauna. *Oceanography and Marine Biology. An Annual Review, 33,* 245–303.

Diaz, S., Settele, J., Brondizio, E., Ngo, H., Guèze, M., Agard, J., et al. (2019). Summary for policymakers of the global assessment report on biodiversity and ecosystem services of the intergovernmental science-policy platform on biodiversity and ecosystem services. *Bonn, Germany.* https://doi.org/10.5281/zenodo.3553579

Dinerstein, E., Vynne, C., Sala, E., Joshi, A., Fernando, S., Lovejoy, T., et al. (2019). A global deal for nature: Guiding principles, milestones, and targets. *Science Advances, 5.* https://doi.org/10.1126/sciadv.aaw2869

Donatti, C., Andrade, A., Cohen-Shacham, E., Fedele, G., Hou-Jones, X., & Robyn, B. (2022). Ensuring that nature-based solutions for climate mitigation address multiple global challenges. *One Earth, 5,* 493–504. https://doi.org/10.1016/j.oneear.2022.04.010

Doney, S., Fabry, V., Feely, R., & Kleypas, J. (2009). Ocean acidification: The other CO_2 problem. *Annual Review of Marine Science, 1,* 169–192. https://doi.org/10.1146/annurev.marine.010908.163834

Dudley, N., Buyck, C., Furuta, N., Pedrot, C., Renaud, F. & Sudmeier-Rieux, K. (2015). Protected areas as tools for disaster risk reduction. *A handbook for practitioners. Ministry of Environment, Japan and IUCN.* https://www.iucn.org/resources/publication/protected-areas-tools-disaster-risk-reduction-handbook-practitioners. Accessed 16 April 2023.

Dudley, N., Stolton, S., Belokurov, A., Krueger, L., Lopoukhine, N., Mackinnon, K., et al. (2010). *Natural Solutions: Protected areas helping people cope with climate change.* IUCN.

Duke, N., Kovacs, J., Griffiths, A., Preece, l., Hill, D., Van Oosterzee, P., et al. (2017). Large-scale dieback of mangroves in Australia's gulf of Carpentaria: A severe ecosystem response, coincidental with an unusually extreme weather event. *Marine and Freshwater Research, 68,* 1816–1829. https://doi.org/10.1071/MF16322

Elsen, P. Monahan, W., Dougherty, E. & Merenlender, A. (2020). Keeping pace with climate change in global terrestrial protected areas. *Science Advances, 6,* eaay0814. https://doi.org/10.1126/sciadv.aay0814

Feeley, K., Bravo-Avila, C., Fadrique, B., Perez, T., & Zuleta, D. (2020). Climate-driven changes in the composition of new world plant communities. *Nature Climate Change, 10,* 965–970. https://doi.org/10.1038/s41558-020-0873-2

Fodrie, F., & Heck jr, K., Powers, S., Graham, W. & Robinson, K. (2010). Climate-related, decadal-scale assemblage changes of seagrass-associated fishes in the northern Gulf of Mexico. *Global Change Biology, 16,* 48–59. https://doi.org/10.1111/j.1365-2486.2009.01889.x

Frölicher, T., Fischer, E., & Gruber, N. (2018). Marine heatwaves under global warming. *Nature, 560*, 360–364. https://doi.org/10.1038/s41586-018-0383-9

Garnett, S., Burgess, N., Fa, J., Fernández-llamazares, A., Molnár, Z., Robinson, C., et al. (2018). A spatial overview of the global importance of indigenous lands for conservation. *Nature Sustainability, 1*, 369. https://doi.org/10.1038/s41893-018-0100-6

Gillingham, P., Alison, J., Roy, D., Fox, R., & Thomas, C. (2015). High abundances of species in protected areas in parts of their geographic distributions colonized during a recent period of climatic change. *Conservation Letters, 8*, 97–106. https://doi.org/10.1111/conl.12118

Gomez-Casanovas, N., Blanc-Betes, E., Moore, C., Bernacchi, C., Kantola, I., & Delucia, E. (2021). A review of transformative strategies for climate mitigation by grasslands. *Science of the Total Environment, 799*, 149466. https://doi.org/10.1016/j.scitotenv.2021.149466

Graham, V., Geldmann, J., Adams, V., Negret, P., Sinovas, P., & Chang, H. (2021). Southeast Asian protected areas are effective in conserving forest cover and forest carbon stocks compared to unprotected areas. *Science Reports, 11*, 23760. https://doi.org/10.1038/s41598-021-03188-w

Griscom, B., Adams, J., Ellis, P., Houghton, R., Lomax, G., Miteva, D., et al. (2017). natural climate solutions. *Proceedings of the National Academy of Sciences, 114*, 11645–11650. https://doi.org/10.1073/pnas.1710465114

Gross, J., Woodley, S., Welling, L., & Watson, J. (2016). *Adapting to Climate Change: guidance for protected area managers and planners. best practice protected area guidelines.* IUCN.

Harrison, I., Green, P., Farrell, T., Juffe-Bignoli, D., Sáenz, L., & Vörösmarty, C. (2016). Protected areas and freshwater provisioning: A global assessment of freshwater provision, threats and management strategies to support human water security. *Aquatic Conservation: Marine and Freshwater Ecosystems, 26*, 103–120. https://doi.org/10.1002/aqc.2652

Hehmeyer, A., Vogel, J., Martin, S. & Bartlett, R. (2019). *Enhancing nationally determined contributions through protected areas.* World Wildlife Fund. https://files.worldwildlife.org/wwfcmsprod/files/Publication/file/3uq3mz 3lhu_05819_WWF_Protected_Areas_Layout_3a_sprds.pdf?_ga=2.23491002.1850305944.1681723127-876922931.1681723127. Accessed 17 April 2023.

Henson, S., Laufkötter, C., Leung, S., Giering, S., Palevsky, H., & Cavan, E. (2022). Uncertain response of ocean biological carbon export in a changing world. *Nature Geoscience, 15*, 248–254. https://doi.org/10.1038/s41561-022-00927-0

Herr, D., Howard, J., Isensee, K., Pidgeon, E. & Ramos, J. (2017). *The blue carbon initiative.* https://www.thebluecarboninitiative.org/about-the-blue-carbon-initiative. Accessed 17 April 2023.

Hilty, J., Worboys, G., Keeley, A., Woodley, S., Lausche, B., Locke, H., et al. (2020). *Guidelines for conserving connectivity through ecological networks and corridors.* IUCN.

Hodgson, E., Kaplan, I., Marshall, K., Leonard, J., Essington, T., Busch, D., et al. (2018). Consequences of spatially variable ocean acidification in the California current: Lower pH drives strongest declines in benthic species in southern regions while greatest economic impacts occur in northern regions. *Ecological Modelling, 383,* 106–117. https://doi.org/10.1016/j.ecolmodel.2018.05.018

Hofmann, G., Barry, J., Edmunds, P., Gates, R., Hutchins, D., Klinger, T., & Sewell, M. (2010). The effect of ocean acidification on calcifying organisms in marine ecosystems: An organism-to-ecosystem perspective. *Annual Review of Ecology, Evolution, and Systematics, 41,* 127–147. https://doi.org/10.1146/annurev.ecolsys.110308.120227

Holsinger, L., Parks, S., Parisien, M., Miller, C., Batllori, E., & Moritz, M. (2019). Climate change likely to reshape vegetation in North America's largest protected areas. *Conservation Science and Practice, 1,* e50. https://doi.org/10.1111/csp2.50

Hook, A., & Laing, T. (2022). the politics and performativity of redd+ reference levels: Examining the Guyana-Norway agreement and its implications for 'offsetting' towards 'net zero.' *Environmental Science and Policy, 132,* 171–180. https://doi.org/10.1016/j.envsci.2022.02.021

IPCC (Intergovernmental Panel on Climate Change) (2019). *Climate Change and Land: summary for policymakers: an IPCC special report on climate change, desertification, land degradation, sustainable land management, food security, and greenhouse gas fluxes in terrestrial ecosystems.* Bonn: WMO and UNEP. https://www.ipcc.ch/site/assets/uploads/2019/11/SRCCL-Full-Report-Compiled-191128.pdf. Accessed 17 April 2023.

IPCC (2021). *Summary for Policymakers. Climate Change 2021: the physical science basis. Contribution of working group to the sixth assessment report of the intergovernmental panel on climate change.* Cambridge: Cambridge University Press.

IPCC (2022). Summary for policymakers. In P. Shukla, J. Skea, R. Slade, Khourdajie, A., Van Diemen, R. & D. McCollum et al. (Eds.), *Mitigation of Climate Change. Contribution of working group iii to the sixth assessment report of the Intergovernmental Panel on Climate Change.* Cambridge and New York: Cambridge University Press. https://www.ipcc.ch/report/ar6/wg3/downloads/report/IPCC_AR6_WGIII_SummaryForPolicymakers.pdf. Accessed 17 April 2023.

Ivanić, K., Stefan, A., Porej, D., & Stolton, S. (2017). Using a participatory assessment of ecosystem services in the Dinaric arc of Europe to support protected area management. *Parks, 23,* 61–74. https://doi.org/10.2305/IUCN.CH.2017.PARKS-23-1K-ZI.en

Jaenicke, J., Englhart, S., & Siegert, F. (2011). Monitoring the effect of restoration measures in Indonesian peatlands by radar satellite imagery. *Journal of Environmental Management, 92,* 630–638. https://doi.org/10.1016/j.jenvman.2010.09.029

Johnston, A., Ausden, M., Dodd, A., Bradbury, R., Chamberlain, D., Jiguet, F., et al. (2013). Observed and predicted effects of climate change on species abundance in protected areas. *Nature Climate Change, 3,* 1055–1061. https://doi.org/10.1038/nclimate2035

Johnston, M., Blakeway, R., O'Connell, K. MacMillan, J., Nuttall, M., Hu, X. et al. (2020). *Long-Term Monitoring at East and West Flower Garden Banks: 2018 Annual Report.* National Marine Sanctuaries Conservation Series ONMS-20–09. U. https://nmssanctuaries.blob.core.windows.net/sanctuaries-prod/media/docs/2018-east-and-west-fgb-monitoringPublication.pdf. Accessed 17 April 2023.

Johnston, M., Sterne, T., Eckert, R., Nuttall, M., Embesi, J., & Walker, R. (2018). *Long-term monitoring at East and West Flower Garden Banks: 2016 Annual Report.* National Marine Sanctuaries Conservation Series ONMS-17–09. https://repository.library.noaa.gov/view/noaa/16332. Accessed 17 April 2023.

Knutson, T., Camargo, S., Chan, J., Emanuel, K., Ho, C., Kossin, J. et al. (2019). Tropical cyclones and climate change assessment: part ii. Projected response to anthropogenic warming. *Bulletin of the American Meteorological Society, 101* (3). https://doi.org/10.1175/BAMS-D-18-0194.1

Kocsis, A., Zhao, Q., Costello, M., & Kiessling, W. (2021). Not all biodiversity rich spots are climate refugia. *Biogeosciences, 18,* 6567–6578. https://doi.org/10.5194/bg-18-6567-2021

Krauss, K., Doyle, T. W., Doyle, T. J., Swarzenski, C., From, A., Day, R., & Conner, W. (2009). Water level observations in mangrove swamps during two hurricanes in Florida. *Wetlands, 29,* 142. https://doi.org/10.1672/07-232.1

Laffoley, D., & Baxter, J. (2019). *Ocean Deoxygenation: Everyone's problem—causes, impacts, consequences and solutions.* IUCN.

Law, B., Hudiburg, T., Berner, L., Kent, J., Buotte, P., & Harmon, M. (2018). Land use strategies to mitigate climate change in carbon dense temperate forests. *Proceedings of the National Academy of Sciences, 115,* 3663–3668.

Lawler, J., Rinnan, D., Michalak, J., Withey, J., Randels, C., & Possingham, H. (2020). Planning for climate change through additions to a national protected area network: Implications for cost and configuration. *Philosophical Transactions of the Royal Society B: Biological Sciences, 375,* 20190117. https://doi.org/10.1098/rstb.2019.0117

Leclère, D., Obersteiner, M., Barrett, M., Butchart, S., Chaudhary, A., De Palma, A., et al. (2020). Bending the curve of terrestrial biodiversity needs an integrated strategy. *Nature, 585,* 551–556. https://doi.org/10.1038/s41586-020-2705-y

Lehikoinen, P., Tiusanen, M., Santangeli, A., Rajasärkkä, A., Jaatinen, K., Valkama, J., et al. (2021). Increasing protected area coverage mitigates climate-driven community changes. *Biological Conservation, 253,* 108892. https://doi.org/10.1016/j.biocon.2020.108892

Lenoir, J., Hattab, T. & Pierre, G. (2017). Climatic microrefugia under anthropogenic climate change: implications for species redistribution. *Ecography, 40,* 253–266. https://doi.org/10.1111/ecog.02788

Lertzman-lepofsky, G., Kissel, A., Sinervo, B., & Palen, W. (2020). water loss and temperature interact to compound amphibian vulnerability to climate change. *Global Change Biology, 26,* 4868–4879. https://doi.org/10.1111/gcb.15231

Littlefield, C., Krosby, M., Michalak, J., & Lawler, J. (2019). Connectivity for species on the move: Supporting climate-driven range shifts. *Frontiers in Ecology and the Environment, 17,* 270–278.

Liu, Y., Stanturf, J., & Goodrick, S. (2010). Trends in global wildfire potential in a changing climate. *Forest Ecology and Management, 259,* 685–697. https://doi.org/10.1016/j.foreco.2009.09.002

Loarie, S., Duffy, P., Hamilton, H., Asner, G., Field, C., & Ackerly, D. (2009). The velocity of climate change. *Nature, 462,* 1052–1055. https://doi.org/10.1002/fee.2043

Lonhart, S., Jeppesen, R., Beas-Luna, R., Crooks, J., & Lorda, J. (2019). Shifts in the distribution and abundance of coastal marine species along the eastern Pacific ocean during marine heatwaves from 2013 to 2018. *Marine Biodiversity Records, 12,* 13. https://doi.org/10.1186/s41200-019-0171-8

Lopoukhine, N., Crawhall, N., Dudley, N., Figgis, P., Karibuhoye, C., Laffoley, D. et al. (2012). Protected areas: Providing natural solutions to 21st century challenges. *Sapiens, 5* (2). http://sapiens.revues.org/1254

Lorenz, K., & Lal, R. (2018). Carbon sequestration in grassland soils. In K. Lorenz & R. Lal (Eds.), *Carbon Sequestration in Agricultural Ecosystems* (pp. 175–209). Springer International Publishing.

Mackey, B., Komos, C., Keith, H., Moomaw, W., Houghton, R., Mittermeier, R., Hole, D., & Hugh, S. (2020). Understanding the importance of primary tropical forest protection as a mitigation strategy. *Mitigation and Adaptation Strategies for Global Change, 25*(5), 763–787.

Mackinnon, K., Smith, R., Dudley, N., Figgis, P., Hockings, M., Keenleyside, K., et al. (2020). Strengthening the global system of protected areas post 2020:

A perspective from the IUCN World Commission on Protected Areas. *Parks Stewardship Forum, 36.* https://doi.org/10.5070/P536248273

Mackinnon, K., Van Ham, C., Reilly, K. & Hopkins, J. (2019). Nature-based solutions and protected areas to improve urban biodiversity and health. In M. Marselle, J. Stadle, H. Korn, K. Irvine & A. Bonn (Eds.), *Biodiversity and Health in the Face of Climate Change* (pp. 363–380). https://doi.org/10. 1007/978-3-030-02318-8_16

Macreadie, P., Costa, M., Atwood, T., Friess, D., Kelleway, J., Kennedy, H., et al. (2021). Blue carbon as a natural climate solution. *Nature Reviews Earth and Environment, 2,* 826–839. https://doi.org/10.1038/s43017-021-00224-1

Maietta, C., Hondula, K., Jones, C., & Palmer, M. (2020). Hydrological conditions influence soil and methane-cycling microbial populations in seasonally saturated wetlands. *Frontiers in Environmental Science, 8.* https://doi.org/10.3389/fenvs.2020.593942

Mcleod, E., Chmura, G., Bouillon, S., Salm, R., Björk, M., Duarte, C., et al. (2011). A blueprint for blue carbon: Toward an improved understanding of the role of vegetated coastal habitats in sequestering CO2. *Frontiers in Ecology and the Environment, 9*(10), 552–560. https://doi.org/10.1890/110004

McMenamin, S., Hadly, E., & Wright, C. (2008). Climatic change and wetland desiccation cause amphibian decline in Yellowstone National Park. *PNAS, 105*(44), 16988–16993. https://doi.org/10.1073/pnas.0809090105

McNeely, J. (2021). Nature and Covid-19: The pandemic, the environment, and the way ahead. *Ambio, 50*(4), 767–781. https://doi.org/10.1007/s13280-020-01447-0

Möller, I., Kudella, M., Rupprecht, F., Spencer, T., Paul, M., Van Wesenbeeck, B., et al. (2014). Wave attenuation over coastal salt marshes under storm surge conditions. *Nature Geoscience, 7,* 727–731. https://doi.org/10.1038/NGEO2251

Moomaw, W., Chmura, G., Davies, G., Finlayson, C., Middleton, B., Natali, S., et al. (2018). Wetlands in a changing climate: Science, policy and management. *Wetlands, 38,* 183–205. https://doi.org/10.1007/s13157-018-1023-8

Moritz, C., Patton, J., Conroy, C., Parra, J., White, G., & Beissinger, S. (2008). Impact of a century of climate change on small-mammal communities in Yosemite National Park, USA. *Science, 322,* 261–264. https://doi.org/10.1126/science.1163428

Myers-Smith, I., Elmendorf, S., Beck, P., Wilmking, M., Hallinger, M., Blok, D., et al. (2015). Climate sensitivity of shrub growth across the tundra biome. *Nature Climate Change, 5,* 887–891. https://doi.org/10.1038/nclimate2697

Nabuurs, G-J., Mrabet, R., Abu Hatab, A., Bustamante, M., Clark, H., Havlík, P. et al. (2022). Agriculture, forestry and other landuses. In P.Shukla, J. Skea,

R. Slade, A. Al Khourdajie, R. van Diemen, D. McCollum et al (Eds.), *IPCC Climate Change 2022: Mitigation of Climate Change. Contribution of Working Group iii to the Sixth Assessment Report of the Intergovernmental Panel on Climate Change*. Cambridge University Press. https://doi.org/10.1017/978 1009157926.009

Nagai, S., Yoshida, G., & Tarutani, K. (2011). Change in species composition and distribution of algae in the coastal waters of western Japan. In S. Casalegno (Ed.), *Global Warming Impacts. Case Studies on the Economy, Human Health, and on Urban and Natural Environments*, (pp. 209–237). Intech Open. https://doi.org/10.5772/25055

Narayan, S., Beck, M., Wilson, P., Thomas, C., Guerrero, A., Shepard, C., et al. (2017). The value of coastal wetlands for flood damage reduction in the north-eastern USA. *Scientific Reports, 7*, 1–12. https://doi.org/10.1038/s41598-017-09269-z

Neugarten, R., Moull, K., Martinez, N., Andriamaro, l., Bernard, C., Bonham, C., et al. (2020). Trends in protected area representation of biodiversity and ecosystem services in five tropical countries. *Ecosystem Services, 42*, 101078. https://doi.org/10.1016/j.ecoser.2020.101078

Newbold, T. (2018). Future effects of climate and land-use change on terrestrial vertebrate community diversity under different scenarios. *Proceedings of the Royal Society b, 285*, 20180792. https://doi.org/10.1098/rspb.2018.0792

Noon, M., Goldstein, A., Ledezma, J., Roehrdanz, P., Cook-Patton, S., Spawn-Lee, S., Wright, T., et al. (2021). Mapping the irrecoverable carbon in earth's ecosystems. *Nature Sustainability, 5*, 37–46. https://doi.org/10.1038/s41 893-021-00803-6

O'Mara, F. (2012). The role of grasslands in food security and climate change. *Annals of Botany, 110*(6), 1263–1270. https://doi.org/10.1093/aob/mcs209

Osipova, E., Emslie-Smith, M., Osti, M., Murai, M., Aberg, U., & Shadie, P. (2020). *IUCN World Heritage Outlook 3: A conservation assessment of all natural world heritage sites, November 2020*. IUCN.

Packalen, M., Finkelstein, S., & McLaughlin, J. (2014). Carbon storage and potential methane production in the Hudson Bay lowlands since mid-Holocene peat initiation. *Nature Communications, 5*, 4078. https://doi.org/10.1038/ncomms5078

Pascoe, K., Fukunaga, A., Kosaki, R., & Burns, J. (2021). 3d assessment of a coral reef at Lalo Atoll reveals varying responses of habitat metrics following a catastrophic hurricane. *Scientific Reports, 11*, 1–9. https://doi.org/10.1038/s41598-021-91509-4

Pascual, U., McElwee, P., Diamond, S., Ngo, H., Bai, X., Cheung, W., et al. (2022). governing for transformative change across the biodiversity–climate–society nexus. *BioScience, 72*(7), 684–704. https://doi.org/10.1093/biosci/biac031

Peng, Q., Xie, S., Wang, D., Huang, R., Chen, G., Shu, Y., Shi, J. & Liu, W. (2022). surface warming–induced global acceleration of upper ocean currents. *Science Advances, 8* (16), eabj8394. https://doi.org/10.1126/sciadv.abj8394

Perkins-Kirkpatrick, S. & Lewis, S. (2020). Increasing trends in regional heatwaves. *Nature Communications, 11*, 1–8. https://doi.org/10.1038/s41467-020-16970-7

Piatt, J., Parrish, J., Renner, H., Schoen, S., Jones, T., Arimitsu, M., et al. (2020). Extreme mortality and reproductive failure of common murres resulting from the northeast Pacific marine heatwave of 2014–2016. *PLoS ONE, 15*, e0226087. https://doi.org/10.1371/journal.pone.0226087

Pinsky, M., Selden, R., & Kitchel, Z. (2020). Climate-driven shifts in marine species ranges: Scaling from organisms to communities. *Annual Review of Marine Science, 12*(12), 153–179. https://doi.org/10.1146/annurev-marine-010419-010916

Poloczanska, E., Brown, C., Sydeman, W., Kiessling, W., Schoeman, D., Moore, P., et al. (2013). Global imprint of climate change on marine life. *Nature Climate Change, 3*(10), 919–925. https://doi.org/10.1038/nclimate1958

Pörtner, H-O., Roberts, D., Adams, H., Adler, C., Aldunce, P., Ali, E., et al. (2022). WGII Contribution to the IPCC Sixth Assessment Report (ar6), Climate Change 2022: Impacts, Adaptation and Vulnerability: Summary for Policymakers. IPCC. https://report.ipcc.ch/ar6/wg2/IPCC_AR6_WGII_FullReport.pdf. Accessed 18 April 2023.

Pörtner, H., Roberts, D., Masson-Delmotte, V., Zhai, P., Tignor, M., Poloczanska, E., et al. (2021). IPBES-IPCC Co-sponsored Workshop Report on Biodiversity and Climate Change. Bonn: IPBES-IPCC. https://www.ipbes.net/sites/default/files/2021-06/20210609_workshop_report_embargo_3pm_CEST_10_june_0.pdf. Accessed 18 April 2023.

Roberts, C., O'Leary, B., & Hawkins, J. (2020). Climate change mitigation and nature conservation both require higher protected area targets. *Philosophical Transactions of the Royal Society b: Biological Sciences, 375*, 20190121. https://doi.org/10.1098/rstb.2019.0121

Roberts, C., O'Leary, B., McCauley, D., Cury, P., Duarte, C., Lubchenco, J., et al. (2017). Marine reserves can mitigate and promote adaptation to climate change. *Proceedings of the National Academy of Science USA, 114*, 6167–6175. https://doi.org/10.1073/pnas.1701262114

Rockström, J., Gaffney, O., Rogelj, J., Meinshausen, M., Nakicenovic, N., & Schellnhuber, H. J. (2017). A roadmap for rapid decarbonization. *Science, 355*(6331), 1269–1271. https://doi.org/10.1126/science.aah3443

Ruthrof, K., Breshears, D., Fontaine, J., Froend, R., Matusick, G., Kala, J., et al. (2018). Subcontinental heat wave triggers terrestrial and marine, multi-taxa responses. *Scientific Reports, 8*, 13094. https://doi.org/10.1038/s41598-018-31236-5

Santora, J., Mantua, N., Schroeder, I., Field, J., Hazen, E., Bograd, S. et al. (2020). Habitat compression and ecosystem shifts as potential links between marine heatwave and record whale entanglements. *Nature Communications, 11* (536), 1–12. https://doi.org/10.1038/s41467-019-14215-w

Schuster, R., Germain, R. R., Bennett, J. R., Reo, N. J., & Arcese, P. (2019). Vertebrate biodiversity on indigenous-managed lands in Australia, Brazil, and Canada equals that in protected areas. *Environmental Science and Policy, 101*, 1–6. https://doi.org/10.1016/j.envsci.2019.07.002

Scridel, D., Brambilla, M., Martin, K., Lehikoinen, A., Iemma, A., Matteo, A., et al. (2018). A review and meta-analysis of the effects of climate change on Holarctic mountain and upland bird populations. *Ibis, 160*, 489–515. https://doi.org/10.1111/ibi.12585

Shi, H., Li, X., Liu, X., Wang, S., Liu, X., Zhang, H., Tang, D., & Li, T. (2020). Global protected areas boost the carbon sequestration capacity: Evidences from econometric causal analysis. *Science of the Total Environment, 715*, 137001. https://doi.org/10.1016/j.scitotenv.2020.137001

Smith, R., Cannizzo, Z., Belle, E.. & Wenzel, L. (2020). Role of protected areas in climate change mitigation, adaptation and disaster risk reduction. climate action. In W. Leal Filho, A. Azul, L. Brandli, P. Özuyar, P & T. Wall (Eds.), *Encylopedia of the UN Sustainable Development Goals*. Springer. https://doi.org/10.1007/978-3-319-71063-1_142-1

Smith, R., Guevara, O., Wenzel, L., Dudley, N., Petrone-Mendoza, V., Cadena, M. & Rhodes, A. (2019). Ensuring co-benefits for biodiversity, climate change and sustainable development. In W. Leal Filho, J. Barbir & R. Preziosi (Eds.), *Handbook of Climate Change and Biodiversity*, (151–166). Springer. https://doi.org/10.1007/978-3-319-98681-4_9

Soto-Navarro, C., Ravilious, C., Arnell, A., de Lamo, X., Harfoot, M., Hill, S., et al. (2020). Mapping co-benefits for carbon storage and biodiversity to inform conservation policy and action. *Philosophical Transactions of the Royal Society B: Biological Sciences, 375*, 20190128. https://doi.org/10.1098/rstb.2019.0128

Sreekar, R., Zeng, Y., Zheng, Q., Lamba, A., Teo, H., Sarira, T., & Koh, L. (2022). Nature-based climate solutions for expanding the global protected area network. *Biological Conservation, 269*, 109529. https://doi.org/10.1016/j.biocon.2022.109529/

Storlazzi, C., Reguero, B., Cole, A., Lowe, E., Shope, J., Gibbs, A. et al. (2019). *Rigorously Valuing the Role of US Coral Reefs in Coastal Hazard Risk Reduction*. US Geological Survey, Open File Report 2019–1027. https://pubs.usgs.gov/of/2019/1027/ofr20191027.pdf. Accessed 18 April 2023.

Strack, M., Davidson, S., Hirano, T., & Dunn, C. (2022). The potential of peatlands as nature-based climate solutions. *Current Climate Change Reports, 1–12,*. https://doi.org/10.1007/s40641-022-00183-9

Stramma, l. & Schmidtko, S. (2019). Global evidence of ocean deoxygenation. In D. Laffoley & J. Baxter (Eds.) *Ocean Deoxygenation: Everyone's Problem* (pp. 23–36). Gland, Switzerland: IUCN. https://portals.iucn.org/library/sites/library/files/documents/02.1%20DEOX.pdf. Accessed 18 April 2023.

Suggitt, A., Gillingham, P., Hill, J., Huntley, B., Kunin, W., Roy, D., & Thomas, C. (2011). Habitat microclimates drive fine-scale variation in extreme temperatures. *Oikos, 120*, 1–8. https://doi.org/10.1111/j.1600-0706.2010.18270.x

Suggitt, A., Wilson, R., Isaac, N., Beale, C., Auffret, A., August, T., et al. (2018). Extinction risk from climate change is reduced by microclimatic buffering. *Nature Climate Change, 8*(8), 713–717. https://doi.org/10.1038/s41558-018-0231-9

Taillardat, P., Friess, D., & Lupascu, M. (2018). Mangrove blue carbon strategies for climate change mitigation are most effective at the national scale. *Biology Letters, 14*, 20180251. https://doi.org/10.1098/rsbl.2018.0251

Taillardat, P., Thompson, B., Garneau, M., Trottier, K., & Friess, D. (2020). Climate change mitigation potential of wetlands and the cost-effectiveness of their restoration. *Interface Focus, 10*, 20190129. https://doi.org/10.1098/rsfs.2019.0129

Terraube, J., & Fernández-Llamazares, A. (2020). Strengthening protected areas to halt biodiversity loss and mitigate pandemic risks. *Current Opinion in Environmental Sustainability, 46*, 35–38. https://doi.org/10.1016/j.cosust.2020.08.014

Terraube, J., Fernández-Llamazares, A., & Cabeza, M. (2017). The role of protected areas in supporting human health: A call to broaden the assessment of conservation outcomes. *Current Opinion in Environmental Sustainability, 25*, 50–58. https://doi.org/10.1016/j.cosust.2017.08.005

Thomas, C., & Gillingham, P. (2015). The performance of protected areas for biodiversity under climate change. *Biological Journal of the Linnean Society, 115*, 718–730. https://doi.org/10.1111/bij.12510

Thomas, C., Gillingham, P., Bradbury, R., Roy, D., Anderson, B., Baxter, J., et al. (2012). Protected areas facilitate species' range expansions. *Proceedings of the National Academy of Sciences, 109*, 14063–14068. https://doi.org/10.1073/pnas.1210251109

UNFCCC (United Nations Framework Convention on Climate Change) (2021). The ocean. https://unfccc.int/topics/ocean. Accessed 15 August 2022.

Vergés, A., Steinberg, P., Hay, M., Poore, C., A., Ballesteros, E., et al. (2014a). The tropicalization of temperate marine ecosystems: Climate-mediated changes in herbivory and community phase shifts. *Proceedings of the Royal Society B: Biological Sciences, 281*, 20140846. https://doi.org/10. 1098/rspb.2014.0846

Vergés, A., Tomas, F., Cebrian, E., Ballesteros, E., Kizilkaya, Z., Dendrinos, P., et al. (2014b). Tropical Rabbitfish and the deforestation of a warming temperate sea. *Journal of Ecology, 102*, 1518–1527. https://doi.org/10. 1111/1365-2745.12324

Vergés, A., Doropoulos, C., Malcolm, H., Skye, M., Garcia-Pizá, M., Marzinelli, E., Campbell, A., et al. (2016). Long-term empirical evidence of ocean warming leading to tropicalization of fish communities, increased herbivory, and loss of kelp. *Proceedings of the National Academy of Sciences, 113*, 13791–13796. https://doi.org/10.1073/pnas.1610725113

Virkkala, R., Heikkinen, R., Kuusela, S., Leikola, N., & Pöyry, J. (2019). Significance of protected area network in preserving biodiversity in a changing northern European climate. In W. Leal-Filho, L. Barbir, & R. Preziosi (Eds.), *Handbook of climate change and biodiversity* (pp. 377–390). Springer. https://doi.org/10.1007/978-3-319-98681-4_23

Wambugu, P., & Henry, R. (2022). Supporting in-situ conservation of the genetic diversity of crop wild relatives using genomic technologies. *Molecular Ecology, 31*, 2207–2222. https://doi.org/10.1111/mec.16402

Watson, J., Evans, T., Venter, O., Williams, B., Tulloch, A., Stewart, C., et al. (2018). the exceptional value of intact forest ecosystems. *Nature Ecology and Evolution, 2*, 599–610. https://doi.org/10.1038/s41559-018-0490-x

West, T., Börner, J., Sills, E., & Kontoleon, A. (2020). Overstated carbon emission reductions from voluntary redd+ projects in the Brazilian Amazon. *Proceedings of the National Academy of Sciences, 117*, 24188–24194. https://doi.org/10.1073/pnas.2004334117

White, E., & Kaplan, D. (2017). Restore or retreat? Saltwater intrusion and water management in coastal wetlands. *Ecosystem Health and Sustainability, 3*, e01258. https://doi.org/10.1002/ehs2.1258

Wigand, C., Ardito, T., Chaffee, C., Ferguson, W., Paton, S., Raposa, K., et al. (2017). A climate change adaptation strategy for management of coastal marsh systems. *Estuaries and Coasts, 40*, 682–693. https://doi.org/10.1007/s12 237-015-0003-y

Wilson, S. & Hebda, R. (2008). *Mitigating and adapting to climate change through the conservation of nature.* Land Trust Alliance of British Columbia, Canada. https://ltabc.ca/wp-content/uploads/2012/02/LTA_ ClimateChangePrint.pdf. Accessed 16 April 2023.

Woods Hole Climate Research Centre, Wild Heritage, Intact, Griffith University, Geos Institute, Australian Rainforest Conservation Society & Frankfurt Zoological Society (2020). *Policy briefs: primary tropical, temperate and boreal forests: critical stores of carbon, biodiversity and freshwater.* https://www.griffith.edu.au/research/research-excellence/griffith-climate-change-response-program/primary-forest-protection-boreal-and-temperate-primary-forests/primary-forests-information. Accessed 17 April 2023.

Xu, X., Huang, A., Belle, E., De Frenne, R. & Jia, G. (2022). Protected areas provide thermal buffer against climate change. *Science Advances, 8* (44). https://doi.org/10.1126/sciadv.abo0119

Zhang, K., Liu, H., Li, Y., Xu, H., Shen, J., Rhome, J., & Smith iii, T. (2012). The role of mangroves in attenuating storm surge. *Estuarine, Coastal and Shelf Science, 102*, 11–23. https://doi.org/10.1016/j.ecss.2012.02.021

Zhu, Z., Middleton, B., Pindilli, E., Johnson, D., Johnson, K., & Covington, S. (2022). Conservation of carbon resources and values on public lands: A case study from the national wildlife refuge system. *PLoS ONE, 17*, e0262218. https://doi.org/10.1371/journal.pone.0262218

The Virtual Wild: Exploring the Intersection of Virtual Reality and Natural Environments

Simone Grassini and Eleanor Ratcliffe

17.1   INTRODUCTION

Mental health is a leading cause of disability globally and there is a growing emphasis on promoting mental well-being (UN General Assembly, 2011; WHO, 2013; 2014). The World Health Organisation (WHO) defines mental health as a state that allows individuals to utilise their abilities, cope with stress, work effectively, and contribute to their community (WHO, 2004). There is evidence linking mental health to experience of nature, including parks and public green spaces in urban areas (Bratman et al., 2012; Douglas 2012; Kaplan, 2001; Maller et al., 2016). While some studies have shown positive associations between parks

S. Grassini (✉)
Department of Psychosocial Science, University of Bergen, Bergen, Norway
e-mail: simone.grassini@uib.no

Cognitive and Behavioral Neuroscience Lab, University of Stavanger, Stavanger, Norway

E. Ratcliffe
School of Psychology, University of Surrey, Guildford, UK
e-mail: eleanor.ratcliffe@surrey.ac.uk

© The Author(s) 2024   327
N. Finneran et al. (eds.), *Managing Protected Areas*,
https://doi.org/10.1007/978-3-031-40783-3_17

and mental health, the effects can vary by gender and age (Annerstedt et al., 2012; Astell-Burt et al., 2014). The quality of parks and green spaces has been shown to be important for mental health, with residents in neighbourhoods with higher-quality parks having lower levels of psychological distress (Francis et al., 2012). Green spaces also play a crucial role in advancing biodiversity and ecological health and are therefore important for both human and planetary well-being.

One of the most significant impacts of urbanisation on the environment is the loss of green spaces. As cities grow, more land is needed to accommodate the growing population, resulting in the destruction of natural habitats. According to a study by the United Nations, the world's urban areas are expected to grow by 2.5 billion people by 2050, which means that more green spaces will be lost to make room for buildings and infrastructure (United Nations, 2018). The scarcity of green spaces in cities has numerous negative effects on people's health and well-being. Studies have shown that access to green spaces can improve mental health and reduce stress levels (Bowler et al., 2010). Without green spaces, people are more likely to experience mental health problems, such as anxiety and depression. Additionally, the lack of green spaces can lead to physical health problems, such as obesity, as people are less likely to engage in outdoor activities.

Virtual reality (VR) has emerged as a potential solution to bridge the gap between people and nature. VR technology can simulate natural environments and provide exposure to a range of settings from scenic landscapes to wildlife habitats, in a controlled and safe environment. VR also has the potential to reach a wide audience, including people who may have difficulty accessing natural environments due to physical or financial limitations. Thus, the integration of VR technology in promoting exposure to natural environments has the potential to offer a new and innovative approach for education, communication, and well-being promotion.

17.2   Overview of Virtual Reality Technology: Definition, Development, and Its Potential Uses

Virtual Reality (VR) has been a popular topic in information technology for the past several years, with the release of the film *Ready Player One* by Steven Spielberg bringing VR to the attention of millions of people. However, VR technology has been around since the 1960s, but its widespread adoption has been hindered by high equipment costs and low quality (Newman et al., 2022; Valmaggia, 2017). VR has been referred to as 'over-hyped' for a long time (Walsh & Pawlowski, 2002) and was even thought to be a 'dead' technology (Slater & Sanchez-Vives, 2016), but the release of affordable, high-quality, and consumer-grade VR headsets for gaming and entertainment has contributed to the revival of the VR technology starting from the mid-2010s.

VR technology simulates a virtual environment that immerses users, giving them a *sense of presence* or of 'being there' (Bowman & McMahan, 2007). Researchers from various disciplines, including computer science, engineering, and the social sciences, have been studying VR technology for decades, but the focus has mostly been on the technology itself, leaving limited understanding of its behavioural and organisational impacts (Walsh & Pawlowski, 2002). As a result, interest in VR research has been increasing among information system researchers (Cavusoglu et al., 2019; Khairunisa et al., 2022; Lee, 2022). The VR market has been estimated to grow from $7.3 billion in 2018 to $120.5 billion in 2026 (Fortune Business Insights, 2019, as reported in Yang & Han, 2021), with a significant portion of the market being consumer software, primarily video games. However, companies such as IKEA, Volkswagen, and Takeda have started to use VR technology for non-gaming purposes. Further development of the idea of using simulated worlds for non-gaming purposes was also embraced by Meta and the development of the idea of the metaverse (Duan, et al., 2021; Knox, 2022). Despite some potentially detrimental side effects, VR technology has also gained uses in the context of health (Grassini & Laumann, 2021).

The ability for VR to promote a high sense of presence in users has been exploited for both clinical uses and well-being promotion (Frost et al., 2022). The technology has been shown to be a promising tool for diagnosing and treating medical conditions, including stress, social anxiety, chronic pain, and Alzheimer's disease (García-Betances et al., 2015; Grassini, 2022a, 2022b; Jones et al., 2016). VR has been effective in treating PTSD and anxiety disorders, as well as distracting patients from physical pain during medical procedures (Indovina et al., 2018). VR-based diagnostic testing could potentially provide objective results for conditions such as schizophrenia, ADHD, and autism, leading to earlier and more accurate diagnoses (Kim et al., 2019a, 2019b).

17.3   Closing the Knowledge-To-Action Gap: Communication Strategies for Climate Change Mitigation and Adaptation

The Intergovernmental Panel on Climate Change (IPCC) has published a special report on the impacts of global climate change on nature and society (Masson-Delmotte et al., 2018). The report highlights the dire consequences of global warming, even if the Earth's temperature increases by just 0.5°C. It predicts that rising temperatures will lead to food shortages, increased wildfires, and the mass die-off of coral reefs by as early as 2040. Despite increasing media attention on the human-induced rate of climate change, research from the behavioural sciences has found that people may perceive climate change as a non-urgent, psychologically distant risk, which has led to delayed action and decision-making about mitigation and adaptation (Van der Linden et al., 2015). However, it is worth noting that an increased awareness of problems related to climate change and concerns about the phenomenon have become more common among the general population (Leiserowitz et al., 2022). This has led to new psychological phenomena of distress, generally referred in current literature as climate anxiety (for a review see Soutar & Wand, 2022).

Traditionally, one of the greatest challenges in communicating climate science to the public is bridging the knowledge-to-action gap (Moser & Dilling, 2011). This gap refers to the general lack of environmental behaviour change by individuals or society at large, even when there is greater communication about environmental problems and heightened public awareness. To bridge this gap, successful communication methods involve cognitive, affective, and behavioural dimensions (Lorenzoni et al., 2007; Moser & Dilling 2011; Nelson et al., 2018a, 2018b; Van der Linden et al., 2015). For example, people need to gain an understanding of the issue, experience an emotional and cognitive response—such as interest or worry—and then actively respond by changing their behaviour or taking political action. Research has shown that the message content (Gifford & Comeau, 2011) and level of visual immersion (Innocenti, 2017; Rosenberg et al., 2013; Shin, 2018) can impact the intensity of emotional responses and influence behaviour (Baberini et al., 2015).

The nature documentary series Our Planet has recently gained significant media attention for using negative framing of human-induced climate change impacts on nature to encourage viewers to visit a conservation website (Lowry, 2019). In addition to framing, virtual reality (VR) technology has also garnered attention for its ability to increase immersion and one's sense of presence, which is believed to impact emotions and improve people's connection to the subject matter (Diemer et al., 2015). In 2017, Fortune magazine published an article claiming that virtual reality (VR) can increase awareness, evoke empathy, and elicit action, citing a Facebook report that '48% of virtual reality charity content viewers were likely to donate to the causes they experienced' and a report by the United Nations stating that their VR production Clouds over Sidra, which depicts the life of a 12-year-old Syrian refugee, helped raise twice the charity's normal rate (Samit, 2017).

However, these studies lack controlled experimental designs, and other uncontrolled factors may have contributed to the supposed increase in donations, such as increased media attention on the Syrian crisis that year, regardless of VR technology's use at the fundraiser. The Facebook report and a study by the market research company Nielsen (2017) found that donation intentions were higher after exposure to a 360° video compared to other media, but these only considered hypothetical donation decisions. The idea of experimentally testing the impact of different levels of immersion (low vs. high immersive environments) on emotional response and prosocial behaviour emerged from the recent surge of media articles suggesting that virtual reality 'can make you a better person' and is 'the ultimate empathy machine' (Chang, 2018; Millar et al., 2023; Samit, 2017; West, 2015). Nelson et al. (2020) further explored this concept in their study reporting promising results.

17.4   Virtual Reality and Prosocial Behaviour: Evidence from Laboratory Studies

Several laboratory studies at Stanford University's Virtual Human Interaction Lab focus on the effects of VR on encouraging prosocial behaviour (Ahn et al., 2014, 2016) and environmental behavioural intentions (Ahn et al., 2014). Evidence suggests that VR interactions can increase a person's sense of presence, improving their connection to the subject matter and affecting behaviour. In one study, Ahn et al. (2014) compared the effects of cutting a tree in VR to reading a written description or watching a video of the tree-cutting process to promote paper conservation. The virtual experience resulted in a 20% decrease in paper use during the experiment compared to participants who read a print description of tree-cutting. In another experiment by Ahn et al. (2016), participants wore a VR headset and experienced what it's like to be a cow raised for dairy and meat. After the experience, participants reported eating less meat. However, preliminary research by the authors of this paper found no statistical differences in university students' charitable giving behaviour for coral reef conservation exposed to varying levels of immersion in an underwater VR world (Nelson et al., 2019). While some scholars believe VR can create empathy with non-human actors, such as simulating the experience of being a cow or coral (Ahn et al., 2016), other scholars are critical of this concept (Ramirez, 2017).

17.5   Emotional Framing of Environmental Communication Using Immersive Systems

In the context of communication, framing refers to the semantic restructuring of identical information (Hallahan, 2008). Valence framing involves presenting information in either a positive or negative light. Positive message valence describes actions that lead to favourable outcomes, while negative message valence describes the consequences of inaction, which lead to adverse conditions (Avineri & Waygood, 2013). Classic economic theory assumes that presenting information in different frames should not affect an individual's preferences. However, research has shown that people's decisions can be influenced by positive or negative semantics, a cognitive bias known as the framing effect (Tversky & Kahneman, 1981).

In the realm of climate change communication, it is crucial to frame messages in a way that encourages environmentally friendly behaviour. While the effects of framing on attitudes are somewhat understood, there is a surprising scarcity of empirical research examining the effects of message framing on pro-environmental behaviour (Cheng et al., 2011; Jacobson et al., 2019). Furthermore, the literature on positive and negative message framing is inconsistent and highly contextual (Rothman et al., 2006). Virtual reality shows promise in enhancing pro-environmental behaviour and charitable giving, despite current limited evidence and the need for more research to understand the underlying mechanisms and address limitations of existing studies.

17.6   Virtual Nature Health Promotion

As noted earlier in this chapter, a substantial body of research has established the connection between nature experience and positive psychological outcomes such as changes in affect, stress recovery, and cognitive restoration (Browning et al., 2020). A significant portion of these studies make use of simulated or virtual nature, which includes visual media, recorded sounds/soundscapes, and virtual reality (VR) in laboratory settings. Prior research has explored the impact of computer-mediated forms of nature, such as images displayed in slide shows (Berman et al., 2008; Gladwell et al., 2012) or videos (Mayer et al., 2009; Van den Berg et al., 2003), on physiological and psychological processes.

For instance, Gladwell et al. (2012) found that viewing nature images led to a significant increase in heart rate variability compared to viewing urban images, and van den Berg et al. (2003) reported improved mood after watching a video of a forest walk compared to a street walk. Grassini et al. (2019) found that people looking at natural images reported to be more relaxed than people looking at urban scenes, and the effect was detectable also from neurophysiological markers. Grassini et al. (2022) found that a similar psychophysiological effect was detectable when participants were exposed to videos of natural environments compared to videos of urban or neutral environments. These studies aimed to control confounding variables by examining the effects of actual nature in controlled experimental settings. However, these studies have all used exposition media that deliver a low level of immersion and therefore sense of presence. Modern VR systems, on the other hand, can be used to stimulate user senses in a more realistic and immersive manner.

Despite not affording the same consistency or strength of outcomes as direct nature experience, VR nature can still have psychological benefits as shown by several studies (Bolouki, 2022; Lee et al., 2022; Ünal et al., 2022). VR nature can provide an alternative for individuals who are unable to experience real nature due to various reasons, including physical limitations, urbanisation, or lack of access to green spaces (Li et al., 2021). Additionally, VR nature can provide a balance between the desire for environmental experience and the need for environmental protection and health and safety, as seen in the use of VR for heritage sites (Godovykh et al., 2022). Such use of VR can help individuals learn about the history, culture, and significance of these sites without putting the sites themselves at risk of damage or degradation.

Systematic reviews and meta-analyses of relevant studies indicate that VR nature is associated with increased positive affective states and reduced negative affective states, self-reported stress, and fatigue (Li et al., 2022). Lee et al. (2022) found in their systematic review that affective recovery from negative emotions was the most consistently observed outcome in VR nature studies. This highlights the potential for VR nature to help individuals cope with stress and negative emotions, and promote positive well-being. Meanwhile, Bolouki's (2022) systematic review and meta-analysis found both positive and negative affect to be consistently affected by VR nature, with greater effects for negative affect. This suggests that VR nature can have a broad impact on individuals' emotional states and may be particularly useful for improving mood and reducing negative affect. However, there are also reports of negative affect reductions in response to VR built environments (Bolouki, 2022). Thus, researchers should consider the psychological benefits of different environments, not just nature, in their studies of VR environments and their impact on psychological well-being (e.g., see Weber & Trojan, 2018, for a review of restorative urban environments).

The use of VR built environments can help individuals experience the benefits of urban green spaces, or the benefits of visiting heritage sites in situations where physical access is not possible. The way VR nature is presented also plays a crucial role in determining its psychological benefits. Li et al. (2022) found that medium immersion in virtual nature was associated with greater positive affect outcomes compared to high or low immersion. This suggests a trade-off between creating a sense of presence in the VR environment and avoiding adverse effects such as VR sickness, which can be attributed to the use of head-mounted devices (HMDs). Lee et al. (2022) reported that exposure duration of 10 minutes or more was associated with more consistent affective benefits but noted the need to balance longer exposure time with the increased possibility of negative user experiences. Bolouki (2022) also highlighted that users' comfort and familiarity with VR technology can impact the psychological benefits achieved in simulated nature. It is essential to consider how VR environments are presented to users in order to maximise their potential to generate positive psychological outcomes.

In conclusion, while VR nature may not have the same psychological benefits as direct nature experience, it can still have positive effects on affective states. Researchers should focus on ensuring that users feel immersed in the VR environment while minimising the occurrence of adverse effects, which may depend on the technology used, administration, and target audience. VR nature can provide an alternative for individuals who are unable to experience real nature and may be particularly useful for improving mood and reducing negative affect. Furthermore, the use of VR nature can provide a balance between the desire for environmental experience and the need for environmental protection and health and safety, making it a valuable tool for promoting well-being and learning about heritage sites. It is important for researchers to continue exploring the potential of VR nature and how it can best be utilised to promote positive psychological outcomes.

17.7 Technology for Ecotourism

The United Nations General Assembly's 2030 Agenda for Sustainable Development includes 17 Sustainable Development Goals, aiming to address environmental, social, and economic challenges (Gue et al., 2020). Among these goals, environmental concerns are particularly challenging, as human activities are often the cause of environmental degradation. The tourism industry, for example, has been associated with environmental degradation due to its contribution to global carbon emissions (Bhutto et al., 2021a, 2021b; Lenzen et al., 2018). However, the tourism sector also contributes significantly to economic growth and cultural enrichment (Movono et al., 2018). As a result, it is crucial to strike a balance between economic growth and environmental sustainability and explore sustainable solutions to reduce the negative impact of tourism on the environment.

One of the proposed solutions to reduce the impact of tourism on the environment is virtual reality tourism (VRT). VRT involves the simulation of tourist destinations through visualisation, immersion, and interactivity, providing individuals with a realistic experience of tourism without the need for physical travel (Gutierrez et al., 2008; Guttentag, 2010). VRT can be a sustainable alternative to traditional tourism, as it does not generate pollution or contribute to environmental degradation. However, it's important to consider that VRT may not always have the same economic benefits as traditional tourism, despite its positive impact on the environment. Therefore, while VRT has the potential to be a sustainable alternative to traditional tourism, it is crucial to assess its full impact on both the environment and the economy before fully embracing it as a solution.

Recent advancements in virtual reality technology, coupled with the COVID-19 pandemic's impact on travel, have made VRT a more viable option for sustainable tourism (Talwar et al., 2022). VRT is not only a solution for sustainable tourism, but it has also been employed to promote awareness of endangered areas such as marine conservation (Koh et al., 2023), the protection of the Great Barrier Reef (Reilly, 2017), and other protected or endangered areas (e.g., Ecoegypt nd.). Organisations such as national parks have also recently adopted VR solutions to allow citizens to experience and explore the great outdoors (some examples: Birtles nd; Lane, 2020; Watson nd). To ensure the long-term acceptance and use of VRT, it is crucial to understand individuals' perceptions and attitudes towards this technology. Research in this area is essential as it can help promote sustainable consumption and drive societal, national, and global sustainability orientation (Anderson & Bows, 2011; Lorek & Fuchs, 2013).

17.8   CHALLENGES AND LIMITATIONS

Virtual Reality (VR) technology has come a long way in terms of its development but still has a long way to go before it can be fully utilised in all its potential applications. One major challenge facing VR technology is the lack of standardisation in its presentation and technology (LaRocco, 2020). Currently, each developer has their own specifications and functionality, making it difficult for applications to be easily transferable between devices. The only standardisation that can be seen is with popular games that are designed to work across different VR platforms. This lack of standardisation also makes it challenging to troubleshoot bugs and receive proper support for any issues. Efforts to standardize VR technology are underway, but they are still in their early stages (LaRocco, 2020).

VR technology is facing other challenges regarding the hardware and software requirements for professional VR development. VR development software tends to consume a large amount of data space on computers and has high power consumption (Fernández & Alonso, 2015). VR headsets can also be heavy and cause physical strain on users, leading to headaches and pain, especially in the neck and shoulders (Kaplan et al., 2020). The long-term effects of VR use on users' eyesight are not yet known, but it is known to cause eye strain, especially with prolonged usage (Hirzle et al., 2022). Long-term effects of prolonged use of VR have not been studied yet, but it has been found that there may be health risks associated with its use (e.g., Grassini & Laumann, 2021), and that individual characteristics of the users may make some of them more susceptible to ill effects related to the use of VR (Grassini & Laumann, 2020; Grassini et al., 2021; Thorp et al., 2022, 2023).

A common issue in VR technology is the lag between a user's movements and the visual display within a VR headset (Laviola, 2000). The headset's tracking often does not keep up with the user's movements, reducing their immersion and causing dizziness or 'cybersickness' which is a phenomenon where users feel symptoms of motion sickness, characterised by, e.g., nausea, dizziness, and light-headedness, because of using a VR device (Hamad, 2021; McCauley & Sharkey, 1992). The exact cause of cybersickness is not yet known, but it is believed to be due to a conflict between the user's visual system and vestibular system or an inability to perceive or react to new dynamic situations (Laviola, 2000; Stanney et al., 2020). Cybersickness can be exacerbated by several factors, including prolonged VR exposure, the user's predisposition to motion sickness, fatigue, or nausea, and how adapted the user is to VR applications (Laviola, 2000). Technical factors, such as noticeable lags, position tracking errors, and visual distortions, can also increase the likelihood of cybersickness (Laviola, 2000). If users continue to experience cybersickness, it can be a hindrance to the widespread development and utilisation of VR applications (Stanney et al., 2020).

Despite these challenges, VR technology is becoming increasingly accessible as it evolves. The cost of VR headsets is still high, but it is comparable to most gaming consoles. Headsets like the Oculus Quest 2, for instance, cost about $300 for the base model and can be fully operated without a computer, making it one of the more accessible VR headsets on the market. Most other VR headsets, however, require a high-end computer with a powerful graphics card that can manage VR applications, making them more expensive overall and out of reach for most people (Kaplan et al., 2020). The cost of VR technology is one of the major barriers to its growth as a household technology, and often to its use in research (Newman et al., 2022).

As VR technology evolves, it is becoming more accessible, especially compared to its earlier stages. The cost of VR headsets on the market is still higher than most people can afford, but it is now on par with most gaming consoles. Consumer-oriented last-generation headsets like the Oculus Quest 2 are becoming increasingly affordable and can be fully operated without a computer. However, most of the 'traditional' VR headsets require a 'VR-ready' computer, which is typically more expensive than most computers, making them more expensive overall and less accessible to most people. Furthermore, the general acceptability of VR among different groups of users may vary depending on various factors, such as age, sex, motivation, and perceived benefits and risks. One of the main factors that may influence the acceptability of VR is age. Older people may have different needs, preferences, and expectations than younger people when using VR. For example, older people may be more concerned about the usability and accessibility of VR devices, as well as the potential negative effects of VR on their health and well-being (e.g., motion sickness, eye strain, isolation). On the other hand, older people may also benefit from using VR for enhancing their cognitive abilities, social engagement, and quality of life (Abdul Rahman et al., 2020). VR has been used for both assessment and training of elderly with impaired episodic memory (La Corte et al., 2019) and as a potential aid for patients with dementia (Hayhurst, 2018).

VR may also pose some risks for communication strategies, as it may be argued that it may provoke disengagement from the real world and potentially reduce the sense of empathy with real entities. However, the current literature on the topic suggests instead a positive effect of VR on empathic feelings towards social situations in the physical world (Schutte & Stilinović, 2017). VR may also raise ethical concerns, as it may expose users to unwanted or harmful content, manipulate their perceptions and emotions, or collect their personal data without their consent (Gonzalez-Franco et al., 2019). A further challenge and potential limitation for the mass adoption of VR systems is how to handle and collect data from the immersive environments. VR systems can generate large amounts of data from various sources, such as user interactions, eye tracking, physiological sensors, and environmental variables (Feltham, 2018).

These data can provide valuable insights into the user's behaviour, preferences, emotions, and cognitive processes. However, they also pose ethical and practical issues regarding data privacy, security, storage, analysis, and presentation (Addis & Kutar, 2018). For example, VR data can reveal sensitive information about the user's identity, health condition, personality traits, or political views (Feltham, 2018). Therefore, researchers need to ensure that they follow appropriate protocols for obtaining informed consent from participants, protecting their anonymity and confidentiality, encrypting, and storing data securely, and complying with relevant regulations and standards, as the European GDPR framework (Henriksson, 2018).

17.9 Conclusion

In conclusion, virtual reality technology has the potential to revolutionise environmental education by providing immersive and interactive learning experiences. The use of VR in environmental education can offer a fresh perspective and a new dimension to learning, making it more engaging and interesting. It can effectively educate individuals on the impact of human actions on the environment and the importance of preserving natural spaces. The use of VR in environmental education can also provide a platform for learners to witness the consequences of their actions and the positive impact of their conservation efforts. Furthermore, VR can be integrated with current tourism strategies, such as virtual guided tours, or guided tours using mixed reality (where virtual elements are added in addition to the real world, see e.g., Futurism, n.d.). This method to promote engaging tourism experiences has been already studied in the context of heritage tourism (see, e.g., Jiang et al., 2023), which can offer a more informative travel experience. This can promote a greater appreciation for the environment and inspire learners to take action to protect it.

Another significant application of VR is improving the health of those who cannot access natural environments by providing nature exposure. This can be especially beneficial for individuals who are unable to travel or have limited mobility. VR can simulate natural environments and offer a calming and restorative experience for individuals who are experiencing stress, anxiety, or depression. However, it is important to note that further research is necessary to explore the full potential of VR in environmental education and its effectiveness in promoting environmental awareness and conservation. The technology is still relatively novel, and there is much to be learned about its impact and effectiveness. Furthermore, it may be questioned whether or not the technology has reached a level of maturity sufficient for a large-scale acceptability. In summary, VR offers a valuable opportunity to inspire individuals to make positive changes to the environment. It can provide a dynamic and engaging learning experience, promote greater appreciation for nature, and improve the health and well-being of individuals. As this technology continues to evolve, it is essential to explore its potential and use it as a tool to create a more sustainable and environmentally conscious society.

REFERENCES

Abdul Rahman, N., Mohamad, A., Ahmad, M., & Yusof, A. (2020). Acceptability of Virtual Reality among Older People. In *Proceedings of International Conference on HCI for Cybersecurity Privacy & Trust HCICPT 2020: HCI International 2020 – Late Breaking Papers: Human Aspects in Cybersecurity Privacy & Trust* (pp. 3–15).

Addis, C., & Kutar, M. (2018, March). The general data protection regulation (GDPR), emerging technologies and UK organisations: Awareness, implementation and readiness. In *UK Academy for Information Systems Conference Proceedings 2018* (p. 29). UKAIS–UK Academy for Information Systems.

Ahn, S. J., & G., Bailenson, J. & Dooyeon Park. (2014). Short- and long-term effects of embodied experiences in immersive virtual environments on environmental locus of control and behavior. *Computers in Human Behavior, 39*, 235–245. https://doi.org/10.1016/j.chb.2014.07.014

Ahn, S. J., & G., Bostick, J., Ogle, E., Nowak, K., McGillicuddy, K. & Bailenson, J. (2016). Experiencing nature: Embodying animals in immersive virtual environments increases inclusion of nature in self and involvement with nature. *Journal of Computer-Mediated Communication, 21*(6), 399–419. https://doi.org/10.1111/jcc4.12172

Anderson, K., & Bows, A. (2011). Beyond 'dangerous' climate change: Emission scenarios for a new world. *Philosophical Transactions of the Royal Society A-Mathematical Physical and Engineering Sciences, 369*(1934), 20–44. https://doi.org/10.1098/rsta.2010.0290

Annerstedt, M., Östergren, P. O., Björk, J., Grahn, P., Skärbäck, E., & Währborg, P. (2012). Green qualities in the neighbourhood and mental health–results from a longitudinal cohort study in Southern Sweden. *BMC Public Health, 12*, 1–13.

Astell-Burt, T., Mitchell, R., & Hartig, T. (2014). The association between green space and mental health varies across the lifecourse. A longitudinal study. *Journal Epidemiol Community Health, 68*(6), 578–583.

Avineri, E., Owen, E., & Waygood, D. (2013). Applying valence framing to enhance the effect of information on transport-related carbon dioxide emissions. *Transportation Research Part a: Policy and Practice, 48*, 31–38. https://doi.org/10.1016/j.tra.2012.10.003

Baberini, M., Coleman, C., Slovic, P., & Västfjäll, D. (2015). Examining the effects of photographic attributes on sympathy, emotions, and donation behavior. *Visual Communication Quarterly, 22*(2), 118–128. https://doi.org/10.1080/15551393.2015.1061433

Berman, M. G., Jonides, J., & Kaplan, S. (2008). The cognitive benefits of interacting with nature. *Psychological Science, 19*(12), 1207–1212.

Bhutto, S., Butt, F., Abbas, Q., Rehman, Z., & Shaikh, F. (2021a). Carbon footprint analysis of global inbound tourist destinations: An empirical study based on panel data estimation techniques. *Journal of Cleaner Production, 279*, 123629.

Bhutto, T., Farooq, R., Talwar, S., Awan, U., & Dhir, A. (2021b). Green inclusive leadership and green creativity in the tourism and hospitality sector: Serial mediation of green psychological climate and work engagement. *Journal of Sustainable Tourism, 29*(10), 1716–1737. https://doi.org/10.1080/096 69582.2020.1867864

Birtles. (n.d.). National parks online: Virtual tours. Trafalgar. https://www.tra falgar.com/real-word/national-parks-online-virtual-tours/. Accessed 9 March 2023.

Bowler, D., Buyung-Ali, L., Knight, T., & Pullin, A. (2010). A systematic review of evidence for the added benefits to health of exposure to natural environments. *BMC Public Health, 10*, 456. https://doi.org/10.1186/1471-2458-10-456

Bratman, G. N., Hamilton, J. P., & Daily, G. C. (2012). The impacts of nature experience on human cognitive function and mental health. *Annals of the New York Academy of Sciences, 1249*(1), 118–136.

Browning, M. H., Shipley, N., McAnirlin, O., Becker, D., Yu, C. P., Hartig, T., & Dzhambov, A. M. (2020). An actual natural setting improves mood better than its virtual counterpart: A meta-analysis of experimental data. *Frontiers in Psychology, 11*, 2200.

Bolouki, A. (2022). The impact of virtual reality natural and built environments on affective responses: A systematic review and metaanalysis. *International Journal of Environmental Health Research*, 1–17.

Bowman, D. A., & McMahan, R. P. (2007). Virtual reality: How much immersion is enough? *Computer, 40*(7), 36–43.

Cavusoglu, H., Dennis, A. R., & Parsons, J. (2019). immersive systems. *Journal of Management Information Systems, 36*(3), 680–682.

Chang, P. (2018). Using virtual reality for increased charity donor outreach and funding. *Journal of Nonprofit and Public Sector Marketing, 30*(4), 434–450. https://doi.org/10.1080/10495142.2017.1399185

Cheng, T., Woon, D., & Lynes, J. (2011). The use of message framing in the promotion of environmentally sustainable behaviors. *Social Marketing Quarterly, 17*(2), 48–62. https://doi.org/10.1080/15245004.2011.570859

Coghlan, A. (2022). Can ecotourism interpretation influence reef protective behaviours? Findings from a quasi-experimental field study involving a virtual reality game. *Journal of Ecotourism, 21*(2), 187–196. https://doi.org/10.1080/14724049.2021.1971240

Diemer, J., Alpers, G., Peperkorn, H., Shiban, Y., & Mühlberger, A. (2015). The impact of perception and presence on emotional reactions: A review of research in virtual reality. *Frontiers in Psychology, 6*, 26. https://doi.org/10.3389/fpsyg.2015.00026

Douglas, I. (2012). Urban ecology and urban ecosystems: Understanding the links to human health and well-being. *Current Opinion in Environmental Sustainability, 4*(4), 385–392.

Duan, H., Li, J., Fan, S., Lin, Z., Wu, X., & Cai, W. (2021, October). Metaverse for social good: A university campus prototype. In *Proceedings of the 29th ACM International Conference on Multimedia* (pp. 153–161). https://doi.org/10.48550/arXiv.2108.08985

Ecoegypt (nd). https://ecoegypt.org/. Accessed 1 March 2023.

Feltham J. (2018, August 6). *Researcher outlines how VR data collection could be used against you.* https://uploadvr.com/report-vr-data-collection-stanford/. Accessed 15 April 2023.

Fernández, R., & Alonso, V. (2015). Virtual Reality in a shipbuilding environment. *Advances in Engineering Software, 81*, 30–40. https://doi.org/10.1016/j.advengsoft.2014.11.001

Francis, J., Wood, L. J., Knuiman, M., & Giles-Corti, B. (2012). Quality or quantity? Exploring the relationship between Public Open Space attributes and mental health in Perth, Western Australia. *Social Science & Medicine, 74*(10), 1570–1577.

Frost, S., Kannis-Dymand, L., Schaffer, V., Millea, P., Allen, A., Stallman, H., et al. (2022). Virtual immersion in nature and psychological well-being: A systematic literature review. *Journal of Environmental Psychology, 101765.* https://doi.org/10.1016/j.jenvp.2022.101765

Futurism. (n.d.). Virtual travel: Google is bringing the national parks to a VR headset near you. *Futurism.* https://futurism.com/virtual-travel-google-is-bringing-the-national-parks-to-a-vr-headset-near-you. Accessed 9 March 2023.

García-Betances, R., Arredondo Waldmeyer, M., Fico, G., & Cabrera-Umpiérrez, M. (2015). A succinct overview of virtual reality technology use in Alzheimer's disease. *Frontiers in Aging Neuroscience, 80.* https://doi.org/10.3389/fnagi.2015.00080

Gifford, R., & Comeau, L. (2011). Message framing influences perceived climate change competence, engagement, and behavioral intentions. *Global Environmental Change, 21*(4), 1301–1307. https://doi.org/10.1016/j.gloenvcha.2011.06.004

Gladwell, V. F., Brown, D. K., Barton, J. L., Tarvainen, M. P., Kuoppa, P., Pretty, J., & Sandercock, G. R. H. (2012). The effects of views of nature on autonomic control. *European Journal of Applied Physiology, 112*, 3379–3386.

Jingjing, G., Yan, Z., Zheng, Y., Yonghua, H., Jun, F., & Weiwei, Z. (2013). The framing effect in medical decision-making: A review of the literature.

Psychology, Health and Medicine, 18(6), 645–653. https://doi.org/10.1080/13548506.2013.766352

Gonzalez-Franco M., Peck T., Rodriguez-Fornells A. & Slater M. (2019). The hidden risk of virtual reality—and what to do about it. World Economic Forum. https://www.weforum.org/agenda/2019/08/the-hidden-risk-of-virtual-reality-and-what-to-do-about-it/. Accessed 15 April 2023.

Grassini, S. (2022a). Virtual reality assisted non-pharmacological treatments in chronic pain management: A systematic review and quantitative meta-analysis. *International Journal of Environmental Research and Public Health, 19*(7), 4071. https://doi.org/10.3390/ijerph19074071

Grassini, S. (2022b). The use of VR natural environments for the reduction of stress: an overview on current research and future prospective. In *Proceedings of the 33rd European Conference on Cognitive Ergonomics* (pp. 1–5). https://doi.org/10.1145/3552327.3552336

Grassini, S., & Laumann, K. (2020). Are modern head-mounted displays sexist? A systematic review on gender differences in HMD-mediated virtual reality. *Frontiers in Psychology, 11*, 1604. https://doi.org/10.3389/fpsyg.2020.01604

Grassini, S. & Laumann, K. (2021). Immersive visual technologies and human health. In *Proceedings of the 32nd European Conference on Cognitive Ergonomics* (pp. 1–6). https://doi.org/10.1145/3452853.3452856

Grassini, S., Laumann, K., de Martin Topranin, V., & Thorp, S. (2021). Evaluating the effect of multi-sensory stimulations on simulator sickness and sense of presence during HMD-mediated VR experience. *Ergonomics, 64*(12), 1532–1542. https://doi.org/10.1080/00140139.2021.1941279

Grassini, S., Revonsuo, A., Castellotti, S., Petrizzo, I., Benedetti, V., & Koivisto, M. (2019). Processing of natural scenery is associated with lower attentional and cognitive load compared with urban ones. *Journal of Environmental Psychology, 62*, 1–11.

Grassini, S., Segurini, G. V., & Koivisto, M. (2022). Watching nature videos promotes physiological restoration: evidence from the modulation of alpha waves in electroencephalography. *Frontiers in Psychology, 13*, 871143.

Godovykh, M., Baker, C., & Fyall, A. (2022). VR in tourism: A new call for virtual tourism experience amid and after the COVID-19 pandemic. *Tourism and Hospitality, 3*(1), 265–275.

Guangzhi, T. Z., Fu., Wu Yuanyuan, & Hongyu, & Wang Jiajing,. (2019). Positive versus negative messaging in discouraging drunken driving: Matching behavior consequences with target groups. *Journal of Advertising Research, 59*(2), 185–195. https://doi.org/10.2501/JAR-2018-029

Gue, I., Ubando, A., Tseng, M., & Tan, R. (2020). Artificial neural networks for sustainable development: A critical review. *Clean Technologies and Environmental Policy, 22,* 1449–1465. https://doi.org/10.1007/s10098-020-01883-2

Gürerk, Ö. & Kasulke, A. (2018). Does virtual reality increase charitable giving? An experimental study. *SSRN Journal* (October 8, 2018). https://doi.org/10.2139/ssrn.3267691.

Gutierrez, M., Vexo, F., & Thalmann, D. (2008). *Stepping into Virtual Reality.*

Guttentag, D. (2010). Virtual reality: Applications and implications for tourism. *Tourism Management, 31*(5), 637–651. https://doi.org/10.1016/j.tourman.2009.07.003

Hakkinen, J., Vuori, T., & Paakka, M. (October 2002). (2002) Postural stability and sickness symptoms after HMD use. *In Proceedings of the IEEE International Conference on Systems, Man and Cybernetics, Yasmine Hammamet, Tunisia,* 6–9, 147–152. https://doi.org/10.1109/ICSMC.2002.1167964

Hallahan, K. (2008). Strategic framing. *The International Encyclopedia of Communication.* https://doi.org/10.1002/9781405186407.wbiecs107

Hamad, A. (2021) Two wheelistic: Development of a high-fidelity virtual reality cycling simulator for transportation safety research. Unpublished Thesis University of Michigan. https://doi.org/10.7302/1035

Hardisty, D., & Weber, E. U. (2009). Discounting future green: Money versus the environment. *Journal of Experimental Psychology: General, 138*(3), 329–340. https://doi.org/10.1037/a0016433

Hayhurst, J. (2018). How augmented reality and virtual reality is being used to support people living with dementia—design challenges and future directions. *Augmented Reality and Virtual Reality: Empowering Human, Place and Business,* 295–305. https://doi.org/10.1007/978-3-319-64027-3_20

Henriksson, E. (2018). Data protection challenges for virtual reality applications. *Interactive Entertainment Law Review, 1*(1), 57–61. https://doi.org/10.4337/ielr.2018.01.05

Hettinger, L. & Riccio, G. (1992). Visually induced motion sickness in virtual environments. *Presence: Teleoperators & Virtual Environments, 1*(3), 306–310.

Hirzle, T., Fischbach, F., Karlbauer, J., Jansen, P., Gugenheimer, J., Rukzio, E., & Bulling, A. (2022). Understanding, addressing, and analysing Digital Eye Strain in Virtual Reality Head-Mounted Displays. *ACM Transactions on Computer-Human Interaction, 29*(4), 1–80. https://doi.org/10.1145/3492802

Indovina, P., Barone, D., Gallo, L., Chirico, A., De Pietro, G., & Antonio, G. (2018). Virtual reality as a distraction intervention to relieve pain and distress during medical rocedures. *Clinical Journal of Pain, 34*(9), 858–877. https://doi.org/10.1097/AJP.0000000000000599

Innocenti, A. (2017). Virtual reality experiments in economics. *Journal of Behavioral and Experimental Economics, 69*, 71–77. https://doi.org/10.1016/j.socec.2017.06.001

Jacobson, S., Morales, N., Chen, B., Soodeen, R., Moulton, M., & Eakta, J. (2019). Love or Loss: Effective message framing to promote environmental conservation. *Applied Environmental Education and Communication, 18*(3), 252–265. https://doi.org/10.1080/1533015X.2018.1456380

Jiang, S., Moyle, B., Yung, R., Tao, L., & Scott, N. (2023). Augmented reality and the enhancement of memorable tourism experiences at heritage sites. *Current Issues in Tourism, 26*(2), 242–257. https://doi.org/10.1080/13683500.2022.2026303

Jones, T., Moore, T., & Choo, J. (2016). The impact of virtual reality on chronic pain. *PLoS ONE, 11*(12), e0167523. https://doi.org/10.1371/journal.pone.0167523

Kandaurova, M., Lee, S. H., & Mark. (2019). The effects of Virtual Reality (VR) on charitable giving: The role of empathy, guilt, responsibility, and social exclusion. *Journal of Business Research, 100*, 571–580. https://doi.org/10.1016/j.jbusres.2019.02.022

Kaplan, A., Cruit, J., Endsley, M., Beers, S., Sawyer, B., & Hancock, P. (2020). The Effects of Virtual Reality, Augmented Reality, and Mixed Reality as Training Enhancement Methods: A Meta-Analysis. *Human Factors: The Journal of the Human Factors and Ergonomics Society, 63*, 706–726. https://doi.org/10.1177/0018720820904229

Kaplan, R. (2001). The nature of the view from home: Psychological benefits. *Environment and Behavior, 33*(4), 507–542.

Khairunisa, Y., Nurhasanah, Y., & Verlaili, R. (2022). Virtual Job Fair Information System Design Based on Augmented Reality/Virtual Reality. *Sinkron: Jurnal Dan Penelitian Teknik Informatika, 7*(4), 2449–2461. https://doi.org/10.33395/sinkron.v7i4.11795

Kim, H., Lim, H.-T., & Ro, Y. (2019a). Deep Virtual Reality Image Quality Assessment with Human Perception Guider for Omnidirectional Image. *IEEE Transactions on Circuits Systems and Video Technology, 30*, 917–928. https://doi.org/10.1109/TCSVT.2019.2898732

Kim, O., Pang, Y., & Kim, J. (2019b). The effectiveness of virtual reality for people with mild cognitive impairment or dementia: A meta-analysis. *BMC Psychiatry, 19*(1), 1–10. https://doi.org/10.1186/s12888-019-2180-x

Knox, J. (2022). The metaverse, or the serious business of tech frontiers. *Postdigital Science and Education, 4*(2), 207–215. https://doi.org/10.1007/s42438-022-00300-9

Koh, L. Y., Wu, M., Wang, X., & Yuen, K. F. (2023). Willingness to participate in virtual reality technologies: Public adoption and policy perspectives for marine conservation. *Journal of Environmental Management, 334*, 117480. https://doi.org/10.1016/j.jenvman.2023.117480

La Corte, V., Sperduti, M., Abichou, K., & Piolino, P. (2019). Episodic memory assessment and remediation in normal and pathological aging using virtual reality: a mini review. *Frontiers in Psychology, 10*, 173. https://doi.org/10.3389/fpsyg.2019.00173. eCollection 2019

Lane, L. (2020, April 12). Travel virtually to U.S. national parks—including ones you probably wouldn't otherwise get to. *Forbes.* https://www.forbes.com/sites/lealane/2020/04/12/travel-virtually-to-us-national-parks-including-ones-you-probably-wouldnt-otherwise-get-to/. Accessed 9 March 2023.

LaRocco, M. (2020). Developing the 'best practices' of virtual reality design: Industry standards at the frontier of emerging media. *Journal of Visual Culture, 19*(1), 96–111. https://doi.org/10.1177/147041292090625

Lateef, F., Chong, Y., Sethi, D., & Loh, C. (2018). Going forward with Pokemon Go. *Journal of Emergencies, Trauma, and Shock, 11*(4), 243–246. https://doi.org/10.4103/JETS.JETS_87_17

Laviola, J., Jr. (2000). A discussion of cybersickness in virtual environments. *ACM SIGCHI Bulletin, 32*(1), 47–56. https://doi.org/10.1145/333329.333344

Lee, U. (2022). Tourism using virtual reality: Media richness and information system successes. *Sustainability, 14*(7), 3975. https://doi.org/10.3390/su14073975

Leiserowitz, A., Carman, J., Buttermore, N., Neyens, L., Rosenthal, S., Marlon, J. et al. (2022). *International Public Opinion on Climate Change, 2022.* Yale Program on Climate Change Communication and Data for Good at Meta.

Lenzen, M., Sun, Y., Faturay, F., Ting, Y., Geschke, A., & Malik, A. (2018). The carbon footprint of global tourism. *Nature Climate Change, 8*(6), 522–528. https://doi.org/10.1038/s41558-018-0141-x

Li, H., Zhang, X., Wang, H., Yang, Z., Liu, H., Cao, Y., & Zhang, G. (2021). Access to nature via virtual reality: A mini-review. *Frontiers in Psychology, 12*, 725288.

Lorek, S., & Fuchs, D. (2013). Strong sustainable consumption governance – precondition for a degrowth path? *Journal of Cleaner Production, 38*, 36–43. https://doi.org/10.1016/j.jclepro.2011.08.008

Lorenzoni, I., Nicholson-Cole, S., & Whitmarsh, L. (2007). Barriers perceived to engaging with climate change among the UK public and their policy implications. *Global Environmental Change, 17*(3–4), 445–459. https://doi.org/10.1016/j.gloenvcha.2007.01.004

Lowry, B. (2019, April 1). 'Our Planet' is visually stunning and deadly serious. CNN. https://edition.cnn.com/2019/04/01/entertainment/nature-documentaries-our-planet/index.html. Accessed 14 April 2023.

Maller, C., Nicholls, L., & Strengers, Y. (2016). Understanding the materiality of neighbourhoods in 'healthy practices': Outdoor exercise practices in a new master-planned estate. *Urban Policy and Research, 34*(1), 55–72.

Marquet, O., Alberico, C., Adlakha, D., & Hipp, J. (2017). Examining motivations to play Pokémon GO and their influence on perceived outcomes and physical activity. *JMIR Serious Games, 5*(4), e21. https://doi.org/10.2196/games.7229

Masson-Delmotte, V., Zhai, P., Pörtner, H., Roberts, D., Skea, J., Shukla, P. et al. (2018). Summary for policymakers. In *Global warming of 1.5°C. An IPCC Special Report on the impacts of global warming of 1.5°C above pre-industrial levels and related global greenhouse gas emission pathways, in the context of strengthening the global response to the threat of climate change, sustainable development, and efforts to eradicate poverty* (p. 32). IPCC. https://www.ipcc.ch/sr15/chapter/spm/. Accessed 15 April 2023.

Mayer, F. S., Frantz, C. M., Bruehlman-Senecal, E., & Dolliver, K. (2009). Why is nature beneficial? The role of connectedness to nature. *Environment and Behavior, 41*(5), 607–643.

McCauley, M. & Sharkey, T. (1992). Cybersickness: Perception of self-motion in virtual environments. *Presence: Teleoperators and Virtual Environments, 1* (3), 311–318. https://doi.org/10.1162/pres.1992.1.3.311

Millar, M., Westfall, R. S., & Fink-Armold, A. (2023). Effects of disease threat and attitude similarity on willingness to help: The mediating role of disgust. Psychological Reports, 126(1), 150-168.

Moser, S. C., & Dilling, L. (2011). COMMUNICATING CHANGE SCIENCE:-CLOSING ACTION CLIMATE. The Oxford handbook of climate change and society, 161. Millar, M., Westfall, R. S., & Fink-Armold

Movono, A., Dahles, H., & Becken, S. (2018). Fijian culture and the environment: A focus on the ecological and social interconnectedness of tourism development. *Journal of Sustainable Tourism, 26*(3), 451–469. https://doi.org/10.1080/09669582.2017.1359280

Nelson, K., Schlüter, A., & Vance, C. (2018a). Distributional preferences and donation behavior among marine resource users in Wakatobi, Indonesia. *Ocean and Coastal Management, 162*, 34–45. https://doi.org/10.1016/j.ocecoaman.2017.09.003

Nelson, K., Schlüter, A., & Vance, C. (2018b). Funding conservation locally: Insights from behavioral experiments in Indonesia. *Conservation Letters, 11*(2), e12378. https://doi.org/10.1111/conl.12378

Nelson, K., Cepok, J., Marnholz, K., Arzaroli, R., Grosse, C., Zachmann, G. (2019). Effects of virtual reality on giving to charity. Experimental Economics for the Environment, Münster, Germany, February 25–26, 2019.

Nelson, K., Anggraini, E., & Schlüter, A. (2020). Virtual reality as a tool for environmental conservation and fundraising. *PLoS ONE, 15*(4), e0223631. https://doi.org/10.1371/journal.pone.0223631

Newman, M., Gatersleben, B., Wyles, K., & Ratcliffe, E. (2022). The use of virtual reality in environment experiences and the importance of realism. *Journal of Environmental Psychology, 79,* 101733. https://doi.org/10.1016/j.jenvp.2021.101733

Nielsen (2017). Virtual empathy: How 360-Degree video can boost the efforts of non-profits. https://www.nielsen.com/us/en/insights/news/2017/how-360-degree-video-can-boost-the-efforts-of-non-profits.html. Accessed 15 April 2023.

O'Keefe, D., & Jensen, J. (2007). The relative persuasiveness of gain-framed loss-framed messages for encouraging disease prevention behaviors: A meta-analytic review. *Journal of Health Communication, 12*(7), 623–644. https://doi.org/10.1080/10810730701615198

Pelletier, L., & Sharp, E. (2008). Persuasive communication and proenvironmental behaviours: How message tailoring and message framing can improve the integration of behaviours through self-determined motivation. *Canadian Psychology/psychologie Canadienne, 49*(3), 210. https://doi.org/10.1037/a0012755

Pounders, K., Seungae, L., & Mackert, M. (2015). Matching temporal frame, self-view, and message frame valence: Improving persuasiveness in health communications. *Journal of Advertising, 44*(4), 388–402. https://doi.org/10.1080/00913367.2015.1071210

Ramirez, E. (2017). Empathy and the limits of thought experiments. *Metaphilosophy, 48*(4), 504–526. https://doi.org/10.1111/meta.12249

Rebenitsch, L. & Owen, C. (2014). Individual variation in susceptibility to cybersickness. In *Proceedings of the 27th Annual ACM Symposium on User Interface Software and Technology* (pp. 309–317). Virtual Event. https://doi.org/10.1145/2642918.2647394.

Reed-Jones, R., Hollands, M., Reed-Jones, J., & Vallis, L. (2009). Visually evoked whole-body turning responses during stepping in place in a virtual environment. *Gait and Posture, 30,* 317–321. https://doi.org/10.1016/j.gaitpost.2009.06.001

Reilly, C. (2017, October 20). Scientists hope building an artificial reef could help save the real one. CNET. Retrieved March 2, 2023 from https://www.cnet.com/science/saving-great-barrier-reef-simulated-oceans-and-3d-coral/. Accessed 15 April 2023.

Rodríguez-Fidalgo, M., & Paíno-Ambrosio, A. (2020). Use of virtual reality and 360° video as narrative resources in the documentary genre: Towards a new immersive social documentary? *Catalan Journal of Communication and Cultural Studies, 12*(2), 239–253. https://doi.org/10.1386/cjcs_00030_1

Rosenberg, R., Baughman, S., & Bailenson, J. (2013). Virtual superheroes: Using superpowers in virtual reality to encourage prosocial behavior. *PLoS ONE, 8*(1), e55003. https://doi.org/10.1371/journal.pone.0055003

Rothman , A., Bartels R., Wlaschin, J & Salovey P. (2006). The strategic use of gain-and loss-framed messages to promote healthy behavior: How theory can inform practice. *Journal of Communication, 56* (suppl_1), S202–S220.

Samit, J. (2017, November 21). How these charities are using virtual reality to reach donors this holiday season. *Fortune.* http://fortune.com/2017/11/21/virtual-reality-charities-donations/. Accessed 15 April 2023.

Schmitt, H., Schmitt, C., & Nelson, K. (2017). Coral reefs: Life below the surface. Youtube https://www.youtube.com/watch?v=2TPG8lcfeDc&t=64s: TheJetlagged. Accessed 15 April 2023.

Schutte, N., & Stilinović, E. (2017). Facilitating empathy through virtual reality. *Motivation and Emotion, 41*, 708–712. https://doi.org/10.1007/s11031-017-9641-7

Shin, D. (2018). Empathy and embodied experience in virtual environment: To what extent can virtual reality stimulate empathy and embodied experience? *Computers in Human Behavior, 78*, 64–73. https://doi.org/10.1016/j.chb.2017.09.012

Slater, M., & Sanchez-Vives, M. (2016). Enhancing our lives with immersive virtual reality. *Frontiers in Robotics and AI, 3*(74), 1–47. https://doi.org/10.3389/frobt.2016.00074

Soutar, C., & Wand, A. (2022). Understanding the spectrum of anxiety responses to climate change: A systematic review of the qualitative literature. *International Journal of Environmental Research and Public Health, 19*(2), 990. https://doi.org/10.3390/ijerph19020990

Spence, A., & Pidgeon, N. (2010). Framing and communicating climate change: The effects of distance and outcome frame manipulations. *Global Environmental Change, 20*(4), 656–667. https://doi.org/10.1016/j.gloenvcha.2010.07.002

Stanney, K., Lawson, B., Rokers, B., Dennison, M., Fidopiastis, C., Stoffregen, T., et al. (2020). Identifying causes of and solutions for cybersickness in immersive technology: Reformulation of a research and development agenda. *International Journal of Human-Computer Interaction, 36*, 1783–1803. https://doi.org/10.1080/10447318.2020.1828535

Sutherland I. (1965) The ultimate display. In: *Proceedings of the International Federation for Information Processing Congress*, New York, pp. 506–508.

Talwar, S., Kaur, P., Escobar, O., & Lan, S. (2022). Virtual reality tourism to satisfy wanderlust without wandering: An unconventional innovation to promote sustainability. *Journal of Business Research, 152*, 128–143. https://doi.org/10.1016/j.jbusres.2022.07.032

Thorp, S., Rimol, L., & Grassini, S. (2023). Association of the big five personality traits with training effectiveness, sense of presence, and cybersickness in virtual Reality. *Multimodal Technologies and Interaction, 7*(2), 11. https://doi.org/10.3390/mti7020011

Thorp, S., Sævild Ree, A., & Grassini, S. (2022). Temporal development of sense of presence and cybersickness during an immersive VR experience. *Multimodal Technologies and Interaction, 6*(5), 31. https://doi.org/10.3390/mti6050031

Tversky, A. & Kahneman, D. (1975). Judgment under Uncertainty: Heuristics and Biases. In D. Wendt & C. Vlek (Eds.), *Utility, probability, and human decision making. Theory and decision library*, vol. 11. https://doi.org/10.1007/978-94-010-1834-0_8

Tversky, A., & Kahneman, D. (1981). *The framing of decisions and the psychology of choice. Science, 211*(4481), 453–458.

Ünal, A. B., Pals, R., Steg, L., Siero, F. W., & van der Zee, K. I. (2022). Is virtual reality a valid tool for restorative environments research? *Urban Forestry & Urban Greening, 74*, 127673.

United Nations (2018). Around 2.5 billion more people will be living in cities by 2050, projects new UN report. https://www.un.org/development/desa/en/news/population/2018-revision-of-world-urbanization-prospects.html. . Accessed 2 March 2023.

Valmaggia, L. (2017). The use of virtual reality in psychosis research and treatment. *World Psychiatry, 16*(3), 246–247. https://doi.org/10.1002/wps.20443

Van den Berg, A. E., Koole, S. L., & van der Wulp, N. Y. (2003). Environmental preference and restoration: (How) are they related? *Journal of Environmental Psychology, 23*(2), 135–146.

Van der Linden, S., Maibach, E., & Leiserowitz, A. (2015). Improving public engagement with climate change: Five "best practice" insights from psychological science. *Perspectives on Psychological Science, 10*(6), 758–763. https://doi.org/10.1177/1745691615598516

Walsh, K., & Pawlowski, S. (2002). Virtual reality: A technology in need of IS research. *Communications for the Association of Information Systems, 8*(20), 297–313. https://doi.org/10.17705/1CAIS.00820

Watson (n.d.). Take a virtual visit to a national park. *National Parks Foundation.* https://www.nationalparks.org/connect/blog/take-virtual-visit-national-park. Accessed 9 March 2023.

Weber, A. M., & Trojan, J. (2018). The restorative value of the urban environment: A systematic review of the existing literature. *Environmental Health Insights, 12*, 1178630218812805.

West, M. (2015). Charities use virtual reality to draw in donors. https://www.wsj.com/articles/charities-use-virtual-reality-to-draw-in-donors-1448663492. Accessed 2 March 2023.

World Health Organization. (2004). *Promoting mental health: Concepts, emerging evidence, practice: Summary report*. World Health Organization.

Yang, H., & Han, S. Y. (2021). Understanding virtual reality continuance: An extended perspective of perceived value. *Online Information Review, 45*(2), 422–439.

INDEX